Neurodevelopmental Disorders in Education:

The Clinical Continuum From Childhood to Adulthood

Irini K. Zerva
School of Medicine, Aristotle University of Thessaloniki, Greece & School of Education, University of Nicosia, Cyprus & Psychology Department, School of Social Sciences, Arts and Humanities, Neapolis University Pafos, Cyprus

IGI Global
Scientific Publishing
Publishing Tomorrow's Research Today

Vice President of Editorial	Melissa Wagner
Director of Acquisitions	Mikaela Felty
Director of Book Development	Jocelynn Hessler
Production Manager	Mike Brehm
Cover Design	Jose Rosado

Published in the United States of America by
IGI Global Scientific Publishing
701 East Chocolate Avenue
Hershey, PA, 17033, USA
Tel: 717-533-8845 | Fax: 717-533-7115
Website: https://www.igi-global.com E-mail: cust@igi-global.com

Library of Congress Cataloging-in-Publication Data

Names: Zerva, Irini K., 1980- author
Title: Neurodevelopmental disorders in education : the clinical continuum from childhood to adulthood / Irini K. Zerva.
Description: Hershey, PA : IGI Global Scientific Publishing, [2026] | Includes bibliographical references and index. | Summary: "This book examines neurodevelopmental disorders across the lifespan, highlighting their impact on learning, behavior, and social development. By integrating clinical insights with educational practice, it offers guidance for early intervention, individualized support, and inclusive strategies to help children and adults with conditions like ADHD, ASD, and learning disorders thrive"-- Provided by publisher.
Identifiers: LCCN 2026002336 (print) | LCCN 2026002337 (ebook) | ISBN 9798369363317 hardcover | ISBN 9798369363324 paperback | ISBN 9798369363331 ebook
Subjects: LCSH: Children with mental disabilities--Education | Children with mental disabilities--Social conditions | Inclusive education | Communicative disorders in children--Treatment | People with mental disabilities--Social conditions
Classification: LCC LC4601 .Z47 2026 (print) | LCC LC4601 (ebook)
LC record available at https://lccn.loc.gov/2026002336
LC ebook record available at https://lccn.loc.gov/2026002337

British Cataloguing in Publication Data
A Cataloguing in Publication record for this book is available from the British Library.

To my daughter, my husband, my family, and my teachers—from childhood through adulthood.

Table of Contents

Foreword

Advances in neuroscience, developmental psychology, and educational research over the past decades have profoundly reshaped how we understand neurodevelopmental disorders (NDDs). Yet, despite this progress, fragmented systems of care and rigid diagnostic boundaries continue to obscure the lived experiences of neurodivergent individuals across their lifespan. *Neurodevelopmental Disorders in Education: The Clinical Continuum From Childhood to Adulthood* arrives at a crucial time, offering a cohesive, interdisciplinary framework that bridges contemporary scientific knowledge with educational and clinical practice. The book draws on recent research results, ensuring that its perspectives reflect current evidence while addressing gaps in understanding and care.

As someone who has spent many years working with children, adolescents, and adults with NDDs in clinical and academic settings, I am struck by the clarity, depth, and coherence of this volume. The author, Dr. Irini Zerva, presents neurodevelopment as a dynamic, lifelong process shaped by interactions among genetic predispositions, brain development, learning environments, cultural contexts, and social experiences. The book's approach is both original and innovative, exploring themes that challenge traditional assumptions and expand the dialogue between research and practice.

The text covers the full spectrum of NDDs, from Autism and ADHD to Learning Disorders, Intellectual Disability, Communication Disorders, and motor-related conditions, emphasizing both shared neurobiological pathways and the unique characteristics of each profile. The discussion of comorbidity and the limitations of rigid diagnostic systems provides a nuanced framework for interpreting complex presentations. Particularly notable is the attention to children with an oncological history and neurodevelopmental imprint, highlighting how early medical experiences can intersect with developmental trajectories to shape learning and psychosocial outcomes.

Equally compelling is the sustained attention to the environments in which children grow and learn. By examining early developmental trajectories, bilingual and twice-exceptional learners, executive-function development, and the socio-emotional consequences of unmet needs, the authors illuminate how educational systems shape developmental outcomes. These chapters offer conceptual clarity alongside practical strategies, emphasizing early identification, culturally grounded assessment, and strengths-based intervention.

The volume extends its lens beyond childhood, addressing adolescence, higher education, and adulthood, domains often overlooked in academic texts. Chapters on university transitions, employment, and adult well-being illustrate how neuro-developmental differences continue to interact with institutional structures, societal expectations, and personal histories. The exploration of neurodivergent experience in urban environments invites readers to consider how cities themselves can become more inclusive, sensory-aware, and participatory.

Additionally, the discussion of artificial intelligence and digital technologies introduces a contemporary perspective on the intersection of innovation and human development. As educational systems adopt AI-driven tools, the book provides guidance for integrating these resources ethically and effectively, without compromising relational connection or developmental integrity.

Altogether, this volume integrates scientific rigor with compassion, theoretical insight with practical relevance, and empirical evidence with novel perspectives. It presents a vision of neurodevelopment that accounts for diversity in learning and life experiences, including children with complex medical histories, and encourages approaches that are holistic, inclusive, and forward-thinking.

For clinicians, educators, researchers, and policymakers, *Neurodevelopmental Disorders in Education: The Clinical Continuum From Childhood to Adulthood* is both a comprehensive resource and an invitation to assess more thoughtfully, listen more attentively, and design environments where neurodivergent individuals of all ages can thrive.

It is with great respect for the author's work and confidence in its contribution to the field that I recommend this volume to readers.

Eleni Bonti
School of Medicine, Aristotle University of Thessaloniki, Greece

Preface

This volume brings together educational, clinical, and psychosocial perspectives to examine neurodevelopmental disorders as dynamic conditions that unfold across the lifespan. Rather than approaching these conditions as isolated diagnostic categories, the book adopts a clinical continuum framework that emphasizes developmental change, contextual influences, and the central role of education in shaping outcomes.

The chapters that follow address neurodevelopmental disorders from early childhood through adulthood, highlighting implications for learning, well-being, inclusion, and participation. The overview below is provided to guide the reader through the structure and focus of each chapter.

CHAPTER OVERVIEW

Chapter 1

Introduces neurodevelopmental disorders within educational contexts, establishing key diagnostic frameworks and their interaction with learning environments.

Chapter 2

Explores the spectrum and comorbidity of neurodevelopmental disorders, highlighting shared mechanisms and challenging rigid categorical approaches.

Chapter 3

Presents the clinical continuum hypothesis, offering a unifying framework for understanding neurodevelopmental disorders across the lifespan.

Chapter 4

Examines neurodevelopmental disorders in early childhood, emphasizing early identification, developmental trajectories, and the impact of early intervention.

Chapter 5

Analyzes the school experience of students with neurodevelopmental disorders, connecting symptom presentation with academic performance and educational support.

Chapter 6

Addresses diagnostic and educational challenges among bilingual and twice-exceptional learners, with a focus on equity, accurate identification, and inclusive assessment practices.

Chapter 7

Investigates executive functions as a cross-cutting mechanism underlying learning difficulties, comorbidity, and functional outcomes across disorders.

Chapter 8

Discusses the psychosocial impact of neurodevelopmental disorders on students and families, highlighting resilience, mental health, and systemic support.

Chapter 9

Situates pediatric cancer survivorship within a neurodevelopmental and educational framework, emphasizing long-term cognitive and academic outcomes.

Chapter 10

Explores the role of technology and artificial intelligence in assessment and intervention, critically examining both innovative potential and ethical considerations.

Chapter 11

Focuses on neurodiversity in the transition to higher education, addressing accessibility, institutional support, and inclusive learning environments.

Chapter 12

Examines employment and labor market participation, highlighting barriers, accommodations, and pathways toward inclusive and sustainable work.

Chapter 13

Explores the imprint of neurodevelopmental disorders in adulthood, emphasizing participation, adaptation, and quality of life across contexts.

Chapter 14

Synthesizes the core themes of the volume and reinforces the importance of a lifespan-oriented, interdisciplinary, and inclusive perspective.

Acknowledgment

I would like to express my heartfelt gratitude to the many individuals who have supported and inspired me throughout the creation of this book.

First, I am deeply thankful to my mentors and teachers, whose guidance, wisdom, and encouragement have been invaluable in shaping both my professional journey and the ideas presented in this volume. Their dedication to knowledge and excellence continues to inspire me.

I am especially grateful to my family, my daughter and my husband, for their unwavering love, patience, and encouragement. Their support has been a constant source of strength, allowing me to pursue this work with dedication and passion.

Most importantly, I wish to acknowledge the children, adolescents, and adults with neurodevelopmental differences who have shared their experiences and perspectives. Their resilience, creativity, and individuality have been at the heart of this book, informing every chapter and guiding my understanding of neurodevelopmental diversity.

Finally, I would like to thank all colleagues, collaborators, and students who contributed their time, insight, and expertise to this project. This book is a product of many minds and generous hearts, and I am deeply grateful to everyone who made it possible.

Introduction

The present volume, *Neurodevelopmental Disorders in Education: The Clinical Continuum From Childhood to Adulthood*, explores the intricate relationship between neurodevelopmental conditions and education through an integrated, longitudinal lens. It invites readers to view these disorders not as isolated medical diagnoses, but as evolving developmental pathways that influence and are shaped by educational experience. The book follows the trajectory of neurodevelopmental disorders across the lifespan, beginning with early childhood identification and intervention, continuing through the school years, and extending into the challenges and support needs that emerge in adulthood. By situating these conditions within the broader context of human growth and learning, the work aims to offer a nuanced understanding of the dilemmas, opportunities, and responsibilities faced by educators, clinicians, and society at large.

Neurodevelopmental disorders comprise a diverse group of conditions that affect brain development and functioning, altering the way individuals process information, regulate emotion, and engage with their environment. In educational settings, these disorders often manifest through academic struggles, behavioral difficulties, or challenges in adapting to the structured demands of schooling. Conditions such as Attention-Deficit/Hyperactivity Disorder (ADHD), Autism Spectrum Disorder (ASD), Speech and Language Disorders, and Specific Learning Disorders (SLD) can disrupt core developmental domains including attention, language, executive functioning, and social communication. These difficulties, in turn, influence a child's self-perception, relationships, and capacity to thrive in learning environments. Because of this interplay, students with neurodevelopmental differences often require individualized instruction and targeted support to access learning on an equal footing with their peers. Early recognition within the school environment is therefore critical, both for initiating appropriate interventions and for promoting inclusive educational practices that value the participation and potential of all learners.

UNDERSTANDING NEURODEVELOPMENTAL DISORDERS THROUGH THE CLINICAL CONTINUUM

Adopting a clinical continuum perspective, one that traces the evolution of disorders from early childhood through adulthood, provides an essential framework for understanding, diagnosing, and supporting individuals across developmental stages. This perspective recognizes that neurodevelopmental disorders are not static conditions; rather, they unfold dynamically over time, shaped by biological maturation, environmental contexts, and educational opportunities.

Difficulties first observed in early childhood, such as language delay or social withdrawal, may transform in character as the individual matures. By examining how these patterns change, educators and clinicians can better understand how early manifestations influence later outcomes, and how timely interventions can alter developmental trajectories in meaningful ways.

A continuum approach also foregrounds the importance of individualized, stage-specific support. During childhood, interventions may emphasize foundational skills such as language development, attention, and social engagement. In adolescence, the focus often shifts toward identity formation, executive functioning, and emotional regulation. Adulthood, in turn, introduces new challenges related to vocational training, independent living, and social participation. Recognizing these evolving needs allows educators and practitioners to design interventions that adapt over time, building not only competence but also resilience and self-efficacy. Early and sustained support can mitigate long-term difficulties, promoting social inclusion, academic achievement, and professional success later in life.

Furthermore, examining neurodevelopmental disorders along a continuum acknowledges their multidimensional nature. These conditions affect multiple domains, cognitive, emotional, social, and physical, and each dimension interacts with the others in complex ways. A comprehensive framework enables the development of interventions that address the person through a holistic perspective than focusing narrowly on isolated symptoms. This holistic understanding also facilitates continuous monitoring and reassessment, ensuring that interventions remain responsive to the individual's growth. A child with ADHD, for instance, may experience entirely different challenges as a teenager or adult, necessitating an evolving strategy that accounts for new contexts and capacities. Continuous evaluation not only refines support but also prevents secondary difficulties such as anxiety, academic disengagement, or social exclusion.

The clinical continuum approach also reinforces the link between sustained support and social equity. True inclusion is not a one-time event but an ongoing process that requires educational systems and communities to adapt continuously. Integrating individuals with neurodevelopmental disorders into society demands more than physical presence; it requires cultivating environments that foster belonging, competence, and agency across the lifespan. Education plays a decisive role in this process, as it shapes both personal development and social participation. Through consistent adaptation and inclusive policy, education becomes a key mechanism for achieving equity and promoting lifelong flourishing.

EDUCATION AS THE CORE FRAMEWORK FOR UNDERSTANDING THE CONTINUUM

Education occupies a unique and central position within this continuum. As one of the earliest and most sustained contexts for human development, it provides the primary setting in which neurodevelopmental differences are first observed, interpreted, and supported. The school environment, with its social interactions and structured demands, often brings neurodevelopmental characteristics to light whether in the form of attention difficulties, language struggles, or challenges in peer relations. Teachers, parents, and psychologists, working collaboratively, play an indispensable role in identifying these signs early and developing individualized educational plans that address both learning and socioemotional needs. The classroom thus becomes both a diagnostic and an intervention space, where observation meets support, and where inclusion begins to take concrete form.

Education is not only instrumental for learning academic skills; it is also a vital arena for social and emotional growth. It offers structured opportunities for developing communication, empathy, self-regulation, and cooperative behavior, all of which are areas of difficulty for many individuals with neurodevelopmental disorders. Through carefully designed pedagogical approaches, education can help these learners strengthen executive functions, manage emotions, and build adaptive social strategies. This dual role, academic and psychosocial, positions education as a uniquely comprehensive framework for understanding and addressing neurodevelopmental disorders.

While medical and therapeutic interventions remain essential, education offers distinct advantages: it is lifelong, developmental, and adaptable. Unlike medical treatments that may occur intermittently, education accompanies the individual through successive stages of growth, continuously adjusting to their changing needs. It provides the structure within which therapeutic principles, such as cognitive training, behavioral modification, and communication enhancement, can be integrated into daily routines. In this sense, the educational process itself becomes therapeutic, fostering not only cognitive learning but also self-awareness, autonomy, and resilience. The collaborative involvement of teachers, psychologists, speech and occupational therapists, and families ensures that interventions are coherent and sustained across contexts.

Furthermore, education provides the mechanisms for ongoing assessment. The continuous observation of students' progress, behaviors, and interactions allows for the early detection of new challenges and the timely modification of support strategies. This cyclical process of assessment, feedback, and adaptation reflects the very essence of the clinical continuum. It enables educators to anticipate developmental transitions, adjust expectations, and sustain progress toward greater independence and inclusion.

Education thus emerges as the most comprehensive and humane context for supporting individuals with neurodevelopmental disorders. It does not merely aim to reduce symptoms but to empower individuals, to help them acquire the skills, confidence, and self-knowledge necessary for meaningful participation in social and professional life. By embracing a continuum approach, education affirms that development is ongoing and that every stage offers opportunities for growth, adaptation, and contribution.

THE PURPOSE AND STRUCTURE OF THE BOOK

This book seeks to bridge the gap between scientific understanding and practical application. It offers an integrated view of neurodevelopmental disorders across all stages of life, illuminating how their manifestations, challenges, and support needs evolve over time. For educators, this longitudinal perspective provides essential insights into identifying and responding to students' learning and socioemotional needs from the earliest stages of schooling. For clinicians and mental health professionals, it highlights the interplay between educational experiences and clinical outcomes, demonstrating how cross-disciplinary collaboration can enhance both learning and well-being.

A major theme running throughout the volume is the value of early diagnosis and sustained intervention. Recognizing neurodevelopmental differences early enables tailored educational planning and prevents the compounding of difficulties. Continuous support across developmental phases further strengthens the individual's capacity to learn, connect, and adapt. Equally important is the book's emphasis on collaboration among professionals, teachers, psychologists, clinicians, and support staff, whose coordinated efforts are fundamental to creating environments of belonging and opportunity.

The chapters that follow explore the full spectrum of neurodevelopmental conditions, integrating scientific research with lived experience. Topics include early childhood development, bilingualism and twice-exceptionality, executive function and learning, resilience, the psychological effects of childhood illness, and the evolving impact of technology and artificial intelligence on diagnosis and intervention. Later sections address neurodiversity in higher education, employment outcomes, and adult adaptation, offering a truly lifespan perspective. Each chapter concludes with a synthesis of key insights and references for further study, providing readers with both theoretical grounding and practical direction.

Underlying this work is a philosophy that values neurodiversity as a natural and enriching dimension of human variation. To study neurodevelopmental disorders is not only to understand challenges, but also to recognize distinct modes of thinking, learning, and experiencing the world. This recognition invites a shift from a deficit-based model toward one of empowerment and inclusion. Embracing neurodiversity encourages empathy and reflection, prompting educators and clinicians alike to reconsider conventional definitions of ability, success, and normalcy.

In doing so, it opens a broader dialogue about what it means to learn and to belong. The process of inclusion, understood as continuous adaptation throughout life, becomes both a moral and a developmental imperative. Education, in this sense, is not merely preparation for participation in society, it is the very medium through which society learns to value difference. By cultivating environments where individuals with neurodevelopmental differences can flourish, we advance not only their well-being but also our collective understanding of human potential.

Chapter 1
Neurodevelopmental Disorders in Educational Contexts

ABSTRACT

This chapter brings together current perspectives on neurodevelopmental disorders, showing how inherited factors, brain development, learning experiences, and context interact to shape children's abilities. It discusses the range of NDDs, Specific Learning Disorders, ADHD, Autism Spectrum Disorder, Communication Disorders, Intellectual Disability, and Motor and Tic Disorders, while noting the shared patterns and unique features within each. The chapter stresses the importance of recognizing difficulties early, using multidimensional assessment, and designing supports that build on strengths rather than deficits. By combining scientific insight with practical guidance, it calls for learning environments that adapt to individual needs and foster meaningful growth.

INTRODUCTION

Neurodevelopmental disorders constitute a group of conditions characterized by early-onset impairments in cognitive, social, and motor development, which can significantly impact educational performance and daily functioning. According to the Diagnostic and Statistical Manual of Mental Disorders, Fifth Edition (DSM-5; American Psychiatric Association, 2013), neurodevelopmental disorders encompass a range of conditions including Intellectual Disability (ID), Communication Disorders (CD), Autism Spectrum Disorder (ASD), Attention-Deficit/Hyperactivity Disorder (ADHD), Specific Learning Disorders (SLDs), and Motor Disorders such as Developmental Coordination Disorder (DCD) and Tic Disorders. These conditions

DOI: 10.4018/979-8-3693-6331-7.ch001

share common features, including onset during the developmental period, deficits in acquiring and using skills appropriate for age, and an interaction between biological, genetic, and environmental factors in shaping their presentation.

Understanding neurodevelopmental disorders in educational contexts is critical because these conditions directly influence how children acquire, process, and apply knowledge. While some students may exhibit specific deficits in areas such as reading, writing, or mathematics (as in SLDs), others may experience broader challenges affecting attention, communication, social interaction, or executive functioning. The heterogeneity of these disorders underscores the need for individualized assessment and intervention strategies that consider the interplay of cognitive, neural, and environmental factors. Contemporary research emphasizes that these disorders are not the product of inadequate instruction or low motivation; rather, they reflect complex differences in neurobiological development, often modulated by genetic predispositions and environmental experiences.

By integrating neurobiological, genetic, and psychoeducational perspectives, educators and clinicians can better understand the mechanisms underlying learning and developmental challenges. This holistic approach facilitates early identification of at-risk students, informs the design of targeted interventions, and promotes educational strategies that are responsive to each learner's unique cognitive and developmental profile. Within this framework, Specific Learning Disabilities exemplify the intricate interplay between neurobiology, genetics, and the learning environment, making them a particularly informative focus for understanding the broader category of neurodevelopmental disorders.

1.SPECIFIC LEARNING DISORDERS

Specific Learning Disorders (SLDs) encompass persistent difficulties in acquiring and using academic skills, manifesting across reading, writing, and mathematics. These difficulties are not explained by intellectual disability, sensory deficits, or insufficient educational opportunity, but rather reflect neurodevelopmental vulnerabilities. Contemporary research frames SLDs as emerging from a complex interplay of genetic, cognitive, and environmental factors, with domain-specific and cross-domain executive function deficits contributing to the severity and persistence of learning challenges (Astle et al., 2022; Thapar et al., 2017). Beyond academics, SLDs affect psychosocial outcomes, motivation, and long-term educational trajectories (Bonti et al., 2024; King, 2022).

Reading Difficulties

Reading challenges are often characterized by difficulties in decoding, word recognition, and comprehension. Neurocognitive studies highlight phonological processing, rapid naming, and working memory deficits as core contributors (Ali et al., 2025; Papadopoulos et al., 2012). These deficits affect the ability to map sounds to letters and recognize word patterns efficiently, impacting fluency and overall comprehension. Cross-linguistic research emphasizes that orthographic transparency modulates the manifestation of reading difficulties, with more consistent orthographies producing specific error patterns in decoding and reading accuracy (Georgiou et al., 2012; Papadopoulos et al., 2021; Parrila et al., 2020).

Executive functions, such as inhibitory control, updating, and cognitive flexibility, are closely linked to reading comprehension. Students with reading difficulties often show inefficiencies in these domains, which can exacerbate comprehension deficits despite adequate decoding skills (Butterfuss & Kendeou, 2018; Capodieci et al., 2023). Additionally, bilingual learners may require adapted assessment methods that account for differential exposure and proficiency across languages to avoid misidentification of reading difficulties (Cowan et al., 2023; Riva et al., 2021).

Educationally, interventions that focus on phonics, decoding practice, vocabulary enrichment, and reading strategies demonstrate the most substantial impact. Longitudinal evidence indicates that early identification and intervention mitigate cumulative academic difficulties, supporting the importance of screening programs in early elementary education (Kendeou et al., 2009; Papadopoulos et al., 2020). Importantly, assessment should include domain-specific tests and comprehensive evaluation of executive function, language, and working memory to tailor interventions effectively (Chalmpe & Vlachos, 2025; Papadopoulos et al., 2020).

Writing Difficulties

Writing difficulties arise from a combination of linguistic, cognitive, and motor processes, affecting spelling, sentence construction, and the organization of ideas. These challenges are not merely reflections of reading deficits but involve metalinguistic awareness, planning, and working memory (Zerva, 2023a, c; Zerva & Koundourou, 2023). Students with writing difficulties may struggle to translate ideas into coherent text, manage spelling conventions, or structure sentences logically, which impacts academic performance across subjects.

Research highlights that executive functions, including planning, sequencing, and self-monitoring, are essential for successful writing. Deficits in these functions often result in fragmented text production, inefficient use of cognitive resources, and increased task-related stress (Capodieci et al., 2023). Assessment of writing abilities should therefore include tasks that evaluate text composition, spelling accuracy, and organization, alongside cognitive measures such as working memory and processing speed (Zerva & Koundourou, 2023).

Effective educational strategies for writing difficulties include scaffolded instruction, graphic organizers, guided writing exercises, and assistive technologies. These strategies reduce cognitive load and allow students to focus on higher-level linguistic processes. Early intervention is critical, as delays in writing can lead to frustration, low motivation, and avoidance behaviors that negatively affect broader academic performance (King, 2022; Zerva, 2023a). Teachers' awareness and training in accommodating neurodiverse learners are also crucial, particularly in multilingual or inclusive classroom settings (Cowan et al., 2023).

Mathematical Difficulties

Mathematical difficulties manifest in number sense, calculation, problem-solving, and mathematical reasoning. Neurocognitive studies link these difficulties to deficits in working memory, visuospatial processing, and processing speed (Papadatou-Pastou et al., 2021; Lievore et al., 2024). Some students also experience mathematics-related anxiety, which exacerbates performance difficulties and reduces engagement (Lievore et al., 2024). Brain lateralization and handedness may play a subtle role in shaping numerical cognition, illustrating the complexity of neurodevelopmental contributions to mathematical learning (Papadatou-Pastou et al., 2021).

Mathematical difficulties often co-occur with reading and writing challenges, reflecting shared executive function deficits such as working memory limitations, planning difficulties, and cognitive flexibility (Crisci et al., 2021; Bonti et al., 2024). This overlap underscores the need for comprehensive, cross-domain assessment, rather than isolated skill evaluations. Evidence-based interventions include explicit instruction in number concepts, structured problem-solving training, and digital adaptive tools to enhance engagement and scaffold learning (Alzahrani & Algahtani, 2025). Teachers should provide contextualized and stepwise instruction to strengthen conceptual understanding and procedural fluency.

Assessment and Intervention Methods

Assessment of SLDs requires a multidimensional approach integrating cognitive, linguistic, and academic evaluations. Standardized tests for reading, writing, and mathematics should be complemented by executive function, working memory, and attention assessments (McDonough et al., 2017; Capodieci et al., 2023). Observational and functional assessments within classroom and home contexts provide critical insights into real-world functioning (Rutherford et al., 2021).

Bilingual students or those with diverse linguistic backgrounds require culturally fair assessment tools that evaluate both languages and account for differential exposure (Cowan et al., 2023; Riva et al., 2021). Longitudinal monitoring helps distinguish developmental lags from persistent SLDs, allowing for more targeted interventions over time (Papadopoulos et al., 2021).

Recent advances in technology, including artificial intelligence and teleassessment platforms, have enabled more precise profiling of neurocognitive strengths and weaknesses, facilitating personalized intervention planning (Ali et al., 2025; Capodieci et al., 2023). These tools can detect subtle patterns in reading fluency, spelling accuracy, and problem-solving performance that may be difficult to identify through traditional methods alone.

Intervention approaches should be evidence-based, individualized, and multimodal, targeting the cognitive and academic domains most affected. In reading, structured phonics programs, vocabulary enrichment, and comprehension strategies are essential. In writing, scaffolding, graphic organizers, and assistive technologies support the planning, organization, and revision processes. In mathematics, explicit instruction, adaptive tools, and anxiety reduction strategies facilitate engagement and skill acquisition (Alzahrani & Algahtani, 2025; King, 2022).

Executive function training, working memory exercises, and metacognitive strategy instruction enhance domain-general skills and support cross-domain learning (Crisci et al., 2021, Zerva, 2023a). Interdisciplinary collaboration between educators, psychologists, and speech-language specialists is critical for creating individualized education plans that address both academic and psychosocial needs (Rutherford et al., 2021; King, 2022).

Psychosocial Implications

SLDs affect more than academic achievement. Students with persistent learning challenges frequently experience low self-esteem, anxiety, and peer difficulties, which can influence motivation and long-term engagement with learning (Bonti et al., 2024; King, 2022). Early identification and supportive interventions can mitigate these effects, emphasizing the importance of integrating social-emotional learning with academic remediation. Promoting resilience, self-efficacy, and participation in classroom activities enhances both learning outcomes and psychosocial development (Zerva, 2023b).

Specific Learning Disorders reflect a complex neurodevelopmental profile affecting reading, writing, and mathematics. Understanding SLDs requires integrating cognitive, linguistic, and environmental perspectives, implementing multidimensional assessment, and delivering targeted, evidence-based interventions. Early identification, individualized support, and attention to executive function and psychosocial development are essential for enabling academic success, adaptive functioning, and long-term well-being. By considering both domain-specific difficulties and cross-domain cognitive skills, educators and clinicians can foster inclusive learning environments and support the developmental potential of every student (Ali et al., 2025; Astle et al., 2022; Zerva, 2023c).

Integrating Neurobiological and Genetic Perspectives

Specific Learning Disorders (SLDs) represent a complex category of neurodevelopmental conditions arising from the dynamic interplay of biological, genetic, cognitive, and environmental factors. They cannot be explained by low intelligence, insufficient effort, or inadequate instruction (Snowling & Hulme, 2025). Instead, SLDs reflect distinct differences in brain organization and cognitive processing that specifically affect skills such as reading, writing, and mathematics. These differences manifest as persistent academic difficulties that are often evident despite age-appropriate educational opportunities. Environmental and educational contexts play a crucial role in either exacerbating or mitigating these challenges, highlighting the need for learning environments that are responsive to each child's unique cognitive profile. Recognizing SLDs as multifactorial phenomena is essential for challenging reductive stereotypes of "laziness" or "inadequacy," and for emphasizing that academic challenges are not indicators of overall intellectual ability.

Neurobiological Underpinnings

Research consistently identifies both structural and functional differences in the brains of individuals with SLDs. Deviations are particularly pronounced in the left hemisphere, especially within the temporoparietal and occipitotemporal regions, which are critical for phonological decoding, visual-word recognition, and the integration of auditory and visual information during reading (Grigorenko et al., 2020; Peters & Ansari, 2019). Functional imaging studies reveal that children with SLDs often exhibit reduced activation in these integrative regions while simultaneously showing hyperactivation in the prefrontal cortex and other regions associated with compensatory cognitive effort. This neural imbalance contributes to characteristic difficulties in reading fluency, spelling, and arithmetic, as children expend more effort to achieve tasks that typically develop automatically in their peers (Grigorenko et al., 2020; Sathyanesan et al., 2019).

At the neurochemical level, variations in synaptic plasticity, myelination, and neurotransmitter systems, including gamma-aminobutyric acid (GABA) and glutamate, affect the efficiency with which the brain processes and consolidates new information (Atasever et al., 2025). Early childhood is a particularly sensitive period for these neural mechanisms, during which the quality and nature of environmental inputs can profoundly influence learning trajectories. These findings collectively reinforce that SLDs are rooted in inherent differences in neural structure and function rather than in deficits of effort or intelligence (Grigorenko et al., 2020; Peters & Ansari, 2019). Interdisciplinary assessments that combine neuroimaging, neuropsychological testing, and psychoeducational evaluation are therefore crucial for accurately identifying SLDs and tailoring individualized interventions (Espinas et al., 2025).

Genetic Contributions

The genetic architecture of SLDs is polygenic, with multiple loci influencing cognitive skills such as phonological processing, reading fluency, and arithmetic ability (Perugini et al., 2024; Georgitsi et al., 2021). Genome-wide association studies (GWAS) have identified numerous variants contributing to susceptibility, complementing earlier linkage and candidate-gene approaches. These genetic differences can shape brain structure, connectivity, and synaptic function, ultimately influencing learning outcomes (Georgitsi et al., 2021). Twin studies suggest that heritability estimates for SLDs range between 50% and 70%, and there is considerable genetic overlap with other neurodevelopmental conditions such as ADHD and developmental coordination disorder, indicating the existence of shared biological pathways (Gialluisi et al., 2021).

Specific genetic variants affecting processes such as Rapid Automatized Naming (RAN) and phonological processing have been linked to genes including MIR924HG, NKAIN3, MRPL19, and KIAA0319L (Müller et al., 2017; Truong et al., 2019). Polygenic risk scores indicate that these inherited factors account for a notable portion of variance in reading ability and can support early identification of children at risk (Müller et al., 2017).

Additionally, genes involved in synaptic plasticity and neuronal maturation, such as MeCP2 and PTEN, demonstrate differential expression patterns in individuals with SLDs, providing insight into the molecular mechanisms underlying learning challenges (Atasever et al., 2025). By integrating neurobiological and genetic findings, researchers can map how inherited factors shape brain development and cognitive function, informing both diagnosis and intervention (Snowling & Hulme, 2025).

Educational Implications

Translating neurobiological and genetic insights into educational practice is essential for supporting children with SLDs effectively. Knowledge of specific cognitive deficits, such as impaired phonological processing or working memory limitations, can guide the design of targeted interventions, including structured literacy programs, scaffolded instruction, and adaptive teaching strategies. Understanding the neural and genetic bases of these difficulties also reinforces the importance of early identification and continuous monitoring, allowing educators to anticipate challenges and intervene proactively rather than reactively. Interventions that leverage a child's strengths while addressing specific weaknesses can improve engagement, motivation, and academic achievement, promoting a more equitable learning experience.

Moreover, integrating SLDs within the broader landscape of neurodevelopmental disorders highlights overlapping cognitive and neural mechanisms. Children with co-occurring ADHD may struggle with attention and executive function, compounding difficulties in reading or math. Similarly, children on the autism spectrum may experience social communication challenges that indirectly affect academic engagement. Recognizing these interactions supports a holistic approach to intervention, emphasizing individualized education plans that address multiple developmental domains simultaneously.

Moving Beyond Misconceptions

A humanized understanding of SLDs emphasizes that academic struggles do not reflect a child's overall intelligence or potential. Children with SLDs often demonstrate high levels of creativity, problem-solving ability, and resilience, particularly when learning strategies are tailored to their cognitive profiles. By focusing on strengths alongside challenges, educators can foster self-efficacy, reduce stigma, and cultivate a growth-oriented mindset. Early and consistent support not only mitigates academic difficulties but also promotes positive social-emotional development, reinforcing the child's sense of competence and agency.

Specific Learning Disorders emerge from the complex interaction of neurobiological, genetic, and environmental factors. Structural, functional, and neurochemical differences in the brain, shaped in part by inherited variation, contribute to the characteristic cognitive profiles observed in affected children. Understanding these mechanisms allows for early identification, tailored interventions, and educational strategies that address both unique and shared developmental pathways. Integrating neurobiological and genetic perspectives with educational practice provides a foundation for individualized support, fostering improved learning outcomes and empowering children with SLDs to achieve their full potential.

2. ATTENTION-DEFICIT/HYPERACTIVITY DISORDER

Understanding and Managing ADHD in Schools

Attention-Deficit/Hyperactivity Disorder (ADHD) is one of the most prevalent neurodevelopmental disorders affecting children and adolescents, with profound implications for academic, social, and behavioral functioning, characterized by persistent patterns of inattention, hyperactivity, and impulsivity, ADHD frequently affects multiple domains of life, including school performance, peer relationships, and family interactions (Bonti et al., 2024). These symptoms often persist into adolescence and adulthood, contributing to long-term difficulties in occupational, interpersonal, and adaptive functioning (APA, 2013; Bonti et al., 2024).

High rates of comorbidity are common among children with ADHD, with co-occurring conditions including mood disorders, conduct disorders, anxiety, learning disabilities, and developmental delays (APA, 2013; Bonti et al., 2024; Cuffe et al., 2020). Evidence suggests a clinical continuum linking ADHD to antisocial, deviant, and delinquent behaviors from childhood through adulthood, highlighting the disorder's broader psychosocial implications (Kuja-Halkola et al., 2015; Moffitt, 2018; Retz et al., 2021).

Within educational settings, ADHD presents unique challenges. Classroom disruptions, difficulties with attention and self-regulation, and repeated academic underachievement are common. Schools often encounter crises stemming from impulsive or oppositional behaviors, which can escalate if interventions are delayed or not informed by a clear understanding of ADHD's neurodevelopmental basis. Effective school management requires proactive strategies that focus on prevention, skill development, and resilience-building, rather than reactive suppression of disruptive behaviors.

Neurobiological Underpinnings of ADHD

ADHD has a well-established neurobiological basis, involving structural and functional variations in brain networks responsible for attention, executive function, and behavioral inhibition (Bonti et al., 2024; Nigg et al., 2020). Neuroimaging studies consistently identify anomalies in the prefrontal cortex, basal ganglia, and cerebellum, regions that contribute to executive functioning, motor regulation, and reward processing (Kofler et al., 2019; Wolfers et al., 2020). Dysfunction in dopaminergic and noradrenergic neurotransmitter systems further contributes to deficits in attentional control, behavioral regulation, and reward-based decision-making. These neurobiological characteristics elucidate why core ADHD symptoms, hyperactivity, impulsivity, and inattention, are not simply issues of motivation or discipline but reflect underlying neural differences.

Genetic research further reinforces the biological basis of ADHD. Heritability estimates range between 60% and 90%, reflecting substantial contributions from both rare copy number variants (CNVs) and common gene polymorphisms (Chen et al., 2017; Faraone & Larsson, 2019; Thapar, 2020). Specific genes implicated in synaptic development, neurotransmitter regulation, and neural connectivity contribute to individual differences in ADHD symptomatology (Brikell et al., 2021; da Silva et al., 2023; Giannoulis et al., 2024). Environmental factors, including prenatal exposure to toxins, maternal lifestyle, dietary influences, and air pollution, interact with genetic predispositions to shape symptom expression and developmental trajectories (Larsson et al., 2013; Mirkovic et al., 2020; Regan et al., 2022). Understanding the neurobiological and genetic mechanisms of ADHD provides a foundation for early identification and intervention, emphasizing that the disorder is inherently neurodevelopmental rather than the result of parenting deficiencies or poor discipline.

Developmental Trajectory and Clinical Continuum

ADHD typically emerges between ages three and seven, coinciding with the onset of formal schooling and increasing demands on attention, behavioral regulation, and rule compliance. The disorder follows a developmental continuum, with symptoms manifesting differently across life stages. In infancy, ADHD may present as excessive crying, restlessness, feeding difficulties, and short sleep duration. During preschool years, hyperactivity, impulsivity, and noncompliance become increasingly apparent. School-aged children often struggle with attention, impulse control, and academic tasks, sometimes accompanied by oppositional behavior. Adolescents frequently experience a reduction in hyperactivity but an increase in risk-taking and conflictual behaviors, including delinquency. In adulthood, hyperactivity often diminishes, while inattention and impulsivity may persist, albeit typically at reduced intensity (Bonti et al., 2024; Cherkasova et al., 2013; Moffitt, 2018).

Cognitive deficits, including impairments in working memory, inhibitory control, planning, and self-monitoring, interact with motivational challenges to create barriers to academic success (Bonti et al., 2024; Mouzaki et al., 2007; Zerva, 2023). Without timely intervention, these deficits can foster frustration, repeated academic failure, and diminished self-esteem, compounding behavioral difficulties and social challenges.

Comorbidity and Behavioral Risk

Comorbidity is highly prevalent in ADHD, with approximately 65% of individuals presenting with one or more co-occurring psychiatric or developmental disorders, including learning disabilities, oppositional defiant disorder, conduct disorder, mood disorders, sleep and tic disorders, and anxiety (APA, 2013; Bonti et al., 2024; Cuffe et al., 2020; Gnanavel et al., 2019). Comorbid conditions exacerbate functional impairments, impacting executive functioning, emotional regulation, and social adaptation, while increasing risks for academic underachievement, peer rejection, and long-term adverse outcomes (Berchiatti et al., 2022; Bőthe et al., 2019; Zerva, 2023c).

Research indicates that ADHD is a significant risk factor for antisocial and delinquent behavior. Approximately 41% of children with ADHD report involvement in criminal activities, including a subset engaged in violent acts (Bonti et al., 2024). Persistent impulsivity, observed in over 50% of adults with ADHD, contributes to rule-breaking and maladaptive social behavior (Faraone & Larsson, 2019; Franke et al., 2018; Retz et al., 2021). These findings underscore the importance of early and effective interventions that target both core ADHD symptoms and secondary behavioral risks, thereby reducing the likelihood of long-term adverse outcomes.

School-Based Management and Intervention

Schools are central to supporting children with ADHD, both academically and behaviorally. Effective management requires early identification, structured behavioral interventions, and, when appropriate, pharmacological treatment. Behavioral classroom management, psychoeducation, and cognitive-behavioral therapy have demonstrated efficacy, particularly when combined with medication in cases of severe symptoms (National Institute of Health and Care Excellence [NICE], 2018; Schatz et al., 2021; Young et al., 2015). Relying solely on suppressive measures such as exclusion or punishment is counterproductive, often exacerbating disruptive behavior and increasing delinquency risk (Ramey & Freelin, 2023).

Educators play a pivotal role in recognizing ADHD symptoms, understanding their neurodevelopmental basis, and implementing proactive strategies. These include classroom accommodations, individualized instruction, reinforcement of self-regulation skills, and social-emotional learning strategies. Collaborative engagement with parents, mental health professionals, and community stakeholders is essential to provide consistent support across school and home contexts (Cuba Bustinza et al., 2022). Preventive and skill-building approaches, rather than reactive suppression, are most effective in mitigating disruptive behaviors and fostering positive academic and social outcomes.

ADHD is a complex, multifactorial neurodevelopmental disorder characterized by persistent inattention, hyperactivity, and impulsivity. Its neurobiological and genetic underpinnings interact with environmental factors to shape behavioral and cognitive profiles, which are further complicated by frequent comorbidity with other psychiatric or developmental conditions. Recognition of ADHD's developmental trajectory and clinical continuum, combined with evidence-based, school-centered interventions, is essential for supporting children's academic, social, and behavioral functioning. By addressing both core symptoms and associated behavioral risks, educators and mental health professionals can foster resilience, enhance engagement, and reduce long-term negative outcomes, enabling children with ADHD to reach their full potential.

3. AUTISM SPECTRUM DISORDER

Autism Spectrum Disorder (ASD) is a complex neurodevelopmental condition characterized by persistent deficits in social communication and interaction, alongside restricted, repetitive patterns of behavior, interests, or activities (Riesco et al., 2025). The term "spectrum" captures the remarkable heterogeneity of ASD, as individuals display varying degrees of functional abilities, cognitive skills, and behavioral presentations. While some individuals may exhibit profound challenges requiring intensive support, others demonstrate subtle social differences with preserved intellectual functioning. Across this diversity, however, certain core characteristics consistently emerge, shaping the developmental trajectories and daily experiences of those with ASD.

Social Communication and Interaction

One of the most salient features of ASD involves difficulties in social communication and interaction. Individuals with ASD often exhibit impairments in understanding social cues, including gestures, facial expressions, and tone of voice, which complicates reciprocal social engagement (Yeung, 2022). Facial emotion recognition, in particular, is frequently disrupted. Yeung's (2022) meta-analysis indicates that deficits in identifying emotions from facial expressions are prevalent among individuals with ASD, although the magnitude of impairment may vary depending on task complexity and methodological factors. These challenges can affect both verbal and nonverbal communication, leading to difficulties in initiating and maintaining conversations, understanding social norms, and forming peer relationships.

Beyond the recognition of emotions, individuals with ASD often demonstrate atypical social behaviors that may appear aloof, rigid, or socially unconventional.

They may struggle to respond appropriately to social contingencies, such as offering comfort or participating in shared attention activities, which can result in misunderstandings and social isolation (Riesco et al., 2025). These social deficits are not attributable to a lack of desire for connection but rather reflect neurodevelopmental differences in processing and interpreting social information.

In addition to social communication difficulties, ASD is defined by restricted and repetitive behaviors, which may manifest as motor stereotypies, insistence on sameness, or highly focused interests (Cummings et al., 2022). Such behaviors are not merely eccentricities; they often serve regulatory or adaptive functions. For instance, repetitive movements or routines can provide a sense of predictability and reduce anxiety in environments perceived as overwhelming. Similarly, intense preoccupations with specific topics allow for the development of deep expertise, though they may limit broader engagement with diverse experiences.

The prevalence and severity of restricted behaviors can differ depending on genetic and neurodevelopmental profiles. Cummings et al. (2022) highlighted that individuals with PTEN mutations, a known genetic contributor to ASD, exhibit high rates of autism-related characteristics, including repetitive behaviors, underscoring the interplay between genetic factors and behavioral phenotypes. Understanding these patterns is crucial for tailoring interventions and support strategies that respect individual preferences while promoting functional flexibility.

Cognitive and Learning Profiles

Cognitive profiles in ASD are heterogeneous, ranging from intellectual disability to above-average intelligence. Many individuals display uneven abilities, excelling in specific domains such as visual-spatial reasoning or rote memory while struggling with executive functioning, planning, and adaptive skills (Blázquez et al., 2025). These cognitive differences interact with behavioral and social characteristics to shape learning experiences. Chen and Jahromi (2025) emphasized the critical role of self-regulation in preschoolers with ASD, demonstrating that difficulties in regulating attention, emotion, and behavior can affect school engagement and academic outcomes. Effective educational strategies must therefore consider both cognitive strengths and regulatory challenges to optimize learning and participation.

Sensory Processing and Emotional Regulation

Sensory processing differences are a frequent feature of ASD, contributing to both behavioral and emotional patterns. Many individuals demonstrate hyper- or hyposensitivity to sensory stimuli, such as light, sound, touch, or taste, which can influence their comfort, attention, and participation in daily activities. Sensory sensitivities are often intertwined with emotional regulation, as overstimulation can trigger anxiety or meltdowns, whereas under-responsiveness may result in seeking intense sensory input (Hustyi et al., 2025). These experiences underscore the importance of individualized support approaches that accommodate sensory profiles while promoting adaptive coping strategies.

Emotional regulation deficits can further exacerbate social and behavioral challenges. Hustyi et al. (2025) reviewed interventions targeting skill-based treatments for challenging behaviors, demonstrating that structured programs emphasizing self-management, functional communication, and reinforcement of adaptive behaviors can improve outcomes. These interventions are most effective when they address underlying regulatory difficulties, rather than focusing solely on overt behaviors, highlighting the intertwined nature of emotional and behavioral characteristics in ASD.

Genetic and Biological Contributions

The diversity of ASD characteristics is partially explained by genetic and biological factors. Advances in genetic sequencing have identified numerous variants associated with ASD, including PTEN mutations and other rare exonic changes, alterations in the DNA sequence within exons (Blázquez et al., 2025). These genetic contributors not only influence susceptibility to ASD but also shape specific behavioral and cognitive profiles, including social deficits, repetitive behaviors, and intellectual variability. Cummings et al. (2022) emphasize that understanding these genetic influences allows for more precise phenotypic characterization and the potential for personalized intervention strategies.

Additionally, neurobiological differences in brain structure and connectivity are thought to underlie many hallmark ASD characteristics. Aberrant connectivity patterns may affect social information processing, executive function, and sensory integration, thereby influencing observable behaviors. While genetic and neurobiological factors provide critical insights, it is important to recognize the role of environmental experiences in shaping the manifestation of ASD traits across development (Riesco et al., 2025).

Early Identification and Developmental Considerations

Early identification of ASD is essential for supporting optimal development. Developmental differences often emerge in infancy or early childhood, including delayed social engagement, atypical gaze patterns, and limited communicative gestures (Riesco et al., 2025). Early intervention programs that target social communication, adaptive skills, and emotional regulation have been shown to improve developmental trajectories, emphasizing the importance of timely recognition and individualized support.

Preschoolers with ASD, in particular, benefit from interventions that promote self-regulation and school engagement. Chen and Jahromi (2025) found that enhancing self-regulatory capacities not only supports learning but also mitigates the impact of core autism characteristics on academic participation. These findings underscore that early tailored approaches can enhance both functional outcomes and quality of life.

Conclusion

Autism Spectrum Disorder is a multifaceted condition defined by core deficits in social communication, repetitive behaviors, and unique cognitive and sensory profiles. Its heterogeneity is shaped by genetic, neurobiological, and environmental factors, resulting in highly individualized patterns of strengths and challenges. Research highlights that while deficits in social reciprocity, facial emotion recognition, and self-regulation are common, interventions tailored to cognitive, behavioral, and sensory profiles can significantly improve functional outcomes (Blázquez et al., 2025; Chen & Jahromi, 2025; Cummings et al., 2022; Hustyi et al., 2025; Riesco et al., 2025; Yeung, 2022). Understanding the general characteristics of ASD is critical for clinicians, educators, and caregivers, as it informs assessment, intervention, and support strategies that respect individual differences while promoting engagement, learning, and well-being.

LINGUISTIC PROFILES IN CHILDREN WITH AUTISM SPECTRUM DISORDER (ASD)

Autism Spectrum Disorder (ASD) is characterized by a heterogeneous neurodevelopmental profile, encompassing variations in social communication, behavior, and cognitive abilities. Among the most complex and variable features of ASD are linguistic and reading skills, which show significant individual differences across children (Åsberg Johnels et al., 2022; Vogindroukas et al., 2022). While some children demonstrate exceptional verbal abilities, including creativity and advanced vocabulary, others experience substantial deficits, with some failing to develop functional speech entirely. These differences highlight the need for a nuanced understanding of language and literacy within ASD, particularly when considering educational interventions and reading development.

Although deficits in language are not a core diagnostic criterion for ASD, difficulties in pragmatic language use, comprehension, and narrative construction are common (Andreou & Raxioni, 2022; Boucher, 2012, as cited in Andreou & Raxioni, 2022). Impaired oral language comprehension, even in children with relatively intact syntax and vocabulary, can constrain reading comprehension and broader academic achievement (Roch et al., 2021). Consequently, literacy acquisition in children with ASD requires careful attention to both underlying language abilities and cognitive processing strategies.

Literacy Skills and Reading Profiles

Literacy in children with ASD depends on many of the same subskills as in typically developing peers, including decoding, fluency, and comprehension (Åsberg Johnels et al., 2022). Nevertheless, deficits across linguistic domains, phonology, vocabulary, syntax, and pragmatics, can limit reading development. Specifically, difficulties in phonological decoding, compounded by impaired oral language comprehension, often contribute to low reading comprehension (Andreou & Raxioni, 2022; Henderson et al., 2014).

Reading profiles in ASD are heterogeneous and may evolve over time. Longitudinal studies indicate that reading comprehension is the most malleable aspect of literacy, whereas decoding and word recognition tend to remain more stable (Solari et al., 2019). Liu et al. (2023) identified distinct reading profiles among autistic children: (a) strong decoders with good listening comprehension, (b) proficient decoders with poor listening comprehension, (c) average decoders and listeners, and (d) poor decoders and poor listeners. Notably, over 50% of children with ASD show uneven profiles, excelling in some reading domains while struggling in others, emphasizing the need for individualized literacy support.

Phonology and Vocabulary

Phonological awareness and vocabulary acquisition are frequently affected in children with ASD, particularly when comorbid developmental disorders are present. Mild, isolated phonological deficits may impact written language acquisition, and broader difficulties with syllabification and phonemic awareness can limit reading fluency (Conner et al., 2022; Henderson et al., 2014; Ricketts et al., 2013; Vogindroukas et al., 2022). Early vocabulary delays are also documented, with children with ASD often demonstrating reduced lexical richness compared to typically developing peers (Papantoniou & Kambourologou, n.d.). These deficits intersect with higher-order language difficulties, such as forming complex sentences and using morphology effectively, which further constrain reading development (Kjelgaard & Tager-Flusberg, 2001; Smith Gabig, 2010; Walenski et al., 2006).

Syntax, Morphology, and Semantics

Reading comprehension in ASD is frequently affected by morphosyntactic deficits, including challenges with verb usage, inflection, and sentence construction (Tek et al., 2014). Morphological and orthographic awareness play critical roles in reading acquisition, and deficits in these domains contribute to slower literacy progress (Liu et al., 2023). Some children with ASD demonstrate delays in semantic development, affecting vocabulary depth and conceptual knowledge (Gilhuber et al., 2023; Kamio et al., 2007; Kelley et al., 2006). Visual imagery-based strategies, which map words onto mental representations, have been shown to support semantic integration and reading comprehension, particularly in children with deficits in oral language (Bednarz et al., 2017; Murdaugh et al., 2017).

Pragmatics and Figurative Language

Pragmatic deficits, including difficulties with conversational turn-taking, over-focus on irrelevant details, and impaired figurative language comprehension, further constrain literacy and reading development (Kelley et al., 2006). Children with ASD may struggle to interpret ambiguous or non-literal sentences, reducing the effectiveness of narrative reading and inferential tasks (Davidson & Ellis Weismer, 2017). Despite these challenges, many children display an early interest in letters and print, sometimes developing advanced word recognition skills, a phenomenon known as "hyperlexia." Although hyperlexia occurs in only a small proportion of autistic children (5–10%), it highlights the potential for advanced decoding abilities even in the presence of broader cognitive or linguistic deficits (Grigorenko et al., 2003; Newman et al., 2007, as cited in Andreou & Raxioni, 2022).

Reading Process and Visual Processing

Children with ASD often exhibit unique reading processes influenced by visual and attentional strategies. Eye-tracking research indicates that they spend more time correcting spelling errors, revisiting previously read text, and resolving semantic ambiguities compared to typically developing peers (Sansosti et al., 2013). Reading performance improves significantly when supported by adults, particularly through systematic teaching methods and software interventions that leverage visual stimuli (Benitez & Domeniconi, 2023; Khoirunnisa & Dewi, 2023). Children with ASD often have superior visual processing abilities, allowing them to benefit from educational approaches that emphasize visual learning and structured visual input (Singh et al., 2021). Environmental factors, such as background noise, appear to have minimal impact on reading performance relative to peers without ASD, suggesting resilience in sensory-mediated reading processes (McClain et al., 2023).

Mathematical Skills in ASD

Mathematical skills in children with Autism Spectrum Disorder (ASD) present a unique profile shaped by both cognitive strengths and challenges inherent to the condition. While ASD is primarily characterized by social communication deficits and repetitive behaviors, these core characteristics interact with learning processes and cognitive abilities, influencing academic outcomes, including numeracy (Chen & Jahromi, 2025). Research indicates that many children with ASD demonstrate particular strengths in systemizing, pattern recognition, and attention to detail, which can facilitate understanding of certain mathematical concepts, especially those involving visual-spatial reasoning or rule-based problem solving (Blázquez et al., 2025). These abilities often allow some children on the spectrum to excel in tasks requiring precise computation, logical sequences, or memorization of mathematical procedures. However, the presence of difficulties in self-regulation, executive function, and flexible thinking can simultaneously pose challenges, particularly in tasks that demand multi-step reasoning, abstract problem solving, or adaptation to novel mathematical contexts (Chen & Jahromi, 2025; Hustyi et al., 2025). For instance, a child with strong rote memorization skills may accurately calculate sums yet struggle to apply the same principles to word problems or unfamiliar scenarios, highlighting the interaction between ASD characteristics and learning outcomes.

Self-regulation, a critical factor in early academic engagement, plays a central role in shaping mathematical performance. Chen and Jahromi (2025) emphasize that preschoolers with stronger self-regulatory abilities exhibit higher engagement in structured learning environments, enabling sustained attention to mathematical tasks and effective management of frustration when encountering challenging problems. Conversely, deficits in regulating attention or emotion can interfere with task persistence, leading to incomplete work or avoidance of mathematically demanding activities. This dynamic illustrates that mathematical skill development in children with ASD is not solely dependent on cognitive aptitude but is deeply intertwined with behavioral and emotional regulation capacities, which may be shaped by both neurodevelopmental and environmental factors.

Genetic and neurodevelopmental influences further contribute to variability in mathematical abilities among children with ASD. Blázquez et al. (2025) note that specific genetic profiles, including rare variants, are associated with distinct cognitive and behavioral patterns, potentially affecting problem-solving strategies and attentional focus during learning. Similarly, Cummings et al. (2022) highlight that children with PTEN mutations often display strong attention to detail and repetitive engagement with preferred activities, which can translate into intensive focus on numerical patterns or mathematical rules, though these tendencies may also manifest as rigidity, limiting flexibility in applying concepts across contexts. These findings underscore that mathematical learning in ASD is best understood through a nuanced lens that accounts for both inherent cognitive capacities and behavioral tendencies characteristic of the disorder.

Interventions tailored to the needs of children with ASD have demonstrated efficacy in enhancing mathematical skills, particularly when they integrate structured learning, visual supports, and reinforcement strategies targeting both cognitive and self-regulatory capacities (Hustyi et al., 2025). By leveraging individual strengths, such as systemizing and pattern recognition, while providing scaffolding to address challenges in executive functioning and flexibility, educators and caregivers can create supportive environments that foster both competence and confidence in mathematics. Additionally, attention to the child's social and emotional context is essential, as engagement and motivation are closely linked to persistence and performance in numeracy tasks. Therefore, mathematical skill development in children with ASD is best conceptualized as the outcome of complex interactions among cognitive strengths, self-regulatory abilities, behavioral patterns, and tailored educational supports, reflecting the broader heterogeneity of the autism spectrum (Riesco et al., 2025).

Cognitive Mechanisms and Theory of Mind

Reading comprehension involves multiple cognitive mechanisms, including central coherence, executive function, and Theory of Mind. Children with ASD frequently exhibit weak central coherence, difficulties with working memory, and challenges in higher-order cognitive processes, which partially explain observed reading comprehension deficits (Lee et al., 2023; Randi et al., 2010). While word recognition and basic decoding skills are often intact, children with ASD demonstrate specific difficulties in interpreting inferential or non-literal text, reflecting Theory of Mind limitations (Lee et al., 2023; Tong et al., 2020).

Moreover, literacy difficulties often co-occur with challenges in mathematics, particularly when executive functioning and working memory deficits are present (Bullen et al., 2022). Cognitive mediation within Theory of Mind appears critical for understanding non-literal language, while literal comprehension remains relatively preserved. Importantly, some pragmatic language abilities remain intact, suggesting that reading interventions can capitalize on preserved skills while addressing specific deficits (Marocchini, 2023; Tong et al., 2020).

The linguistic profiles of children with ASD are highly heterogeneous, reflecting complex interactions between neurodevelopmental, cognitive, and environmental factors. While some children demonstrate exceptional decoding abilities or creativity in language, many experience deficits across phonology, syntax, semantics, morphology, and pragmatics, which constrain reading comprehension and literacy development. Educational strategies must be individualized, leveraging visual strengths, structured teaching, and targeted interventions to address both preserved and impaired skills. By understanding the multifaceted linguistic and cognitive profiles of children with ASD, educators and clinicians can promote effective literacy acquisition, enhance comprehension, and support broader academic and social development.

4.COMMUNICATION DISORDERS

Communication is a complex and dynamic human process that integrates linguistic, cognitive, affective, and social functions. It enables individuals to exchange information, express inner thoughts and emotions, and build relationships within their environment. As defined in linguistic and neurocognitive terms, language represents a symbolic system of sounds and structures governed by rules that allow for the creation and interpretation of meaning. Speech, in turn, constitutes the motor mechanism by which linguistic symbols are realized audibly, while communication encompasses both verbal and non-verbal forms of interaction, including gestures, facial expressions, prosody, and contextual cues.

From a semiotic perspective, communication extends beyond the verbal domain. As Umberto Eco argued, all cultural and symbolic systems, from traffic signs to theatrical gestures, constitute semiotic codes that mediate meaning and social understanding. Therefore, communication is not merely a linguistic phenomenon but a multimodal, socially embedded activity dependent on shared symbols, interpretive skills, and contextual knowledge. This socio-cultural dimension was emphasized by Bernstein's theory of linguistic codes and Bourdieu's concept of *cultural capital*, which highlight how environmental and experiential factors shape communicative competence and access to linguistic resources.

Language development unfolds as an ongoing, interactive process shaped by biological maturation, cognitive growth, and environmental exposure. Typically developing children acquire speech and language skills in a predictable sequence without explicit instruction. Successful language acquisition requires the integration of *form* (phonology, morphology, syntax), *content* (semantics), and *use* (pragmatics), which together enable effective expression and comprehension. Disruptions in any of these components can lead to communication disorders that affect not only linguistic performance but also learning, emotional regulation, and social participation (Bishop et al., 2016; Bruinsma, Wijnen, & Gerrits, 2024).

Neurodevelopmentally, language is supported by a distributed network involving perisylvian regions such as Broca's and Wernicke's areas, as well as associative cortical and subcortical structures involved in auditory processing, working memory, and executive function. The left hemisphere is typically dominant for linguistic processing, but right hemispheric and cerebellar regions also contribute to prosody and pragmatic inference. Early neuropsychological studies, such as those of Broca (1861) and Wernicke (1876), demonstrated the anatomical specialization of language, while contemporary neuroimaging has confirmed that communication is underpinned by dynamic, plastic networks rather than discrete centers (Neef & Chang, 2024).

Crucially, communication disorders (CDs) are not isolated impairments of speech or language, but multifactorial conditions reflecting atypical neurodevelopmental trajectories. They can interfere with educational attainment, emotional well-being, and interpersonal functioning. As highlighted in recent research (Bertollo et al., 2025), neurodevelopmental disorders, including communication disorders, share overlapping neurobiological pathways with mood and cognitive dysregulation, suggesting a continuum rather than discrete categorical boundaries. Early identification and targeted intervention are therefore essential to mitigate cascading effects on literacy, academic performance, and psychosocial outcomes (Burnley et al., 2024).

Communication Disorders within the DSM-5 Framework

The *Diagnostic and Statistical Manual of Mental Disorders, Fifth Edition* (DSM-5; APA, 2013) reclassified communication-related difficulties under the broader umbrella of *Neurodevelopmental Disorders* (NDDs), acknowledging their early onset, neurobiological etiology, and developmental course. This represents a conceptual shift from earlier DSM editions, where "Speech and Language Disorders" were treated as separate or peripheral clinical entities. The DSM-5 (APA, 2013) framework now includes the following categories: 1) *Language Disorder (315.32)*, 2) *Speech Sound Disorder (315.39)*, 3) *Childhood-Onset Fluency Disorder (Stuttering) (315.35)*, 4) *Social (Pragmatic) Communication Disorder (315.39)*, and 5) *Unspecified Communication Disorder* (307.9). This classification reflects a developmental continuum, with distinctions made according to the linguistic, articulatory, fluency, or pragmatic dimensions affected. Importantly, the DSM-5 conceptualization situates these disorders within the same etiological domain as Autism Spectrum Disorder (ASD), Attention-Deficit/Hyperactivity Disorder (ADHD), and Specific Learning Disorder (SLD), acknowledging their frequent comorbidity and shared neurocognitive mechanisms (Bishop et al., 2017; Jain et al., 2025).

Developmental *Language Disorder*, previously termed *Specific Language Impairment (SLI)*, involves persistent deficits in comprehension and/or production of spoken language that cannot be attributed to sensory, motor, or intellectual impairment. The shift in terminology, advocated by the CATALISE Consortium (Bishop et al., 2016; 2017), emphasizes that these difficulties reflect a broader developmental language disorder (DLD) rather than a discrete or "specific" impairment.

Speech Sound Disorder (SSD) encompasses difficulties in articulating or producing phonemes appropriately for one's developmental level. While often transient in preschool years, persistent SSD may indicate underlying deficits in phonological processing or motor planning.

Childhood-Onset Fluency Disorder (COFD), or stuttering, involves disturbances in the normal fluency and time patterning of speech, typically emerging between ages two and seven. Recent research identifies stuttering as a disorder of language encoding rather than merely motor output (Bernstein Ratner & Brundage, 2024). Neuroimaging findings point to atypical connectivity in cortico-basal ganglia-thalamocortical circuits involved in timing and speech initiation (Neef & Chang, 2024).

Social (Pragmatic) Communication Disorder (SCD), a new diagnostic entity introduced in DSM-5, captures individuals who exhibit significant pragmatic deficits, such as difficulties using language for social purposes, following conversational rules, or interpreting non-literal expressions, without the repetitive behaviors' characteristic of ASD (Lampri et al., 2024; Jain et al., 2025).

Lastly, *Unspecified Communication Disorder* serves as a residual category when symptoms cause significant impairment but do not meet full criteria for any specific subtype.

From a clinical standpoint, the DSM-5 conceptualization underscores the *neurodevelopmental nature* of communication disorders: early onset, cross-domain impact, and often persistent course into adolescence and adulthood. Comorbidity is the rule rather than the exception, with many children exhibiting overlapping features of DLD, fluency difficulties, or pragmatic impairments, sometimes alongside ASD or ADHD. These complexities necessitate multidimensional assessment and individualized intervention planning (Bertollo et al., 2025; Bruinsma et al., 2024). Furthermore, emerging evidence highlights the psychosocial dimension of communication disorders. Qualitative work with children with DLD, for example, reveals long-term consequences on self-esteem, social relationships, and identity formation (Burnley et al., 2024). The DSM-5 model, by embedding CDs within the neurodevelopmental framework, aligns diagnostic practice with the growing understanding that language, cognition, and emotion are interdependent developmental systems.

Neurodevelopmental Underpinnings of Communication Disorders

Communication disorders (CDs) are fundamentally neurodevelopmental in origin, reflecting alterations in the maturation of brain systems involved in language, motor control, and social cognition. Neurodevelopmental disorders (NDDs) such as autism spectrum disorder (ASD), attention-deficit/hyperactivity disorder (ADHD), and developmental language disorder (DLD) frequently share overlapping neurobiological mechanisms. According to Bertollo et al. (2025), these disorders involve atypical neurobiological trajectories, including dysregulation of synaptic plasticity, neurotransmitter systems, and neural connectivity especially within cortico-striatal and fronto-temporal circuits.

During early development, genetic and environmental factors interact to shape the architecture of neural networks responsible for auditory processing, motor planning, and linguistic encoding. Studies have demonstrated that children with CDs often exhibit atypical activation patterns in perisylvian regions, as well as in subcortical and cerebellar structures implicated in timing and coordination (Neef & Chang, 2024). These findings support a distributed model of communication, in which functional integration among neural networks is crucial for fluent speech and pragmatic communication.

From a neurobiological perspective, language and communication rely on both domain-specific and domain-general processes. Domain-specific regions, such as Broca's and Wernicke's areas, support phonological and syntactic processing, whereas domain-general networks, such as the dorsal attention system and executive control networks, contribute to planning, inhibition, and cognitive flexibility. Altered connectivity among these systems may result in the linguistic, fluency, and pragmatic deficits observed in CDs (Bernstein Ratner & Brundage, 2024).

Genetic studies further underscore the heritability of communication disorders (Magielski et al., 2025). Family and twin studies have identified significant genetic contributions to language and fluency impairments, with several candidate genes implicated in neuronal migration and synaptic transmission. Nevertheless, gene–environment interactions play a critical role; adverse early environments, reduced linguistic input, and limited social interaction can exacerbate underlying vulnerabilities (Bishop et al., 2017; Newbury & Monaco, 2010).

Importantly, CDs are often comorbid with emotional and behavioral dysregulation. Bertollo et al. (2025) emphasize neurobiological overlaps between NDDs and mood disorders, suggesting shared disruptions in stress-regulatory and dopaminergic pathways. This intersection explains why children with DLD or stuttering frequently present with anxiety, social withdrawal, or depressive symptoms (Burnley et al.,

2024). A comprehensive understanding of CDs thus requires integrating neurobiological, cognitive, and psychosocial dimensions.

i) Language Disorder (Developmental Language Disorder)

Language Disorder (LD), previously referred to as Specific Language Impairment (SLI), is characterized by persistent difficulties in the acquisition and use of language across modalities, spoken, written, or gestural, due to deficits in comprehension or production (APA, 2013). The CATALISE consortium (Bishop et al., 2016; 2017) proposed the term *Developmental Language Disorder (DLD)* to replace SLI, emphasizing that linguistic deficits often co-occur with other developmental challenges rather than existing in isolation.

Children with DLD exhibit significant limitations in vocabulary, morphosyntax, and narrative organization, often accompanied by working memory and auditory processing deficits. Epidemiological studies estimate a prevalence of 6–8% among school-aged children, with higher incidence in boys (Bishop et al., 2017). DLD frequently persists into adolescence and adulthood, manifesting as difficulties with academic performance, literacy, and social communication (Bruinsma, Wijnen, & Gerrits, 2024).

Cognitive and Linguistic Characteristics

DLD encompasses both expressive and receptive deficits. Expressive impairments involve limited sentence complexity and word retrieval problems, while receptive difficulties affect comprehension of abstract and syntactically complex sentences. Children with DLD often struggle with phonological short-term memory, impacting vocabulary learning and reading development. These cognitive-linguistic challenges indicate impairments in the integration of form, content, and use, the three pillars of language identified in linguistic theory.

Neurobiological Basis

Neuroimaging studies reveal atypical asymmetry in perisylvian regions and altered white matter connectivity between frontal and temporal lobes. These findings point to disruptions in auditory-motor integration and temporal processing. Furthermore, electrophysiological studies show reduced mismatch negativity responses, suggesting impaired neural sensitivity to speech sound contrasts. Such neurofunctional abnormalities reinforce the conceptualization of DLD as a neurodevelopmental disorder rooted in brain network atypicalities (Bertollo et al., 2025).

Psychosocial Impact

Beyond linguistic impairments, DLD has profound social and emotional consequences. Burnley et al. (2024) conducted qualitative research showing that individuals with DLD often experience social exclusion, low self-esteem, and identity challenges stemming from early communicative difficulties. These psychosocial outcomes highlight the need for holistic interventions that address both linguistic and emotional domains.

Comorbidity and Intervention

DLD frequently co-occurs with reading disabilities, ADHD, and speech sound disorders. Intervention strategies include individualized speech-language therapy, parent-mediated communication training, and educational accommodations. Early intervention has been shown to significantly improve long-term academic outcomes (Bruinsma et al., 2024).

ii) Speech Sound Disorder (SSD)

Speech Sound Disorder (SSD) refers to difficulties with articulation and/or phonological representation of speech sounds that interfere with speech intelligibility (APA, 2013). SSD affects approximately 3–5% of children, with higher rates among preschoolers due to developmental phonological immaturity.

Linguistic and Motor Components

SSD may involve errors such as omissions, substitutions, or distortions of phonemes. These can result from deficits in articulatory planning, phonological processing, or both. SSD is often linked to atypical development of the phonological system rather than solely to motor dysfunction, distinguishing it from dysarthria or apraxia of speech.

Neurodevelopmental and Genetic Basis

Functional neuroimaging reveals reduced activation in the left inferior frontal gyrus and supplementary motor area during speech production tasks, suggesting immature or inefficient cortical motor planning. Moreover, SSD shares genetic overlap with DLD, with twin studies indicating substantial heritability (Bishop et al., 2017). Early phonological processing deficits may predict later literacy difficulties, emphasizing the importance of early detection and intervention.

Comorbidity and Prognosis

Approximately half of children with SSD present with additional developmental disorders, such as language impairment or stuttering. Prognosis is generally favorable with early intervention, though persistent cases may be associated with reading and spelling difficulties. Intervention typically focuses on phonological awareness, articulation drills, and motor sequencing exercises.

iii) Childhood-Onset Fluency Disorder (Stuttering)

Childhood-Onset Fluency Disorder (COFD), or stuttering, is characterized by disruptions in the flow and timing of speech, including repetitions, prolongations, and blocks that interfere with communication. The onset typically occurs between 2 and 7 years of age, coinciding with a period of rapid language development.

Neurobiological and Cognitive Mechanisms

Traditionally viewed as a motor speech disorder, recent evidence redefines stuttering as a disorder of language encoding and timing (Bernstein Ratner & Brundage, 2024). Neurobiological studies identify abnormalities in left-hemispheric speech-motor networks and overactivation of right-hemispheric homologues, reflecting compensatory mechanisms. Diffusion tensor imaging indicates disrupted white matter integrity in the arcuate fasciculus and basal ganglia circuits responsible for motor sequencing (Neef & Chang, 2024).

Genetic and Environmental Factors

Genetic studies have identified mutations in genes associated with lysosomal metabolism and neuronal connectivity. However, environmental factors, such as stress and parental communication style, can influence the persistence and severity of stuttering. Family aggregation data indicate heritability rates of 60–70%, suggesting a strong genetic component.

Developmental Trajectories

Stuttering often follows a fluctuating course, with spontaneous recovery occurring in up to 80% of affected children. Chronic stuttering is associated with heightened anxiety and avoidance behaviors, leading to reduced participation in academic and social contexts. Early intervention focusing on fluency modification and anxiety reduction yields the best outcomes (Bernstein Ratner & Brundage, 2024).

Therapeutic Approaches

Evidence-based treatments combine behavioral and cognitive-behavioral techniques. Fluency shaping and stuttering modification programs aim to enhance speech timing and reduce emotional reactivity. Family-centered interventions support parental communication strategies, promoting acceptance and confidence in the child's communicative competence.

iv) Social (Pragmatic) Communication Disorder (SCD)

Social (Pragmatic) Communication Disorder (SCD) was introduced in DSM-5 to describe individuals with significant difficulties using verbal and non-verbal language for social interaction, in the absence of restricted or repetitive behaviors seen in ASD (APA, 2013).

Core Features

SCD involves deficits in adapting language to context, following conversational rules, understanding nonliteral language (e.g., idioms, sarcasm), and interpreting paralinguistic cues such as facial expressions or tone of voice (Lampri, Peristeri, Marinis, & Andreou, 2024). Children with SCD often demonstrate adequate grammar and vocabulary but fail to use language effectively for social purposes.

Differential Diagnosis

The distinction between SCD and ASD remains a diagnostic challenge due to overlapping symptoms. However, the absence of restricted interests and repetitive behaviors is a defining criterion for SCD. Pragmatic difficulties may also overlap with ADHD or social anxiety disorder, underscoring the need for multidisciplinary assessment (Jain et al., 2025).

Neurocognitive Basis

SCD reflects disruptions in the neural systems underlying social cognition, including the temporo-parietal junction, medial prefrontal cortex, and superior temporal sulcus. These regions are implicated in theory-of-mind and inferencing processes. Functional MRI studies suggest atypical activation during tasks requiring interpretation of figurative language and irony (Lampri et al., 2024).

Clinical Implications

Pragmatic impairments can significantly affect academic achievement, peer relationships, and mental health. Children with SCD may experience loneliness, misunderstanding in social interactions, and difficulty participating in group learning contexts. Interventions target three domains: (a) social understanding, (b) pragmatic communication skills, and (c) inferential language comprehension. Social communication training and group therapy have demonstrated effectiveness in improving functional communication outcomes (Lampri et al., 2024; Bruinsma et al., 2024).

v) Unspecified Communication Disorder

The DSM-5 includes *Unspecified Communication Disorder (UCD)* to account for clinically significant communication difficulties that do not fully meet criteria for any of the other specified subtypes, Language Disorder, Speech Sound Disorder, Childhood-Onset Fluency Disorder, or Social (Pragmatic) Communication Disorder (APA, 2013). This category recognizes that communication impairments may manifest heterogeneously or with insufficient information for a definitive diagnosis, especially in early childhood or in the presence of co-occurring conditions.

Clinical Features

UCD may be applied when an individual presents with persistent breakdowns in expressive or receptive communication that interfere with social participation, academic performance, or occupational functioning, yet the symptom pattern is *atypical or subthreshold.* For example, a child may exhibit mild pragmatic difficulties combined with inconsistent articulation errors, or an adult may show language retrieval problems secondary to developmental factors but not attributable to acquired brain injury.

Diagnostic Challenges

Clinicians use this diagnosis when assessment data are incomplete, when comorbidity obscures the primary deficit, or when cultural-linguistic factors complicate interpretation. As Bishop et al. (2016, 2017) emphasize, many individuals fall within a "grey zone" of communication competence that does not align neatly with categorical frameworks. The unspecified category thus provides diagnostic flexibility, preventing under-identification of individuals who nevertheless require support.

Research and Conceptual Considerations

Although empirical research on UCD is limited, scholars have argued that it reflects the dimensional nature of communicative abilities and deficits (Bertollo et al., 2025). Rather than a stable diagnostic entity, UCD serves as a *placeholder* for emerging or mixed profiles that may later crystallize into a specific disorder. This perspective aligns with transdiagnostic models of neurodevelopmental disorders, which emphasize overlapping cognitive and neurobiological mechanisms across conditions (Bertollo et al., 2025; Bruinsma et al., 2024).

From a service-delivery standpoint, individuals labeled with UCD should still receive intervention proportional to their functional needs. Assessment should be ongoing to determine whether the communication difficulties evolve into a more defined pattern over time. The unspecified classification reminds clinicians to prioritize *functional communication outcomes* rather than categorical precision alone.

Comorbidity and Differential Diagnosis among CDs

Communication disorders seldom occur in isolation. Rather, they exist along a continuum with other neurodevelopmental and psychiatric conditions. The DSM-5 explicitly acknowledges that comorbidity is the norm rather than the exception. Developmental language disorder (DLD) frequently co-occurs with speech sound disorder, specific learning disorder (particularly dyslexia), and attention-deficit/hyperactivity disorder (Bishop et al., 2017). These overlapping conditions suggest shared vulnerabilities in phonological processing, working memory, and executive function.

Autism spectrum disorder (ASD) represents another domain of overlap, especially with social (pragmatic) communication disorder (SCD). Jain et al. (2025) found that children with ASD and SCD exhibit comparable difficulties in adaptive social behavior, although ASD involves additional repetitive behaviors and restricted interests. The demarcation between the two conditions remains clinically challenging and requires careful assessment across developmental history, pragmatic competence, and behavioral repertoire.

Bertollo et al. (2025) propose that many neurodevelopmental disorders, including communication disorders, ASD, and mood dysregulation share dysregulation within cortico-limbic circuits governing emotional processing. This neurobiological convergence may account for elevated rates of anxiety and depression among individuals with DLD or stuttering (Burnley et al., 2024). Moreover, comorbid emotional disturbances can exacerbate communicative impairments by increasing avoidance, cognitive load, and attentional biases.

Differential diagnosis thus demands a multidimensional evaluation integrating linguistic, cognitive, behavioral, and neuropsychological data. Standardized assessments (e.g., CELF-5, CCC-2) should be supplemented by naturalistic observations and input from teachers and parents (Bruinsma et al., 2024). The aim is not only categorical classification but a functional understanding of each individual's communicative profile.

Clinical Implications and Intervention Approaches

Given their neurodevelopmental foundation, communication disorders require early, individualized, and contextually sensitive intervention. Evidence-based practices emphasize a multimodal and family-centered approach, recognizing that communicative competence emerges from dynamic interactions among biological, cognitive, and environmental systems.

Early Identification and Assessment

Early screening in preschool years enables timely intervention and prevention of cascading effects on literacy and academic outcomes. Risk indicators include delayed vocabulary growth, limited babbling, and poor speech intelligibility. For bilingual children, assessment must distinguish language difference from disorder by considering proficiency in both languages (Bishop et al., 2017).

Intervention for Developmental Language Disorder

Effective DLD interventions target specific linguistic domains while fostering functional communication. Explicit grammar teaching, narrative-based therapy, and phonological awareness training have strong empirical support (Bruinsma et al., 2024). School-based collaboration between speech-language pathologists (SLPs) and educators enhances generalization across settings. Burnley et al. (2024) stress the importance of addressing emotional well-being through counseling or peer-support groups, as low self-esteem and social anxiety often accompany linguistic difficulties.

Intervention for Speech Sound and Fluency Disorders

For SSD, phonological contrast therapy and stimulability approaches aim to reorganize the child's phonological system, while articulatory-motor exercises refine accuracy and consistency. In stuttering, integrated treatment models combine fluency shaping, stuttering modification, and cognitive-behavioral therapy to manage anxiety and improve communicative confidence (Bernstein Ratner & Brundage, 2024). Parent-mediated programs, such as the Lidcombe approach, demonstrate high efficacy when applied early.

Intervention for Social (Pragmatic) Communication Disorder

SCD interventions emphasize pragmatic language use within natural contexts. Social-communication groups employ role-play, video modeling, and peer-mediated learning to enhance inferencing and conversational repair strategies (Lampri et al., 2024). Collaboration with teachers supports generalization of pragmatic skills in the classroom. Digital and virtual-reality platforms are emerging as tools for practicing social scenarios in safe, controlled environments.

Neurobiological and Technological Innovations

Advances in neuroimaging and neurostimulation hold promise for precision interventions. Although still experimental, transcranial magnetic stimulation and neurofeedback are being explored as adjuncts to behavioral therapy in fluency and language disorders (Neef & Chang, 2024). Assistive communication technologies, including speech-generating devices and AI-based language training applications, offer new opportunities for individuals with persistent communication difficulties.

Communication disorders represent a heterogeneous cluster of neurodevelopmental conditions that disrupt the human capacity for linguistic and social exchange. The DSM-5 classification reflects an important paradigm shift, from viewing these disorders as isolated speech or language deficits to understanding them as manifestations of broader neurocognitive and social developmental processes.

The synthesis of recent research underscores three major themes. First, neurobiological integration highlights that communication disorders arise from atypical development of distributed neural networks responsible for auditory-motor integration, timing, and social cognition (Bertollo et al., 2025; Neef & Chang, 2024). Second, dimensional continuity emphasizes that the boundaries among developmental language disorder, stuttering, social (pragmatic) communication disorder, and other neurodevelopmental disorders are fluid, with overlapping genetic and neurocognitive substrates that necessitate transdiagnostic research approaches. Third, psychosocial interdependence illustrates that communicative competence is closely linked to emotional regulation, identity formation, and social participation (Burnley et al., 2024).

Future directions call for integrative frameworks that link neurobiological mechanisms with ecological and psychosocial perspectives. Longitudinal studies should examine developmental trajectories across the lifespan, identifying protective factors that promote resilience and compensatory strategies. Clinically, the emphasis must shift toward personalized, strength-based interventions that leverage each individual's communicative potential. Ultimately, communication is the foundation of human connection. Understanding its disorders through the lens of neurodevelopment not only advances scientific knowledge but also promotes inclusivity, empathy, and effective support for individuals navigating these lifelong challenges.

5.INTELLECTUAL DISABILITY

Intellectual Disability (ID) is a neurodevelopmental condition characterized by significant limitations in both intellectual functioning and adaptive behavior that manifest during the developmental period. Rooted in diverse genetic, neurobiological, and environmental mechanisms, ID impacts cognition, communication, and social participation throughout the lifespan. The DSM-5 conceptualizes ID not as a static cognitive deficit but as a disorder of neurodevelopment, emphasizing adaptive skills and contextual functioning (APA, 2013). Recent advances in neurogenetics, developmental neuroscience, and longitudinal research have refined understanding of ID's etiology, comorbidity patterns, and prognosis. This chapter synthesizes evidence from epidemiological, genetic, cognitive, and psychosocial studies, illustrating how neurogenetic syndromes such as Down syndrome, Fragile X syndrome, and Williams syndrome exemplify the complex relationships between brain development and behavioral outcomes. Additionally, the discussion addresses assessment strategies, mental health comorbidities, and evidence-based interventions aimed at promoting inclusion, quality of life, and autonomy for individuals with ID in educational and community settings.

Intellectual Disability represents one of the most prevalent and multifaceted neurodevelopmental disorders, affecting approximately 1–3% of the global population (Patel et al., 2020). It is defined by deficits in intellectual functioning, including reasoning, problem-solving, and abstract thinking, alongside impairments in adaptive behavior encompassing conceptual, social, and practical skills (APA, 2013). Historically, ID was viewed as a static, lifelong cognitive impairment; however, contemporary conceptualizations emphasize developmental trajectories, functional potential, and the importance of contextual adaptation (Rodan et al., 2025). This shift toward a developmental framework aligns ID with other neurodevelopmental disorders, such as autism spectrum disorder (ASD), communication disorders, and specific learning disorders, highlighting shared neurobiological and genetic substrates (Jansen et al., 2023). Consequently, ID is increasingly understood as a dynamic interplay between genetic vulnerability, brain maturation, environmental influences, and social participation, rather than a simple measure of reduced IQ.

The DSM-5 defines ID as deficits in both intellectual and adaptive functioning that originate before the age of 18 (APA, 2013). The severity of ID is determined not solely by IQ scores but by the degree of impairment in conceptual, social, and practical domains. Similarly, the ICD-11 classifies ID under neurodevelopmental disorders, emphasizing functional context over rigid intelligence thresholds. Diagnosis typically involves standardized intelligence testing, such as the Wechsler scales, combined with adaptive behavior assessment using validated instruments like the Vineland Adaptive Behavior Scales. Rodan et al. (2025) stress the importance of incorporating genetic and metabolic evaluations into the diagnostic process, as 40–60% of ID cases have identifiable genetic origins. The shift from the term "mental retardation" to "intellectual disability" reflects a strengths-based, person-first perspective, consistent with broader societal movements toward social inclusion and human rights for individuals with disabilities (Patel et al., 2020).

Epidemiology and Clinical Phenotypes

Prevalence rates of ID vary across regions and socioeconomic contexts, with higher rates observed in low-income populations where prenatal and perinatal risk factors remain prevalent. Individuals with ID experience markedly increased multimorbidity throughout life compared to the general population, including elevated risks of metabolic, cardiovascular, and psychiatric conditions (Baksh et al., 2023). Clinically, ID presents along a spectrum ranging from mild to profound severity. Those with mild ID often achieve partial independence with support, while individuals with severe or profound ID typically require lifelong assistance. Phenotypic heterogeneity arises from diverse genetic and environmental etiologies, resulting in distinct cognitive and behavioral profiles (Jansen et al., 2023).

Etiology and Neurogenetic Syndromes

Genetic factors contribute to approximately two-thirds of ID cases (Rodan et al., 2025). Advances in genomic technologies, including chromosomal microarray and whole-exome sequencing, have identified hundreds of causative genes. Jansen et al. (2023) emphasize the involvement of genes regulating synaptic plasticity, neuronal migration, and chromatin remodeling, demonstrating that ID often results from disruptions in critical neurodevelopmental signaling pathways.

Certain neurogenetic syndromes serve as illustrative models for understanding ID. Down syndrome, caused by trisomy 21, typically results in mild-to-moderate intellectual impairment, characteristic facial morphology, and increased susceptibility to Alzheimer-like neuropathology (Santoro et al., 2021). Fragile X syndrome, the most common inherited form of ID, arises from FMR1 gene expansion and is associated with high seizure prevalence and longitudinal variability in cognitive function, reflecting developmental instability within neural networks (Berry-Kravis et al., 2021). Williams syndrome, caused by a microdeletion on chromosome 7q11.23, demonstrates relative strengths in verbal and social domains, coupled with deficits in visuospatial construction (Miezah et al., 2021).

Environmental and epigenetic factors also contribute to ID. Prenatal exposure to toxins such as alcohol and lead, maternal infections, malnutrition, and perinatal hypoxia can exacerbate developmental delays. While these factors rarely act independently of biological susceptibility, environmental deprivation and psychosocial neglect may further compound cognitive and adaptive impairments.

Cognitive and Adaptive Functioning

The hallmark of ID is reduced intellectual efficiency combined with impaired adaptive behavior. Cognitive deficits frequently include limitations in working memory, processing speed, and executive functioning (Patel et al., 2020). Adaptive functioning deficits manifest across conceptual, social, and practical domains, encompassing skills such as language, literacy, numeracy, empathy, social judgment, communication, self-care, money management, and vocational competence. Leonard et al. (2022) demonstrated through latent-profile analysis that quality of life among individuals with ID is influenced more by social connectedness, autonomy, and emotional well-being than by cognitive ability, underscoring the centrality of adaptive functioning in both diagnosis and intervention.

Comorbidity and Differential Diagnosis

Comorbidity is pervasive in ID, encompassing neurological, psychiatric, and medical conditions. Individuals with ID frequently present with anxiety, depression, and behavioral dysregulation, particularly when comorbid with ASD or epilepsy (Glasson et al., 2020; Totsika et al., 2022). Differential diagnosis requires careful distinction between ID and conditions such as global developmental delay, specific learning disorder, or communication disorders. Rodan et al. (2025) note that persistent global developmental delay beyond age five may indicate evolving ID, emphasizing the need for comprehensive evaluation across cognitive, linguistic, and adaptive domains.

Assessment combines standardized testing, functional observation, and developmental history. Genetic evaluations are increasingly recommended for children with unexplained ID (Rodan et al., 2025), and neuroimaging can reveal structural anomalies, such as cortical malformations or corpus callosum agenesis, aiding etiological classification. Multidisciplinary collaboration among psychologists, geneticists, pediatric neurologists, and speech-language therapists ensures accurate diagnosis and intervention planning. Culturally fair testing and dynamic assessment approaches are critical to avoid over- or under-diagnosis in linguistically diverse populations.

Psychosocial, Educational, and Therapeutic Approaches

Early intervention integrating cognitive stimulation, parent training, and communication enrichment produces optimal developmental outcomes (Patel et al., 2020). Family empowerment and psychoeducation reduce stress and enhance coping strategies. Inclusive education prioritizes differentiated instruction, individualized education plans, and assistive technologies, while teacher training ensures curriculum adaptation and socio-emotional support. Collaboration between educators and therapists facilitates skill generalization across home and school contexts.

Behavioral interventions, grounded in applied behavior analysis (ABA) and cognitive-behavioral therapy (CBT), effectively reduce maladaptive behaviors and enhance adaptive functioning. Psychopharmacological treatments, when indicated, are combined with psychosocial support to address co-occurring psychiatric symptoms (Totsika et al., 2022).

Quality of Life and Lifespan Outcomes

Individuals with ID experience earlier onset of chronic physical diseases and shorter life expectancy (Baksh et al., 2023). Preventive health monitoring, accessible medical care, and interdisciplinary management are crucial for improving longevity and quality of life. Leonard et al. (2022) highlight that subjective well-being depends on autonomy, participation, and supportive relationships, emphasizing empowerment and community inclusion over mere symptom reduction.

As individuals with ID age, they may encounter new challenges, including neurodegenerative conditions (Santoro et al., 2021). Lifespan research suggests that targeted health surveillance and adaptive skill training can mitigate these risks and support sustained functioning. Advances in genomics and neuroimaging offer opportunities for precision medicine and personalized education, facilitating tailored interventions based on individual brain-behavior profiles (Jansen et al., 2023; Rodan et al., 2025).

Future priorities in ID research and practice include developing integrative models linking genetic, cognitive, and environmental factors, conducting cross-cultural studies of adaptive behavior and social participation, performing longitudinal analyses to map developmental trajectories, and advocating for inclusive education, employment, and healthcare policies. Intellectual disability should be understood not merely as a deficit but as a diverse expression of human neurodevelopment. Comprehensive understanding of its biological bases and social dimensions is essential for promoting dignity, equality, and full participation for individuals with ID.

6.MOTOR DISORDERS AND TIC DISORDERS

Motor coordination and tic disorders represent a continuum of neurodevelopmental conditions characterized by atypical motor control, timing, and sensorimotor integration. Within DSM-5, Developmental Coordination Disorder (DCD) and Tic Disorders, including Tourette Syndrome (TS), are classified under motor disorders. Both are increasingly conceptualized as disorders of neural network connectivity, implicating cerebellar, basal ganglia, and frontostriatal systems. Emerging research highlights overlapping genetic risk factors, shared cognitive profiles, and comorbid psychiatric symptoms, particularly anxiety, obsessive–compulsive behaviors, and attention deficits. This chapter integrates neurobiological, cognitive, and psychosocial perspectives on DCD and TS, drawing on multidimensional models that consider genetic, neurophysiological, and environmental contributions (Indelicato et al., 2025; Ke et al., 2023). Evidence-based interventions, including cognitive–behavioral therapy, motor training, and novel neuromodulation approaches, are discussed alongside inclusive educational strategies for neurodiverse learners. The synthesis emphasizes developmental trajectories, early detection, and lifelong support within a biopsychosocial framework.

Motor coordination and tic disorders are classified as neurodevelopmental disorders in the Diagnostic and Statistical Manual of Mental Disorders, Fifth Edition (DSM-5) (APA, 2013). This category includes *Developmental Coordination Disorder* (DCD), *Stereotypic Movement Disorder*, and the spectrum of *Tic Disorders*, encompassing *Tourette Syndrome* (TS), *Persistent (Chronic) Motor or Vocal Tic Disorder*, and *Provisional Tic Disorder*. These disorders are unified by early-onset impairments in motor planning, timing, and control, which interfere with daily functioning and social participation. Their developmental nature distinguishes them from acquired movement disorders, such as Parkinson's disease or dystonia, although shared neurobiological pathways are increasingly recognized (Ferrante et al., 2025; Tarrano et al., 2024). DCD affects approximately 5–6% of school-aged children and is characterized by clumsiness, slow or inaccurate motor performance, and difficulty learning coordinated movements (Wilson et al., 2020). Tourette Syndrome, with a prevalence of about 0.5–1% in youth, involves multiple motor and at least one vocal tic persisting for over one year (Ramteke & Lamture, 2022).

Neurodevelopmental Basis of Motor Disorders

Motor coordination relies on the interaction among cerebellar, basal ganglia, and cortical circuits. Functional MRI studies show that children with DCD exhibit hypoactivation in cerebellar–parietal–frontal pathways responsible for predictive motor control and sensorimotor timing (Wilson et al., 2020). Similarly, neuroimaging in TS reveals abnormal activity within cortico-striato-thalamo-cortical (CSTC) circuits and reduced inhibition in sensorimotor cortices (Graff, 2024).

These findings support the concept of neural network dysfunction rather than localized lesions as the underlying mechanism of motor neurodevelopmental disorders (Indelicato et al., 2025). Cerebellar abnormalities appear to contribute to deficits in both fine motor coordination and tic suppression, indicating a shared mechanism involving error monitoring and timing regulation (Tarrano et al., 2024). Recent genomic research demonstrates overlapping polygenic influences between motor coordination and tic disorders. Indelicato et al. (2025) identified convergence in genes related to synaptic signaling, dopamine modulation, and neurodevelopmental patterning, suggesting a continuum between movement and neuropsychiatric disorders. Family and twin studies further indicate high heritability for both DCD and TS, although environmental modulation and epigenetic regulation remain critical for phenotypic expression (Ke et al., 2023; Peall et al., 2024).

i)Developmental Coordination Disorder

According to DSM-5, DCD is characterized by a marked impairment in the acquisition and execution of coordinated motor skills, which interferes with academic performance and activities of daily living, and is not better explained by intellectual disability or neurological conditions (APA, 2013). Deb et al. (2022) classify DCD as a specific developmental motor disorder within the neurodevelopmental spectrum, often co-occurring with specific learning and communication disorders.

Research has clarified that DCD is not merely a disorder of movement but also involves cognitive and perceptual dysfunction. Wilson et al. (2020) found deficits in executive functions, visuospatial processing, and motor imagery, indicating broader neural network inefficiencies. Ke et al. (2023) emphasize a "multimodal" approach to understanding DCD, integrating neurocognitive, motor, and emotional domains rather than treating it as an isolated motor deficit. Children with DCD often display poor automaticity in motor learning, reduced proprioceptive feedback integration, and slower processing speed, which affect both fine and gross motor skills (Ke et al., 2023).

Psychosocial and Emotional Implications of DCD

Beyond motor symptoms, DCD significantly affects psychosocial development. Pimenta et al. (2023) identified distinct mental health profiles among children with DCD, including subgroups with elevated anxiety and social withdrawal. Poor motor competence can lead to low self-esteem, peer exclusion, and avoidance of physical activities, contributing to sedentary lifestyles and obesity risk. DCD interferes with handwriting, sports, and self-care, often resulting in academic difficulties and reduced participation in school activities. Early identification and occupational therapy interventions can improve function and confidence, though many children remain undiagnosed (Deb et al., 2022). Inclusive classroom strategies, as emphasized by Ndou-Chikwena and Sefotho (2025), support participation and promote neurodiversity in mainstream education.

ii)Tic Disorders and Tourette Syndrome

Tic disorders are characterized by sudden, rapid, recurrent, nonrhythmic motor movements or vocalizations. Tourette Syndrome (TS), the most severe form, requires multiple motor tics and at least one vocal tic persisting for more than one year (APA, 2013). Tics typically begin between ages 5 and 7, peak in early adolescence, and may diminish in adulthood (Ramteke & Lamture, 2022).

TS is associated with dopaminergic dysregulation and reduced cortical inhibition within the CSTC loop. Graff (2024) argues that Tourette Syndrome provides a valuable model for studying the neural basis of motor inhibition and cognitive control, as many individuals with "pure" TS (without comorbid OCD or ADHD) exhibit distinctive cognitive profiles rather than generalized impairment. Tarrano et al. (2024) further demonstrated cerebellar modulation abnormalities that influence cortical excitability, linking motor and psychiatric manifestations. Cerebellar–striatal–cortical interactions may explain the co-occurrence of motor tics and emotional dysregulation.

Leclerc et al. (2024) conducted a randomized controlled trial comparing cognitive-behavioral treatments for tics in youth and adults, confirming that cognitive factors, particularly premonitory urges and attention to tics, play central roles in symptom expression. Comorbidities such as ADHD, obsessive–compulsive disorder, and anxiety are present in up to 80% of individuals with TS, complicating diagnosis and intervention (Ramteke & Lamture, 2022).

Shared Mechanisms and Overlapping Pathophysiology

Both DCD and TS reflect atypical sensorimotor integration and timing dysfunction. Cerebellar and basal ganglia networks underpin motor learning and inhibitory control, while dopaminergic imbalance modulates movement precision (Indelicato et al., 2025). Peall et al. (2024) describe rare genetic brain disorders that blur the boundary between movement and neuropsychiatric phenotypes, supporting a shared neurodevelopmental origin. López-Rodríguez et al. (2025) note that noninvasive brain stimulation studies reveal overlapping cortical excitability patterns across neurodevelopmental disorders, including DCD and tic disorders, highlighting transdiagnostic mechanisms.

Accurate diagnosis requires comprehensive clinical assessment and exclusion of neurological disease. For DCD, standardized motor tests such as the Movement Assessment Battery for Children (MABC-2) and Developmental Coordination Disorder Questionnaire (DCDQ) are recommended (Ke et al., 2023). For TS, diagnostic interviews and tic severity scales, such as the Yale Global Tic Severity Scale, are standard tools (Leclerc et al., 2024). Clinicians must differentiate DCD from mild cerebral palsy or ASD, and TS from functional tic-like behaviors. A developmental history emphasizing onset, context, and impact of symptoms is critical (Deb et al., 2022).

Psychosocial and Educational Implications

Motor disorders profoundly affect social inclusion and quality of life. Pimenta et al. (2023) report that children with DCD frequently experience peer victimization and low participation in physical activities, while individuals with TS often face stigma and social anxiety due to visible tics (Leclerc et al., 2024). Inclusive education and teacher awareness are critical, yet Ndou-Chikwena and Sefotho (2025) found that educators often lack training to accommodate neurodiverse learners with motor coordination difficulties, underscoring the need for systemic inclusion policies.

For DCD, task-specific motor training and cognitive-motor approaches, such as the CO-OP model, improve coordination and self-efficacy (Ke et al., 2023). For TS, Comprehensive Behavioral Intervention for Tics (CBIT) and habit-reversal training show strong efficacy across age groups (Leclerc et al., 2024). In moderate-to-severe TS, dopamine-blocking agents, such as aripiprazole, are often used alongside behavioral therapy (Ramteke & Lamture, 2022). Emerging noninvasive brain stimulation techniques, including transcranial magnetic and direct-current stimulation, offer promising adjuncts for both DCD and tic disorders by modulating cortical excitability and motor inhibition (López-Rodríguez et al., 2025).

Given the persistence of coordination and tic symptoms into adulthood, interventions should target motor, cognitive, and emotional domains simultaneously. Ke et al. (2023) propose a multivariate, multimodal, and multidimensional treatment framework integrating neuropsychology, physiotherapy, and psychosocial support. Recent research converges on the view that DCD and TS exist along a motor neurodevelopmental spectrum, where shared genetic and neural substrates interact with environmental and cognitive factors (Indelicato et al., 2025). Longitudinal studies are essential to trace developmental trajectories and identify protective mechanisms promoting adaptive outcomes. Future work should prioritize cross-disciplinary research integrating genetics, neuroscience, and education, while precision interventions using neuroimaging biomarkers and inclusive educational frameworks are examined (Ndou-Chikwena & Sefotho, 2025). Understanding motor coordination and tic disorders through a neurodevelopmental lens advance both scientific knowledge and quality of life for affected individuals.

CONCLUSION

As we draw the basic presentation of NDDs in school context to a close, it becomes evident that neurodevelopmental disorders represent a diverse and interconnected landscape of conditions that shape the early developmental experiences of children and adolescents. The DSM-5 identifies a spectrum of these disorders, including Autism Spectrum Disorder, Attention-Deficit/Hyperactivity Disorder, Intellectual Disability, Communication Disorders, Specific Learning Disorders, and Motor Disorders such as Developmental Coordination Disorder and Tic Disorders. While each disorder carries its own defining characteristics, whether social communication deficits, attentional and executive function difficulties, learning challenges, or motor impairments, they all converge in their potential to influence the ways children engage with the world, process information, and form relationships. These conditions underscore the importance of viewing development through a lens that appreciates both the uniqueness and the commonality of neurodiverse profiles.

Understanding and supporting children with these disorders requires more than a checklist of behaviors; it demands attention to the interplay between cognitive capacities, emotional regulation, and environmental contexts. A child with ASD, for example, may exhibit remarkable pattern recognition or problem-solving abilities alongside challenges in social reciprocity, just as a child with a learning disorder may demonstrate profound creativity in certain domains while struggling to decode written language. Recognizing these strengths and challenges in tandem allows for interventions that are not only effective but also empowering, fostering engagement, motivation, and confidence. Early identification, tailored educational strategies, and supportive therapeutic approaches are therefore critical in promoting both functional skills and overall well-being, emphasizing the potential for growth that exists within each child.

Ultimately, the DSM-5 provides a structured framework for conceptualizing neurodevelopmental disorders, but the lived experience of each child extends far beyond diagnostic criteria. Appreciating the complexity, diversity, and individuality inherent in these conditions allows clinicians, educators, and caregivers to approach intervention holistically, creating environments that cultivate learning, adaptive functioning, and personal development. In doing so, we move beyond viewing these disorders solely through the lens of deficits, instead embracing a perspective that values potential, nurtures strengths, and supports children in achieving meaningful participation, resilience, and fulfillment in their communities.

Table 1. Neurodevelopmental disorders: Core neurocognitive domains and educational implications

Disorder	Primary Neurocognitive Domains	Core Developmental Characteristics	Educational Implications and Support Priorities
Specific Learning Disorders (SLD)	Phonological processing, working memory, executive functions	Persistent difficulties in reading, writing, and/or mathematics that are unexpected in relation to age, intelligence, and educational opportunity	Early identification, individualized instruction, structured literacy approaches, executive function support, and ongoing monitoring
Attention-Deficit/ Hyperactivity Disorder (ADHD)	Attention regulation, inhibitory control, executive functions	Developmentally inappropriate levels of inattention, impulsivity, and/or hyperactivity with variability across the lifespan	Classroom accommodations, behavioral and self-regulation strategies, structured learning environments, collaboration with mental health professionals
Autism Spectrum Disorder (ASD)	Social cognition, communication, executive functioning, sensory processing	Atypical social communication, restricted or repetitive behaviors, sensory sensitivities, heterogeneous cognitive profiles	Inclusive educational practices, visual supports, predictable routines, individualized learning pathways
Communication Disorders	Language comprehension and/or production, speech motor planning, pragmatic skills	Difficulties in expressive and/or receptive language, speech sound production, fluency, or pragmatic language use	Speech and language therapy, language-rich classroom environments, targeted support for literacy and social communication
Intellectual Disability (ID)	Global cognitive functioning, adaptive behavior	Limitations in intellectual functioning and adaptive skills with onset during the developmental period	Functional and adaptive curricula, emphasis on daily living skills, interdisciplinary educational planning
Motor and Tic Disorders	Motor coordination, motor planning, inhibitory control	Difficulties in fine and/or gross motor coordination and/or presence of tics or involuntary movements	Accommodations for motor demands, occupational therapy input, reduction of performance-related stress

REFERENCES

Ali, S. A., Nisar, H., Ab Aziz, N. A., Fadzil, N. A., Zaber, N. S. M., & Ismail, L. I. (2025). Understanding dyslexia and the potential of artificial intelligence in detecting neurocognitive impairment in dyslexia. In *Artificial Intelligence and Multimodal Signal Processing in Human-Machine Interaction* (pp. 151–170). Academic Press. DOI: 10.1016/B978-0-443-29150-0.00017-2

Alzahrani, S., & Algahtani, F. (2025). Implications and Identification of Specific Learning Disability Using Weighted Ensemble Learning Model. *Child: Care, Health and Development*, *51*(1), e70026. DOI: 10.1111/cch.70026

American Psychiatric Association. (2013). *Diagnostic and statistical manual of mental disorders* (5th ed.). Author.

Andreou, G., & Raxioni, K. (2022). Language development, reading and word learning in Autism Spectrum Disorder (ASD): A review on eye tracking studies. *International Journal of Developmental Disabilities*, •••, 1–18. DOI: 10.1080/20473869.2021.2024404

Åsberg Johnels, J., Fernell, E., Kjellmer, L., Gillberg, C., & Norrelgen, F. (2022). Language/cognitive predictors of literacy skills in 12-year-old children on the autism spectrum. *Logopedics, Phoniatrics, Vocology*, *47*(3), 166–170. DOI: 10.1080/14015439.2021.1884897

Astle, D. E., Holmes, J., Kievit, R., & Gathercole, S. E. (2022). Annual Research Review: The transdiagnostic revolution in neurodevelopmental disorders. *Journal of Child Psychology and Psychiatry, and Allied Disciplines*, *63*(4), 397–417. DOI: 10.1111/jcpp.13481

Atasever, F., Şahin, N. Ö., & Işık, C. M. (2025). Altered expression of MeCP2 and PTEN genes in the molecular basis of specific learning disorder. *Journal of Molecular Neuroscience*, *75*(2), 74. Advance online publication. DOI: 10.1007/s12031-025-02370-3

Baksh, R. A., Pape, S. E., Chan, L. F., Aslam, A. A., Gulliford, M. C., & Strydom, A. (2023). Multiple morbidity across the lifespan in people with Down syndrome or intellectual disabilities: A population-based cohort study using electronic health records. *The Lancet. Public Health*, *8*(6), e453–e462. DOI: 10.1016/S2468-2667(23)00057-9

Bednarz, H. M., Maximo, J. O., Murdaugh, D. L., O'Kelley, S., & Kana, R. K. (2017). "Decoding versus comprehension": Brain responses underlying reading comprehension in children with autism. *Brain and Language*, *169*, 39–47. DOI: 10.1016/j.bandl.2017.01.002

Benitez, P., & Domeniconi, C. (2023). Equivalence-based instruction to teaching reading by families and teachers' students with autism and/or intellectual disabilities. *Behavioral Interventions*, *38*(3), 861–880. Advance online publication. DOI: 10.1002/bin.1932

Berchiatti, M., Ferrer, A., Badenes-Ribera, L., & Longobardi, C. (2022). School Adjustments in Children with Attention Deficit Hyperactivity Disorder (ADHD): Peer Relationships, the Quality of the Student-Teacher Relationship, and Children's Academic and Behavioral Competencies. *Journal of Applied School Psychology*, *38*(3), 241–261. DOI: 10.1080/15377903.2021.1941471

Bernstein Ratner, N., & Brundage, S. B. (2024). Advances in understanding stuttering as a disorder of language encoding. *Annual Review of Linguistics*, *10*(1), 127–143. DOI: 10.1146/annurev-linguistics-030521-044754

Berry-Kravis, E., Filipink, R. A., Frye, R. E., Golla, S., Morris, S. M., Andrews, H., Choo, T.-H., & Kaufmann, W. E. (2021). Seizures in fragile X syndrome: Associations and longitudinal analysis of a large clinic-based cohort. *Frontiers in Pediatrics*, *9*, 736255. DOI: 10.3389/fped.2021.736255

Bertollo, A. G., Puntel, C. F., da Silva, B. V., Martins, M., Bagatini, M. D., & Ignácio, Z. M. (2025). Neurobiological relationships between neurodevelopmental disorders and mood disorders. *Brain Sciences*, *15*(3), 307. DOI: 10.3390/brainsci15030307

Bishop, D. V. M., Snowling, M. J., Thompson, P. A., Greenhalgh, T., & the CATALISE-2 Consortium. (2017). CATALISE: A multinational and multidisciplinary Delphi consensus study of problems with language development. Phase 2: Terminology. *PeerJ* Preprints, *5*, e2484v2. DOI: 10.7287/peerj.preprints.2484v2

Bishop, D. V. M., Snowling, M. J., Thompson, P. A., & Greenhalgh, T.CATALISE Consortium. (2016). CATALISE: A multinational and multidisciplinary Delphi consensus study, Identifying language impairments in children. *PLoS One*, *11*(7), e0158753. DOI: 10.1371/journal.pone.0158753

Blázquez, A., Rodriguez-Revenga, L., Alvarez-Mora, M. I., & Calvo, R. (2025). Clinical and genetic findings in autism spectrum disorders analyzed using exome sequencing. *Frontiers in Psychiatry*, *16*, 1515793. DOI: 10.3389/fpsyt.2025.1515793

Bonti, E., Zerva, I. K., Koundourou, C., & Sofologi, M. (2024). The High Rates of Comorbidity among Neurodevelopmental Disorders: Reconsidering the Clinical Utility of Distinct Diagnostic Categories. *Journal of Personalized Medicine*, *14*(3), 300. DOI: 10.3390/jpm14030300

Bőthe, B., Koós, M., Tóth-Király, I., Orosz, G., & Demetrovics, Z. (2019). Investigating the Associations Of Adult ADHD Symptoms, Hypersexuality, and Problematic Pornography Use Among Men and Women on a Largescale, Non-Clinical Sample. *Journal of Sexual Medicine*, *16*(4), 489–499. DOI: 10.1016/j.jsxm.2019.01.312

Brikell, I., Burton, C., Mota, N. R., & Martin, J. (2021). Insights into attention-deficit/hyperactivity disorder from recent genetic studies. *Psychological Medicine*, *51*(13), 2274–2286. DOI: 10.1017/S0033291721000982

Bruinsma, G. I., Wijnen, F., & Gerrits, E. (2024). Communication in daily life of children with developmental language disorder: Parents' and teachers' perspectives. *Language, Speech, and Hearing Services in Schools*, *55*(1), 105–129. DOI: 10.1044/2023_LSHSS-23-00051

Bullen, J. C., Zajic, M. C., McIntyre, N., Solari, E., & Mundy, P. (2022). Patterns of math and reading achievement in children and adolescents with autism spectrum disorder. *Research in Autism Spectrum Disorders*, *92*, 101933. DOI: 10.1016/j.rasd.2022.101933

Burnley, A., St Clair, M., Dack, C., Thompson, H., & Wren, Y. (2024). Exploring the psychosocial experiences of individuals with developmental language disorder during childhood: A qualitative investigation. *Journal of Autism and Developmental Disorders*, *54*(8), 3008–3027. DOI: 10.1007/s10803-023-05946-3

Butterfuss, R., & Kendeou, P. (2018). The role of executive functions in reading comprehension. *Educational Psychology Review*, *30*(3), 801–826. DOI: 10.1007/s10648-017-9422-6

Capodieci, A., Ruffini, C., Frascari, A., Rivella, C., Bombonato, C., Giaccherini, S., Scali, V., Luccherino, L., Viterbori, P., Traverso, L., Usai, M. C., Marzocchi, G. M., & Pecini, C. (2023). Executive functions in children with specific learning disorders: Shedding light on a complex profile through teleassessment. *Research in Developmental Disabilities*, *142*, 104621. DOI: 10.1016/j.ridd.2023.104621

Chalmpe, M., & Vlachos, F. (2025). Are there distinct subtypes of developmental dyslexia? *Frontiers in Behavioral Neuroscience*, *18*, 1512892. DOI: 10.3389/fnbeh.2024.1512892

Chen, H., Gu, X. H., Zhou, Y., Ge, Z., Wang, B., Siok, W. T., Wang, G., Huen, M., Jiang, Y., Tan, L.-H., & Sun, Y. (2017). A genome-wide association study identifies genetic variants associated with mathematics ability. *Scientific Reports*, *7*(1), 40365. DOI: 10.1038/srep40365

Chen, Y., & Jahromi, L. B. (2025). Self-Regulation and Academic Learning in Preschoolers with Autism Spectrum Disorder: Links to School Engagement and Levels of Autism Characteristics. *Journal of Autism and Developmental Disorders*, *55*(4), 1286–1301. DOI: 10.1007/s10803-024-06288-4

Cherkasova, M., Sulla, E. M., Dalena, K. L., Pondé, M. P., & Hechtman, L. (2013). Developmental course of attention deficit hyperactivity disorder and its predictors. *Journal of the Canadian Academy of Child and Adolescent Psychiatry*, 47, *22*(1), 47-54. PMID: 23390433; PMCID: PMC3565715.

Conner, C., Allor, J. H., Al Otaiba, S., Yovanoff, P., & LeJeune, L. (2022). Early Reading Outcomes in Response to a Comprehensive Reading Curriculum for Students With Autism Spectrum Disorder and Intellectual Disability. *Focus on Autism and Other Developmental Disabilities*, *10883576221137905*. Advance online publication. DOI: 10.1177/10883576221137905

Cowan, T. M., Galley, K. O., & Miller, C. A. (2023). Clinical bilingual evaluations: Exploring ways to overcome barriers in the direct assessment of both languages when there is a suspicion of language disorder. *Perspectives of the ASHA Special Interest Groups*, *8*(3), 607–617. DOI: 10.1044/2023_PERSP-22-00097

Crisci, G., Caviola, S., Cardillo, R., & Mammarella, I. C. (2021). Executive functions in neurodevelopmental disorders: Comorbidity overlaps between attention deficit and hyperactivity disorder and specific learning disorders. *Frontiers in Human Neuroscience*, *15*, 594234. DOI: 10.3389/fnhum.2021.594234

Cuba Bustinza, C., Adams, R. E., Claussen, A. H., Vitucci, D., Danielson, M. L., Holbrook, J. R., Charania, S. N., Yamamoto, K., Nidey, N., & Froehlich, T. E. (2022). Factors associated with bullying victimization and bullying perpetration in children and adolescents with ADHD: 2016 to 2017 national survey of children's health. *Journal of Attention Disorders*, *26*(12), 1535–1548. DOI: 10.1177/10870547221085502

Cuffe, S. P., Visser, S. N., Holbrook, J. R., Danielson, M. L., Geryk, L. L., Wolraich, M. L., & McKeown, R. E. (2020). ADHD and Psychiatric Comorbidity: Functional Outcomes in a School-Based Sample of Children. *Journal of Attention Disorders*, *24*(9), 1345–1354. DOI: 10.1177/1087054715613437

Cummings, K., Watkins, A., Jones, C., Dias, R., & Welham, A. (2022). Behavioural and psychological features of PTEN mutations: A systematic review of the literature and meta-analysis of the prevalence of autism spectrum disorder characteristics. *Journal of Neurodevelopmental Disorders*, *14*(1), 1. DOI: 10.1186/s11689-021-09406-w

Da Silva, B. S., Grevet, E. H., Silva, L. C. F., Ramos, J. K. N., Rovaris, D. L., & Bau, C. H. D. (2023). An overview on neurobiology and therapeutics of attention-deficit/hyperactivity disorder. *Discover Mental Health*, *3*(1), 2. DOI: 10.1007/s44192-022-00030-1

Davidson, M. M., & Ellis Weismer, S. (2017). Reading comprehension of ambiguous sentences by school-age children with autism spectrum disorder. *Autism Research*, *10*(12), 2002–2022. DOI: 10.1002/aur.1850

Deb, S. S., Roy, M., Bachmann, C., & Bertelli, M. O. (2022). Specific learning disorders, motor disorders, and communication disorders. In *Textbook of Psychiatry for Intellectual Disability and Autism Spectrum Disorder* (pp. 483–511). Springer. DOI: 10.1007/978-3-319-95720-3_18

Espinas, D. R., Ramos, M. R., & Fernández, P. (2025). Interventions for children and adolescents with specific learning disorders: A review and future directions. *Pediatric Research*. Advance online publication. DOI: 10.1038/s41390-025-04261-0

Faraone, S. V., & Larsson, H. (2019). Genetics of attention deficit hyperactivity disorder. *Molecular Psychiatry*, *24*(4), 562–575. DOI: 10.1038/s41380-018-0070-0

Ferrante, F. J., Escobar Grisales, D., López, M. F., Lopes da Cunha, P., Sterpin, L. F., Vonk, J. M., Chaná Cuevas, P., Estienne, C., Hesse, E., Amoruso, L., Orozco Arroyave, J. R., & García, A. M. (2025). Cognitive phenotyping of Parkinson's disease patients via digital analysis of spoken word properties. *Movement Disorders*, *40*(11), 2354–2366. Advance online publication. DOI: 10.1002/mds.70005

Franke, B., Michelini, G., Asherson, P., Banaschewski, T., Bilbow, A., Buitelaar, J. K., Cormand, B., Faraone, S. V., Ginsberg, Y., Haavik, J., Kuntsi, J., Larsson, H., Lesch, K.-P., Ramos-Quiroga, J. A., Réthelyi, J. M., Ribases, M., & Reif, A. (2018). Live fast, die young? A review on the developmental trajectories of ADHD across the lifespan. *European Neuropsychopharmacology*, *28*(10), 1059–1088. DOI: 10.1016/j.euroneuro.2018.08.001

Georgitsi, M., Dermitzakis, I., Soumelidou, E., & Bonti, E. (2021). The polygenic nature and complex genetic architecture of specific learning disorder. *Brain Sciences*, *11*(5), 631. DOI: 10.3390/brainsci11050631

Gialluisi, A., Andlauer, T. F. M., Mirza-Schreiber, N., Moll, K., Becker, J., Hoffmann, P., Ludwig, K. U., Czamara, D., Pourcain, B. S., Honbolygó, F., Tóth, D., Csépe, V., Huguet, G., Chaix, Y., Iannuzzi, S., Demonet, J.-F., Morris, A. P., Hulslander, J., Willcutt, E. G., & Schulte-Körne, G. (2021). Genome-wide association study reveals new insights into the heritability and genetic correlates of developmental dyslexia. *Molecular Psychiatry*, *26*(7), 3004–3017. DOI: 10.1038/s41380-020-00898-x

Giannoulis, S. V., Müller, D., Kennedy, J. L., & Gonçalves, V. (2024). Systematic review of mitochondrial genetic variation in attention-deficit/hyperactivity disorder. *European Child & Adolescent Psychiatry*, *33*(6), 1675–1685. DOI: 10.1007/s00787-022-02030-6

Gilhuber, C. S., Raulston, T. J., & Galley, K. (2023). Language and communication skills in multilingual children on the autism spectrum: A systematic review. *Autism*, *13623613221147780*(6), 1516–1531. Advance online publication. DOI: 10.1177/13623613221147780

Glasson, E. J., Buckley, N., Chen, W., Leonard, H., Epstein, A., Skoss, R., Jacoby, P., Blackmore, A. M., Bourke, J., & Downs, J. (2020). Systematic review and meta-analysis: Mental health in children with neurogenetic disorders associated with intellectual disability. *Journal of the American Academy of Child and Adolescent Psychiatry*, *59*(9), 1036–1048. DOI: 10.1016/j.jaac.2020.01.006

Gnanavel, S., Sharma, P., Kaushal, P., & Hussain, S. (2019). Attention deficit hyperactivity disorder and comorbidity: A review of literature. *World Journal of Clinical Cases*, *7*(17), 2420–2426. DOI: 10.12998/wjcc.v7.i17.2420

Graff, M. (2024). Cognitive characteristics of "pure" Tourette Syndrome and what we can learn from neurological study. *Journal of Cognitive Psychology*, *36*(6), 673–687. DOI: 10.1080/20445911.2024.2361958

Grigorenko, E. L., Compton, D. L., Fuchs, L. S., Wagner, R. K., Willcutt, E. G., Grigorenko, E. L., Klin, A., & Volkmar, F. (2003). Annotation: Hyperlexia: disability or superability? *Journal of Child Psychology and Psychiatry, and Allied Disciplines*, *44*(8), 1079–1091. DOI: 10.1111/1469-7610.00193

Henderson, L. M., Clarke, P. J., & Snowling, M. J. (2014). Reading comprehension impairments in autism spectrum disorders. *L'Année Psychologique*, *114*(4), 779–797.

Hustyi, K. M., Logue, J. J., & Hall, S. S. (2025). Skill-based treatment for challenging behavior in autism spectrum disorder: A scoping review of treatment characteristics and outcomes. *Research in Autism Spectrum Disorders*, *119*, 102523. DOI: 10.1016/j.rasd.2024.102523

Indelicato, E., Zech, M., Eberl, A., & Boesch, S. (2025). Insights on the shared genetic landscape of neurodevelopmental and movement disorders. *Current Neurology and Neuroscience Reports*, *25*(1), 24. DOI: 10.1007/s11910-025-01414-w

Jain, D., Multani, K. S., Dodiya, A., Benani, U., & Iyer, A. (2025). Adaptive behavior and its differences between children with autism spectrum disorder and social communication disorder. *Autism*, *29*(4), 829–837. DOI: 10.1177/13623613251317787

Jansen, S., Vissers, L. E., & de Vries, B. B. (2023). The genetics of intellectual disability. *Brain Sciences*, *13*(2), 231. DOI: 10.3390/brainsci13020231

Kamio, Y., Robins, D., Kelley, E., Swainson, B., & Fein, D. (2007). Atypical lexical/semantic processing in high-functioning autism spectrum disorders without early language delay. *Journal of Autism and Developmental Disorders*, *37*(6), 1116–1122. DOI: 10.1007/s10803-006-0254-3

Ke, L., Su, X., Yang, S., Du, Z., Huang, S., & Wang, Y. (2023). New trends in developmental coordination disorder: Multivariate, multidimensional and multimodal. *Frontiers in Psychiatry*, *14*, 1116369. DOI: 10.3389/fpsyt.2023.1116369

Kelley, E., Paul, J. J., Fein, D., & Naigles, L. R. (2006). Residual language deficits in optimal outcome children with a history of autism. *Journal of Autism and Developmental Disorders*, *36*(6), 807–828. DOI: 10.1007/s10803-006-0111-4

Khoirunnisa, A. N., & Dewi, L. (2023). Design and Prototype Development of Augmented Reality in Reading Learning for Autism. *Computers*, *12*(3), 55. DOI: 10.3390/computers12030055

King, S. (2022). The education context for twice-exceptional students: An overview of issues in special and gifted education. *Neurobiology of Learning and Memory*, *193*, 107659. DOI: 10.1016/j.nlm.2022.107659

Kjelgaard, M. M., & Tager-Flusberg, H. (2001). An investigation of language impairment in autism: Implications for genetic subgroups. *Language and Cognitive Processes*, *16*(2-3), 287–308. DOI: 10.1080/01690960042000058

Kofler, M. J., Irwin, L. N., Soto, E. F., Groves, N. B., Harmon, S. L., & Sarver, D. E. (2019). Executive functioning heterogeneity in pediatric ADHD. *Journal of Abnormal Child Psychology*, *47*(2), 273–286. DOI: 10.1007/s10802-018-0438-2

Kuja-Halkola, R., Lichtenstein, P., D'Onofrio, B. M., & Larsson, H. (2015). Codevelopment of ADHD and externalizing behaviour from childhood to adulthood. *Journal of Child Psychology and Psychiatry, and Allied Disciplines*, *56*(6), 640–647. DOI: 10.1111/jcpp.12340

Lampri, S., Peristeri, E., Marinis, T., & Andreou, M. (2024). Figurative language processing in autism spectrum disorders: A review. *Autism Research, 17*(4), 674–689. DOI: 10.1002/aur.3069

Larsson, H., Asherson, P., Chang, Z., Ljung, T., Friedrichs, B., Larsson, J.-O., & Lichtenstein, P. (2013). Genetic and environmental influences on adult attention deficit hyperactivity disorder symptoms: A large Swedish population-based study of twins. *Psychological Medicine, 43*(1), 197–207. DOI: 10.1017/S0033291712001067

Leclerc, J. B., O'Connor, K. P., Gauthier, B., Singer, I., Woods, D. W., Blanchet, P., & Lavoie, M. E. (2024). Comparison of cognitive-behavioral treatments for tics and Tourette syndrome in youth and adults: A randomized controlled trial. *Journal of Behavioral and Cognitive Therapy, 100491*(2), 100491. Advance online publication. DOI: 10.1016/j.jbct.2024.100491

Lee, H. K., Chan, W. S., & Tong, S. X. (2023). The heterogeneity and interrelationships among theory of mind, executive function, and reading comprehension deficits in Hong Kong Chinese children with Autism. *Read Writ 36*, 1539–1565. https://doi.org/10.1007/s11145-022-10298-y

Leonard, H., Whitehouse, A., Jacoby, P., Benke, T., Demarest, S., Saldaris, J., Wong, K., Reddihough, D., Williams, K., & Downs, J. (2022). Quality of life beyond diagnosis in intellectual disability – latent profiling. *Research in Developmental Disabilities, 129*, 104322. DOI: 10.1016/j.ridd.2022.104322

Lievore, R., Caviola, S., & Mammarella, I. C. (2024). How trait and state mathematics anxiety could affect performance: Evidence from children with and without Specific Learning Disorders. *Learning and Individual Differences, 112*, 102459. DOI: 10.1016/j.lindif.2024.102459

Liu, J., Wang, Y., & Yi, L. (2023). Heterogeneity and imbalance of reading profiles in Autism Spectrum Disorders. *Research in Autism Spectrum Disorders, 100*, 102088. DOI: 10.1016/j.rasd.2022.102088

López-Rodríguez, S., Coelho, D. R., Renet, C., Vieira, W. F., Tural, Ü., Cassano, P., & Camprodon, J. A. (2025). Noninvasive brain stimulation for neurodevelopmental disorders: A systematic review. *The Journal of Neuropsychiatry and Clinical Neurosciences, 37*(4), 297–312. Advance online publication. DOI: 10.1176/appi.neuropsych.20240127

Magielski, J. H., Ruggiero, S. M., Xian, J., Parthasarathy, S., Galer, P. D., Ganesan, S., Back, A., McKee, J. L., McSalley, I., Gonzalez, A. K., Morgan, A., Donaher, J., & Helbig, I. (2025). The clinical and genetic spectrum of paediatric speech and language disorders. *Brain, 148*(2), 663–674. DOI: 10.1093/brain/awae264

Marocchini, E. (2023). Impairment or difference? The case of Theory of Mind abilities and pragmatic competence in the Autism Spectrum. *Applied Psycholinguistics, 44*(3), 1–19. DOI: 10.1017/S0142716423000024

McClain, M. B., Yoho, S. E., Drill, R. B., Haverkamp, C. R., Schwartz, S. E., Barker, B. A., & Upton, S. R. (2023). Reading Skills and Background Noise in Autistic and Non-autistic Children: A Pilot Study. *Contemporary School Psychology*, •••, 1–13. DOI: 10.1007/s40688-023-00450-y

McDonough, E. M., Flanagan, D. P., Sy, M., & Alfonso, V. C. (2017). Specific Learning Disorder. In Goldstein, S., & DeVries, M. (Eds.), *Handbook of DSM-5 Disorders in Children and Adolescents*. Springer. DOI: 10.1007/978-3-319-57196-6_4

Miezah, D., Porter, M., Rossi, A., Kazzi, C., Batchelor, J., & Reeve, J. (2021). Cognitive profile of young children with Williams syndrome. *Journal of Intellectual Disability Research, 65*(8), 784–794. DOI: 10.1111/jir.12860

Mirkovic, B., Chagraoui, A., Gerardin, P., & Cohen, D. (2020). Epigenetics and attention-deficit/hyperactivity disorder: New perspectives? *Frontiers in Psychiatry, 11*, 579. DOI: 10.3389/fpsyt.2020.00579

Moffitt, T. E. (2018). Male antisocial behavior in adolescence and beyond. *Nature Human Behaviour, 2*(3), 177–186. DOI: 10.1038/s41562-018-0309-4

Mouzaki, A., Spantidakis, I., Vamvoukas, M., Liva, E., & Ralli, M. (2007). Development of cognitive and academic skills in children with learning difficulties in early primary school grades. Paper presented at the Conference of the Hellenic Association of Speech-Language Pathologists, *Language Difficulties and Written Expression in the Context of School Learning*, Athens, Greece, 23–25.

Müller, B., Schaadt, G., Boltze, J., Emmrich, F., Skeide, M. A., Neef, N. E., Kraft, I., Brauer, J., Friederici, A. D., Kirsten, H., & Wilcke, A.LEGASCREEN Consortium. (2017). ATP 2C2 and DYX 1C1 are putative modulators of dyslexia-related MMR. *Brain and Behavior, 7*(11), e00851. DOI: 10.1002/brb3.851

Murdaugh, D. L., Maximo, J. O., Cordes, C. E., O'Kelley, S. E., & Kana, R. K. (2017). From word reading to multisentence comprehension: Improvements in brain activity in children with autism after reading intervention. *NeuroImage. Clinical, 16*, 303–312. DOI: 10.1016/j.nicl.2017.08.012

National Institute for Health and Care Excellence. (2018). Attention deficit hyperactivity disorder: diagnosis and management (update), NICE guideline: methods. www.nice.org.uk/guidance/ng87/evidence (Retrieved: March 6th 2024).

Ndou-Chikwena, N. N., & Sefotho, M. M. (2025). Stakeholders' perspectives on the inclusion of neurodiverse learners in the mainstream curricula. *South African Journal of Childhood Education, 15*(1), 1622. https://hdl.handle.net/10520/ejc-sajce_v15_n1_a1622. DOI: 10.4102/sajce.v15i1.1622

Neef, N. E., & Chang, S. E. (2024). Knowns and unknowns about the neurobiology of stuttering. *PLoS Biology, 22*(2), e3002492. DOI: 10.1371/journal.pbio.3002492

Newbury, D. F., & Monaco, A. P. (2010). Genetic advances in the study of speech and language disorders. *Neuron, 68*(2), 309–320. DOI: 10.1016/j.neuron.2010.10.001

Nigg, J. T., Sibley, M. H., Thapar, A., & Karalunas, S. L. (2020). Development of ADHD: Etiology, heterogeneity, and early life course. *Annual Review of Developmental Psychology, 2*(1), 559–583. DOI: 10.1146/annurev-devpsych-060320-093413

Papadatou-Pastou, M., Panagiotidou, D. A., Abbondanza, F., Fischer, U., Paracchini, S., & Karagiannakis, G. (2021). Hand preference and Mathematical Learning Difficulties: New data from Greece, the United Kingdom, and Germany and two meta-analyses of the literature. *Laterality, 26*(5), 485–538. DOI: 10.1080/1357650X.2021.1906693

Papadopoulos, T. C., Csépe, V., Aro, M., Caravolas, M., Diakidoy, I. A., & Olive, T. (2021). Methodological issues in literacy research across languages: Evidence from alphabetic orthographies. *Reading Research Quarterly, 56*(S1), S351–S370. https://ila.onlinelibrary.wiley.com/doi/epdf/10.1002/rrq.407. DOI: 10.1002/rrq.407

Papadopoulos, T. C., Kendeou, P., & Spanoudis, G. (2012). Investigating the factor structure and measurement invariance of phonological abilities in a sufficiently transparent language. *Journal of Educational Psychology, 104*(2), 321–336. https://psycnet.apa.org/record/2012-00883-001. DOI: 10.1037/a0026446

Papadopoulos, T. C., Spanoudis, G., & Chatzoudi, D. (2020). A longitudinal investigation of the double dissociation between reading and spelling deficits: The role of linguistic and executive function skills. *Reading and Writing, 33*(4), 1075–1104. https://link.springer.com/article/10.1007/s11145-020-10029-1. DOI: 10.1007/s11145-020-10029-1

Papadopoulos, T. C., Spanoudis, G., Ktisti, C., & Fella, A. (2021). Precocious readers: A cognitive or a linguistic advantage? *European Journal of Psychology of Education, 36*(1), 63–90. https://pure.unic.ac.cy/en/publications/precocious-readers-a-cognitive-or-a-linguistic-advantage. DOI: 10.1007/s10212-020-00470-9

Papantoniou, M., & Kambourologou, M. (n.d.). *Development and disorders of communication and language in autism.* EPEAEK: Access for All. Institute of Educational Policy. Retrieved March 20, 2023, from https://prosvasimo.iep.edu.gr/docs/pdf/epimorfwtiko-uliko-autismos/Autismos1.pdf

Parrila, R. K., Georgiou, G. K., & Papadopoulos, T. C. (2020). Dyslexia in a consistent orthography: Evidence from reading-level match design. *Dyslexia (Chichester, England), 26*(4), 343–358. https://onlinelibrary.wiley.com/doi/epdf/10.1002/dys.1650. DOI: 10.1002/dys.1650

Patel, D. R., Cabral, M. D., Ho, A., & Merrick, J. (2020). A clinical primer on intellectual disability. *Translational Pediatrics, 9*(S1, Suppl 1), S23–S35. DOI: 10.21037/tp.2020.02.02

Peall, K. J., Owen, M. J., & Hall, J. (2024). Rare genetic brain disorders with overlapping neurological and psychiatric phenotypes. *Nature Reviews. Neurology, 20*(1), 7–21. DOI: 10.1038/s41582-023-00896-x

Perugini, A., Fontanillas, P., Gordon, S. D., Fisher, S. E., Martin, N. G., Bates, T. C., & Luciano, M. (2024). Dyslexia polygenic scores show heightened prediction of verbal working memory and arithmetic. *Scientific Studies of Reading, 28*(5), 549–563. DOI: 10.1080/10888438.2024.2365697

Peters, L., & Ansari, D. (2019). Are specific learning disorders truly specific, and are they disorders? *Trends in Neuroscience and Education, 17*, 100115. DOI: 10.1016/j.tine.2019.100115

Pimenta, R. A., Fuchs, C., Fears, N. E., Mariano, M., & Tamplain, P. (2023). Distinct mental health profiles in children with developmental coordination disorder: A latent class analysis and associations. *Research in Developmental Disabilities, 132*, 104377. DOI: 10.1016/j.ridd.2022.104377

Ramey, D. M., & Freelin, B. N. (2023). Exploring the relationships between school suspension, ADHD diagnoses, and delinquency across different school punitive and special education climates. *Children and Youth Services Review, 148*, 106849. DOI: 10.1016/j.childyouth.2023.106849

Ramteke, A., & Lamture, Y. (2022). Tics and Tourette Syndrome: A literature review of etiological, clinical, and pathophysiological aspects. *Cureus, 14*(8), e28575. DOI: 10.7759/cureus.28575

Randi, J., Newman, T., & Grigorenko, E. L. (2010). Teaching children with autism to read for meaning: Challenges and possibilities. *Journal of Autism and Developmental Disorders, 40*(7), 890–902. DOI: 10.1007/s10803-010-0938-6

Regan, S. L., Williams, M. T., & Vorhees, C. V. (2022). Review of rodent models of attention deficit hyperactivity disorder. *Neuroscience and Biobehavioral Reviews, 132*, 621–637. DOI: 10.1016/j.neubiorev.2021.11.041

Retz, W., Ginsberg, Y., Turner, D., Barra, S., Retz-Junginger, P., Larsson, H., & Asherson, P. (2021). Attention-Deficit/Hyperactivity Disorder (ADHD), antisociality and delinquent behavior over the lifespan. *Neuroscience and Biobehavioral Reviews, 120*, 236–248. DOI: 10.1016/j.neubiorev.2020.11.025

Ricketts, J., Jones, C. R., Happé, F., & Charman, T. (2013). Reading comprehension in autism spectrum disorders: The role of oral language and social functioning. *Journal of Autism and Developmental Disorders, 43*(4), 807–816. DOI: 10.1007/s10803-012-1619-4

Riesco, K., Ukawa, Y., Alfreh, P., & Abdelfattah, E. (2025, March). *Analyzing Autism Spectrum Disorder in Children. In 2025 Northeast Section Conference.*

Riva, A., Musetti, A., Bomba, M., Milani, L., Montrasi, V., & Nacinovich, R. (2021). Language-related skills in bilingual children with specific learning disorders. *Frontiers in Psychology, 11*, 564047. DOI: 10.3389/fpsyg.2020.564047

Roch, M., Cain, K., & Jarrold, C. (2021). Reading for comprehension in individuals with Down syndrome, autism spectrum disorder and typical development: Similar or different patterns of ability? *Brain Sciences, 11*(7), 828. DOI: 10.3390/brainsci11070828

Rodan, L. H., Stoler, J., Chen, E., Geleske, T., Vergano, S., Holm, I. A., Jones, K., Kalish, J., Monteil, D. C., Pritchard, A. B., Rasmussen, S. A., Russell, B., Santoro, S. L., Trapane, P., & Weaver, K. N. (2025). Genetic evaluation of the child with intellectual disability or global developmental delay: Clinical report. *Pediatrics, 156*(1), e2025072219. DOI: 10.1542/peds.2025-072219

Rutherford, M., Maciver, D., Johnston, L., Prior, S., & Forsyth, K. (2021). Development of a pathway for multidisciplinary neurodevelopmental assessment and diagnosis in children and young people. *Children (Basel, Switzerland), 8*(11), 1033. DOI: 10.3390/children8111033

Sansosti, F. J., Was, C., Rawson, K. A., & Remaklus, B. L. (2013). Eye movements during processing of text requiring bridging inferences in adolescents with higher functioning autism spectrum disorders: A preliminary investigation. *Research in Autism Spectrum Disorders, 7*(12), 1535–1542. DOI: 10.1016/j.rasd.2013.09.001

Santoro, J. D., Pagarkar, D., Chu, D. T., Rosso, M., Paulsen, K. C., Levitt, P., & Rafii, M. S. (2021). Neurologic complications of Down syndrome: A systematic review. *Journal of Neurology, 268*(12), 4495–4509. DOI: 10.1007/s00415-020-10179-w

Sathyanesan, A., Zhou, J., Scafidi, J., Heck, D. H., Sillitoe, R. V., & Gallo, V. (2019). Emerging connections between cerebellar development, behaviour and complex brain disorders. *Nature Reviews. Neuroscience, 20*(5), 298–313. DOI: 10.1038/s41583-019-0152-2

Schatz, N. K., Fabiano, G. A., Raiker, J. S., Hayes, T. B., & Pelham, W. E.Jr. (2021). Twenty-year trends in elementary teachers' beliefs about best practices for students with ADHD. *School Psychology, 36*(4), 203–213. DOI: 10.1037/spq0000442

Singh, B. D., Moore, D. W., Furlonger, B. E., Anderson, A., Fall, R., & Howorth, S. (2021). Reading comprehension and autism spectrum disorder: A systematic review of interventions involving single-case experimental designs. *Review Journal of Autism and Developmental Disorders, 8*(1), 3–21. DOI: 10.1007/s40489-020-00200-3

Smith Gabig, C. (2010). Phonological awareness and word recognition in reading by children with autism. *Communication Disorders Quarterly, 31*(2), 67–85. DOI: 10.1177/1525740108328410

Snowling, M. J., & Hulme, C. (2025). Risk factors for dyslexia: Addressing oral language deficits. *Mind, Brain and Education : the Official Journal of the International Mind, Brain, and Education Society, 19*(3), 168–175. DOI: 10.1111/mbe.70009

Solari, E. J., Grimm, R. P., McIntyre, N. S., Zajic, M., & Mundy, P. C. (2019). Longitudinal stability of reading profiles in individuals with higher functioning autism. *Autism, 23*(8), 1911–1926. DOI: 10.1177/1362361318812423

Tarrano, C., Galléa, C., Delorme, C., McGovern, E. M., Atkinson-Clement, C., Brochard, V., Thobois, S., Tranchant, C., Grabli, D., Degos, B., Corvol, J. C., Pedespan, J.-M., Krystkowiak, P., Houeto, J.-L., Degardin, A., Defebvre, L., Beranger, B., Martino, D., Apartis, E., & Worbe, Y. (2024). Psychiatric phenotype in neurodevelopmental myoclonus-dystonia is underpinned by abnormality of cerebellar modulation on the cerebral cortex. *Scientific Reports, 14*(1), 22341. DOI: 10.1038/s41598-024-73386-9

Tek, S., Mesite, L., Fein, D., & Naigles, L. (2014). Longitudinal analyses of expressive language development reveal two distinct language profiles among young children with autism spectrum disorders. *Journal of Autism and Developmental Disorders, 44*(1), 75–89. DOI: 10.1007/s10803-013-1853-4

Thapar, A. (2020). ADHD: Progressing from genetic discoveries to biological insights. *The American Journal of Psychiatry*, *177*(9), 802–804. DOI: 10.1176/appi.ajp.2020.20070961

Thapar, A., Cooper, M., & Rutter, M. (2017). Neurodevelopmental disorders. *The Lancet. Psychiatry*, *4*(4), 339–346. DOI: 10.1016/S2215-0366(16)30376-5

Tong, S. X., Wong, R. W. Y., Kwan, J. L. Y., & Arciuli, J. (2020). Theory of mind as a mediator of reading comprehension differences between Chinese school-age children with autism and typically developing peers. *Scientific Studies of Reading*, *24*(4), 292–306. DOI: 10.1080/10888438.2019.1666133

Totsika, V., Liew, A., Absoud, M., Adnams, C., & Emerson, E. (2022). Mental health problems in children with intellectual disability. *The Lancet. Child & Adolescent Health*, *6*(6), 432–444. DOI: 10.1016/S2352-4642(22)00067-0

Truong, D. T., Adams, A. K., Paniagua, S., Frijters, J. C., Boada, R., Hill, D. E., Lovett, M. W., Mahone, E. M., Willcutt, E. G., Wolf, M., Defries, J. C., Gialluisi, A., Francks, C., Fisher, S. E., Olson, R. K., Pennington, B. F., Smith, S. D., Bosson-Heenan, J., & Gruen, J. R. (2019). Multivariate genome-wide association study of rapid automatised naming and rapid alternating stimulus in Hispanic American and African American youth. *Journal of Medical Genetics*, *56*(8), 557–566. https://jmg.bmj.com/content/56/8/557.info. DOI: 10.1136/jmedgenet-2018-105874

Vogindroukas, I., Stankova, M., Chelas, E. N., & Proedrou, A. (2022). Language and speech characteristics in Autism. *Neuropsychiatric Disease and Treatment*, *18*, 2367–2377. DOI: 10.2147/NDT.S331987

Walenski, M., Tager-Flusberg, H., & Ullman, M. T. (2006). Language in Autism. In S. O. Moldin & J. L. R. Rubenstein (Eds.), *Understanding autism: From basic neuroscience to treatment* (pp. 175–203). CRC Press/Routledge/Taylor & Francis Group. DOI: 10.1201/9781420004205.ch9

Wiener, J., & Mak, M. (2009). Peer victimization in children with attention-deficit/hyperactivity disorder. *Psychology in the Schools*, *46*(2), 116–131. DOI: 10.1002/pits.20358

Williams, N. M., Zaharieva, I., Martin, A., Langley, K., Mantripragada, K., Fossdal, R., Stefansson, H., Stefansson, K., Magnusson, P., Gudmundsson, O. O., Gustafsson, O., Holmans, P., Owen, M. J., O'Donovan, M., & Thapar, A. (2010). Rare chromosomal deletions and duplications in attention-deficit hyperactivity disorder: A genome-wide analysis. *Lancet*, *376*(9750), 1401–1408. DOI: 10.1016/S0140-6736(10)61109-9

Wilson, P., Ruddock, S., Rahimi-Golkhandan, S., Piek, J., Sugden, D., Green, D., & Steenbergen, B. (2020). Cognitive and motor function in developmental coordination disorder. *Developmental Medicine and Child Neurology*, *62*(11), 1317–1323. DOI: 10.1111/dmcn.14646

Yeung, M. K. (2022). A systematic review and meta-analysis of facial emotion recognition in autism spectrum disorder: The specificity of deficits and the role of task characteristics. *Neuroscience and Biobehavioral Reviews*, *133*, 104518. DOI: 10.1016/j.neubiorev.2021.104518

Young, S., Sedgwick, O., Fridman, M., Gudjonsson, G., Hodgkins, P., Lantigua, M., & González, R. (2015). Co-morbid psychiatric disorders among incarcerated ADHD populations: A meta-analysis. *Psychological Medicine*, *45*(12), 2499–2510. DOI: 10.1017/S0033291715000598

Zerva, I. K. (2023a). Metalinguistic abilities: The contribution to writing. *Social Sciences & Humanities Open*, *8*(1), 100609. DOI: 10.1016/j.ssaho.2023.100609

Zerva, I. K. (2023c). Early Signs of Reading Difficulties and Phonological Awareness in Early School Age. In *Perspectives of Cognitive, Psychosocial, and Learning Difficulties From Childhood to Adulthood: Practical Counseling Strategies* (pp. 21-36). IGI Global. DOI: 10.4018/978-1-6684-8203-2.ch002

Zerva, I. K., & Koundourou, C. (2023). Neurodevelopmental Language Disorders in School Context: Specific Learning Disorders in Reading and Writing. In *Developmental Language Disorders in Childhood and Adolescence* (pp. 195-206). IGI Global. DOI: 10.4018/979-8-3693-0644-4.ch015

Zerva, K. I. (2023b). The imprint of specific learning difficulties and ADHD on individual and environmental mental health. In Katsarou, D., Koundourou, Ch., & Reppa, G. (Eds.), *New trends and approaches in psychopathology and mental health of children and adults* (pp. 185–204). Gutenberg. (In Greek)

Chapter 2
Understanding and Exploring the Spectrum of Neurodevelopmental Disorders Comorbidity

ABSTRACT

This chapter examines how comorbidity shapes the landscape of NDDs, showing that overlapping traits and shared developmental pathways are the rule rather than the exception. Drawing on findings from genetics, neurobiology, cognition, and epidemiology, it explains why conditions such as ASD, ADHD, SLDs, ID, CD, and DCD often present together. The chapter also discusses how these intersections complicate diagnosis, particularly when secondary difficulties are overlooked through diagnostic overshadowing. It argues for flexible, dimensional assessment approaches and interventions that respond to functional needs, build on strengths, and reflect neurodiversity principles. Together, these perspectives support more precise, inclusive, and person-centered practice.

INTRODUCTION

Neurodevelopmental disorders (NDDs) represent a complex, heterogeneous, and interrelated set of conditions characterized by atypical brain development, manifesting across cognition, behavior, motor coordination, social communication, and adaptive functioning. Traditional categorical models delineated conditions such as autism spectrum disorder (ASD), attention-deficit/hyperactivity disorder (ADHD), intellectual disability (ID), specific learning disorders (SLDs), communication disorders, and developmental coordination disorder (DCD) as separate entities (APA, 2013).

DOI: 10.4018/979-8-3693-6331-7.ch002

However, a substantial body of contemporary research emphasizes high rates of comorbidity, shared genetic etiology, and overlapping neurocognitive profiles across NDDs, challenging these rigid classifications (Astle et al., 2022; Michelini et al., 2024; Morris-Rosendahl & Crocq, 2020). Comorbidity is increasingly recognized not as an incidental phenomenon but as a defining feature of NDDs, with significant implications for diagnosis, intervention, and educational support.

The conceptualization of NDDs has evolved considerably over the past decades. Early models emphasized discrete syndromes with clear diagnostic boundaries. Yet, longitudinal and cross-sectional studies demonstrated that individuals frequently meet criteria for multiple conditions, prompting reconsideration of categorical models (Bonti et al., 2024; Roccella & Vetri, 2021). DSM-5 and ICD-11 classifications reflect a partial acknowledgment of these overlaps, grouping NDDs under a single developmental framework characterized by early onset, persistence across the lifespan, and deficits in adaptive, social, and executive functions (APA, 2013; Stein et al., 2020). This evolution reflects both clinical observation and empirical findings highlighting shared developmental pathways and the limitations of treating each NDD as fully independent.

Transdiagnostic approaches, which focus on dimensional traits rather than strictly defined categories, are particularly relevant for understanding comorbidity. Astle et al. (2022) argue for a "transdiagnostic revolution," emphasizing that executive dysfunction, attentional dysregulation, and cognitive inflexibility often underlie multiple NDDs. Similarly, Michelini et al. (2024) propose a general neurodevelopmental factor, which captures variance across different conditions and interacts with environmental risk factors, such as perinatal complications or psychosocial adversity, to influence phenotypic expression. This perspective aligns with the notion of overlapping continua rather than discrete, isolated disorders, with comorbidity emerging naturally from shared neurodevelopmental vulnerabilities.

Epidemiological studies consistently report high rates of co-occurrence among NDDs. Approximately 50% of children with ASD meet criteria for ADHD, and a substantial proportion of children with ADHD exhibit learning difficulties or motor impairments (Bonti et al., 2024; Francés et al., 2022). Liu et al. (2025) report that in a Chinese national sample of school-attending children, ADHD frequently co-occurs with learning challenges and emotional dysregulation, demonstrating that comorbidity is globally prevalent and culturally consistent. Njardvik et al. (2025) conducted a systematic review and meta-analysis, finding that children and adolescents with ADHD have elevated rates of anxiety, depression, and behavioral disorders, further illustrating the pervasive nature of psychiatric comorbidity within NDD populations.

Somatic comorbidities are also noteworthy. Agnafors et al. (2019) highlight that children with psychiatric disorders often experience somatic illnesses, including gastrointestinal, metabolic, and neurological conditions, which may exacerbate functional impairment and complicate clinical management. The interplay between somatic and neurodevelopmental comorbidity underscores the importance of comprehensive health monitoring, as noted by Gnanavel and Hussain (2018), who emphasize that children receiving antipsychotic medications for NDDs require careful surveillance to prevent adverse outcomes.

Koyuncu et al. (2022) further illustrate the complexity of comorbidity in ADHD, emphasizing that anxiety disorders frequently co-occur with ADHD in both children and adults. This comorbidity is associated with diagnostic and therapeutic challenges, including overlapping symptom presentations, heightened functional impairment, and treatment resistance. One crucial clinical phenomenon emerging in this context is diagnostic overshadowing, wherein the presence of a prominent diagnosis (e.g., ADHD or ASD) leads clinicians to overlook co-occurring conditions. For instance, anxiety or learning difficulties may be misattributed to the primary NDD, delaying appropriate interventions (Koyuncu et al., 2022; Njardvik et al., 2025). Recognizing diagnostic overshadowing is essential for improving outcomes and ensuring that secondary but functionally significant disorders are not neglected.

GENETIC AND NEUROBIOLOGICAL OVERLAP

Genetic investigations provide strong evidence for shared etiology across NDDs. Genome-wide association studies and sequencing analyses reveal overlapping genetic variants across ASD, ADHD, ID, and language disorders, with key pleiotropic genes including SHANK3, CNTNAP2, and CHD8 (Kelleher & Corvin, 2015; Rovira et al., 2020). These genetic intersections contribute to variable phenotypic expression, whereby individuals carrying the same variant may exhibit divergent clinical presentations, a phenomenon termed variable expressivity (Morris-Rosendahl & Crocq, 2020). The presence of shared genetic risk factors helps explain why comorbidity is not only common but expected within neurodevelopmental populations.

Neuroimaging studies further corroborate overlapping pathophysiology. Disruptions in fronto-striatal, default mode, and cerebellar networks are observed across ADHD, ASD, and DCD, with differential modulation of these circuits contributing to distinct behavioral profiles (Roccella & Vetri, 2021; Thapar et al., 2017). Such convergence supports the conceptualization of NDDs as a spectrum of interrelated conditions, rather than strictly independent disorders. Cognitive studies highlight that deficits in executive functions, including working memory, inhibitory control, and cognitive flexibility, often co-occur in children with ADHD and SLDs, serving as transdiagnostic endophenotypes that underpin comorbid presentations (Crisci et al., 2021; Tsujii et al., 2021). Importantly, diagnostic overshadowing can also arise at the neurobiological level, as overlapping neural patterns may obscure the detection of distinct, co-existing conditions, leading to under-recognition of specific deficits (Michelini et al., 2024).

The developmental cascade framework provides a valuable lens for understanding how early neurocognitive or sensorimotor deficits propagate secondary difficulties across the lifespan. For instance, attentional dysregulation or early language impairments may precipitate learning difficulties, social challenges, or emotional dysregulation (Roccella & Vetri, 2021). Ter-Stepanian et al. (2017) emphasize that co-occurring attentional and motor dysfunctions underlie many comorbid presentations, while Sokolova et al. (2017) demonstrate overlapping neuropsychological origins of ASD and ADHD traits. These insights underscore the importance of holistic assessment models that capture the cascading effects of early developmental vulnerabilities, rather than focusing solely on categorical diagnoses.

Phenotypically, comorbidity can manifest in complex behavioral profiles. Vigil-Pérez et al. (2025) describe repetitive and restrictive behaviors in multiple NDDs, revealing that these traits are not exclusive to ASD but may also appear in ADHD or SLD populations, contributing to functional impairment. Similarly, Albayrak et al. (2025) document that autistic traits within ADHD and SLD cohorts are linked to internalizing and externalizing symptoms, illustrating the intricate interplay of social, cognitive, and emotional domains. Lopez-Villatoro et al. (2025) extend this understanding by demonstrating that ADHD-related traits can intersect with broader psychiatric conditions, such as borderline personality disorder subtypes, leading to social cognition deficits. Diagnostic overshadowing is particularly relevant here, as complex behavioral profiles can mask underlying disorders, delaying or misdirecting therapeutic intervention.

NEURODIVERSITY, CLINICAL ASSESSMENT AND TRANSLATIONAL PRACTICE

The neurodiversity framework offers a paradigm shift in conceptualizing comorbidity. Miranda-Ojeda et al. (2025) advocate for understanding NDDs as natural variations in neurocognitive profiles rather than solely as deficits. This approach emphasizes individual strengths and adaptive capacities, supporting inclusive, person-centered interventions that reduce stigma and enhance participation. By integrating neurodiversity with transdiagnostic evidence, clinicians and educators can move beyond categorical labeling and mitigate the risks of diagnostic overshadowing, ensuring that functional needs across overlapping disorders are appropriately addressed.

Longitudinal studies indicate that children with multiple NDDs are at increased risk for subsequent mental health challenges. Zhou et al. (2025) report that co-occurring NDDs significantly elevate the likelihood of anxiety and depression compared to single NDDs, highlighting the compounded impact of comorbidity on psychological well-being. Diagnostic overshadowing can exacerbate these outcomes by obscuring secondary disorders, delaying intervention, and intensifying functional impairment (Koyuncu et al., 2022; Njardvik et al., 2025). These findings reinforce the necessity of early identification, preventive strategies, and integrated care pathways that address overlapping cognitive, behavioral, and emotional vulnerabilities.

Accurate diagnosis in the context of comorbidity is challenging. Standard categorical frameworks may fail to capture subthreshold or overlapping features, potentially delaying intervention (Francés et al., 2022; Stein et al., 2020). Diagnostic overshadowing further complicates clinical evaluation, as the presence of a prominent NDD may lead to misattribution of secondary symptoms to the primary diagnosis (Koyuncu et al., 2022; Njardvik et al., 2025). Multidisciplinary assessment pathways, encompassing developmental pediatricians, psychologists, speech-language pathologists, occupational therapists, and geneticists, improve diagnostic precision, reduce redundant testing, and support individualized care planning (Gnanavel & Hussain, 2018; Rutherford et al., 2021; Simon et al., 2022).

Emerging dimensional assessment tools, such as the Behavior Rating Inventory of Executive Function and the Social Responsiveness Scale, quantify symptom expression across domains, capturing the nuanced impact of comorbidity and mitigating the effects of diagnostic overshadowing (Michelini et al., 2024). Digital phenotyping via wearables or mobile platforms provides ecological validity, monitoring real-world fluctuations in attention, social engagement, and motor performance. Integrating genetic, neuroimaging, and behavioral data facilitates precision psychiatry approaches, optimizing individualized interventions and prognostic understanding.

EDUCATIONAL AND INTERVENTION IMPLICATIONS

Comorbidity significantly influences educational outcomes. Children with overlapping NDDs require flexible curricula, multimodal instruction, and assistive technologies to optimize learning and participation (Ndou-Chikwena & Sefotho, 2025). Close collaboration among educators, clinicians, and families is critical to identify overlapping deficits, leverage strengths, and promote adaptive development across cognitive, social, and emotional domains. Awareness of diagnostic overshadowing is essential in educational settings, as the presence of a primary diagnosis may lead to under-identification of learning or attentional difficulties, reducing the effectiveness of support strategies (Francés et al., 2022; Lamsal et al., 2020).

Intervention strategies increasingly target shared mechanisms across comorbid conditions. Cognitive training for executive control, social skills programs, mindfulness-based approaches, and parent-mediated interventions have shown efficacy in enhancing attention, cognitive flexibility, and emotional regulation across ADHD, ASD, and SLD populations (Astle et al., 2022; Crisci et al., 2021). Environmental enrichment during critical developmental windows harnesses neuroplasticity to redirect trajectories and reduce the functional impact of comorbidity.

Spectrum-based service models advocate for eligibility criteria grounded in functional need rather than categorical diagnosis, ensuring equitable access to interventions and educational support (Lamsal et al., 2020). Family-centered approaches recognize the intergenerational nature of NDDs, addressing parental mental health, sibling adjustment, and broader social supports, ultimately fostering resilience and adaptive functioning. Ethical considerations include reducing diagnostic labeling stigma, promoting neurodiversity-aligned policies, and integrating lived experience into care planning. Addressing diagnostic overshadowing at the policy level ensures that all functional impairments are recognized and supported, irrespective of the prominence of the primary diagnosis.

Comorbidity is a central, defining feature of neurodevelopmental disorders, reflecting shared genetic, neurobiological, and cognitive mechanisms. Diagnostic overshadowing represents a key challenge in this context, as overlapping disorders may be under-recognized or misattributed to a primary condition. Integrating dimensional, transdiagnostic, and neurodiversity-informed frameworks provides a nuanced understanding of overlapping presentations, facilitates early identification, and guides holistic interventions. Recognition of comorbidity has profound implications for clinical practice, education, and policy, enabling person-centered, strengths-based approaches that promote functional outcomes and inclusive societal participation. By synthesizing empirical evidence with lived experience and addressing diagnostic overshadowing, researchers, clinicians, educators, and policymakers can support individuals across the neurodevelopmental spectrum, ultimately advancing equitable, effective, and ethical care.

REFERENCES

Agnafors, S., Norman Kjellström, A., Torgerson, J., & Rusner, M. (2019). Somatic comorbidity in children and adolescents with psychiatric disorders. *European Child & Adolescent Psychiatry*, *28*(11), 1517–1525. DOI: 10.1007/s00787-019-01313-9 PMID: 30895480

Albayrak, Z. S., Serdengeçti, N., Ateş, B. Ö., Seçinti, D. D., Kadak, M. T., & Doğangün, B. (2025). *Autistic traits and their relationship with internalization-externalization symptoms in ADHD and SLD*. Turkish Journal of Child and Adolescent Mental Health. DOI: 10.4274/tjcamh.galenos.2024.60252

American Psychiatric Association. (2013). *Diagnostic and statistical manual of mental disorders* (5th ed.). American Psychiatric Publishing.

Astle, D. E., Holmes, J., Kievit, R., & Gathercole, S. E. (2022). Annual research review: The transdiagnostic revolution in neurodevelopmental disorders. *Journal of Child Psychology and Psychiatry, and Allied Disciplines*, *63*(4), 397–417. DOI: 10.1111/jcpp.13481 PMID: 34296774

Bonti, E., Zerva, I. K., Koundourou, C., & Sofologi, M. (2024). The high rates of comorbidity among neurodevelopmental disorders: Reconsidering the clinical utility of distinct diagnostic categories. *Journal of Personalized Medicine*, *14*(3), 300. DOI: 10.3390/jpm14030300 PMID: 38541042

Crisci, G., Caviola, S., Cardillo, R., & Mammarella, I. C. (2021). Executive functions in neurodevelopmental disorders: Comorbidity overlaps between attention deficit and hyperactivity disorder and specific learning disorders. *Frontiers in Human Neuroscience*, *15*, 594234. DOI: 10.3389/fnhum.2021.594234 PMID: 33732121

Francés, L., Quintero, J., Fernández, A., Ruiz, A., Caules, J., Fillon, G., Hervás, A., & Soler, C. V. (2022). Current state of knowledge on the prevalence of neurodevelopmental disorders in childhood according to the DSM-5: A systematic review in accordance with the PRISMA criteria. *Child and Adolescent Psychiatry and Mental Health*, *16*(1), 27. DOI: 10.1186/s13034-022-00462-1 PMID: 35361232

Gnanavel, S., & Hussain, S. (2018). Audit of physical health monitoring in children and adolescents receiving antipsychotics in neurodevelopmental clinics in Northumberland. *World Journal of Psychiatry*, *8*(1), 27–32. DOI: 10.5498/wjp.v8.i1.27 PMID: 29568729

Kelleher, E., & Corvin, A. (2015). Overlapping etiology of neurodevelopmental disorders. In *The genetics of neurodevelopmental disorders* (pp. 29–48). Wiley. DOI: 10.1002/9781118524947.ch2

Koyuncu, A., Ayan, T., İnce Guliyev, E., Erbilgin, S., & Deveci, E. (2022). ADHD and anxiety disorder comorbidity in children and adults: Diagnostic and therapeutic challenges. *Current Psychiatry Reports, 24*(2), 129–140. DOI: 10.1007/s11920-022-01324-5 PMID: 35076887

Lamsal, R., Finlay, B., Whitehurst, D. G. T., & Zwicker, J. D. (2020). Generic preference-based health-related quality of life in children with neurodevelopmental disorders: A scoping review. *Developmental Medicine and Child Neurology, 62*(2), 169–177. DOI: 10.1111/dmcn.14301 PMID: 31225644

Liu, J., Jiang, Z., Li, F., Zheng, Y., Cui, Y., Xu, H., & Li, Y. (2025). Prevalence and comorbidity of attention deficit hyperactivity disorder in Chinese school-attending students aged 6–16: A national survey. *Annals of General Psychiatry, 24*(1), 23. DOI: 10.1186/s12991-025-00558-w PMID: 40223102

Lopez-Villatoro, J. M., Galvez-Merlin, A., Diaz-Marsá, M., & Carrasco, J. L. (2025). Social cognition impairment associated to a borderline personality disorder subtype linked to attention deficit hyperactivity disorder. *Journal of Psychiatric Research, 185*, 26–30. DOI: 10.1016/j.jpsychires.2025.03.031 PMID: 40147151

Michelini, G., Carlisi, C. O., Eaton, N. R., Elison, J. T., Haltigan, J. D., Kotov, R., Krueger, R. F., Latzman, R. D., Li, J. J., Levin-Aspenson, H. F., Salum, G. A., South, S. C., Stanton, K., Waldman, I. D., & Wilson, S. (2024). Where do neurodevelopmental conditions fit in transdiagnostic psychiatric frameworks? Incorporating a new neurodevelopmental spectrum. *World Psychiatry; Official Journal of the World Psychiatric Association (WPA), 23*(3), 333–357. DOI: 10.1002/wps.21225 PMID: 39279404

Miranda-Ojeda, R., Wickramasinghe, A., Ntolkeras, G., Castanho, I., & Yassin, W. (2025). The neurodiversity framework in medicine: On the spectrum. *Developmental Neurobiology, 85*(1), e22960. DOI: 10.1002/dneu.22960 PMID: 39874176

Morris-Rosendahl, D. J., & Crocq, M. A. (2020). Neurodevelopmental disorders—The history and future of a diagnostic concept. *Dialogues in Clinical Neuroscience, 22*(1), 65–72. DOI: 10.31887/DCNS.2020.22.1/macrocq PMID: 32699506

Njardvik, U., Wergeland, G. J., Riise, E. N., Hannesdottir, D. K., & Öst, L. G. (2025). Psychiatric comorbidity in children and adolescents with ADHD: A systematic review and meta-analysis. *Clinical Psychology Review, 102571*, 102571. Advance online publication. DOI: 10.1016/j.cpr.2025.102571 PMID: 40245462

Roccella, M., & Vetri, L. (2021). Embracing the complexity of neurodevelopmental disorders. *Brain Sciences, 11*(11), 1385. DOI: 10.3390/brainsci11111385 PMID: 34827384

Rovira, P., Demontis, D., Sánchez-Mora, C., Zayats, T., Klein, M., Mota, N. R., Weber, H., Garcia-Martínez, I., Pagerols, M., Vilar-Ribó, L., Arribas, L., Richarte, V., Corrales, M., Fadeuilhe, C., Bosch, R., Martin, G. E., Almos, P., Doyle, A. E., Grevet, E. H., & Ribasés, M. (2020). Shared genetic background between children and adults with attention deficit/hyperactivity disorder. *Neuropsychopharmacology : Official Publication of the American College of Neuropsychopharmacology, 45*(10), 1617–1626. DOI: 10.1038/s41386-020-0664-5 PMID: 32279069

Rutherford, M., Maciver, D., Johnston, L., Prior, S., & Forsyth, K. (2021). Development of a pathway for multidisciplinary neurodevelopmental assessment and diagnosis in children and young people. *Children (Basel, Switzerland), 8*(11), 1033. DOI: 10.3390/children8111033 PMID: 34828746

Simon, J., Hyde, C., Saravanapandian, V., Wilson, R., Distefano, C., Besterman, A., & Jeste, S. (2022). The diagnostic journey of genetically defined neurodevelopmental disorders. *Journal of Neurodevelopmental Disorders, 14*(1), 27. DOI: 10.1186/s11689-022-09439-9 PMID: 35501684

Sokolova, E., Oerlemans, A. M., Rommelse, N. N., Groot, P., Hartman, C. A., Glennon, J. C., Claassen, T., Heskes, T., & Buitelaar, J. K. (2017). A causal and mediation analysis of the comorbidity between attention deficit hyperactivity disorder (ADHD) and autism spectrum disorder (ASD). *Journal of Autism and Developmental Disorders, 47*(6), 1595–1604. DOI: 10.1007/s10803-017-3083-7 PMID: 28255761

Ter-Stepanian, M., Grizenko, N., Cornish, K., Talwar, V., Mbekou, V., Schmitz, N., & Joober, R. (2017). Attention and executive function in children diagnosed with attention deficit hyperactivity disorder and comorbid disorders. *Journal of the Canadian Academy of Child and Adolescent Psychiatry = Journal de l'Académie Canadienne de Psychiatrie de l'Enfant et de l'Adolescent, 26*(1), 21–30. PMID: 28331500

Tsujii, N., Usami, M., Naya, N., Tsuji, T., Mishima, H., Horie, J., Fujiwara, M., & Iida, J. (2021). Efficacy and safety of medication for attention-deficit hyperactivity disorder in children and adolescents with common comorbidities: A systematic review. *Neurology and Therapy, 10*(2), 499–522. DOI: 10.1007/s40120-021-00249-0 PMID: 34089145

Vigil-Pérez, A., Blázquez, A., Garcia-Delgar, B., Ortiz, A. E., Borràs, R., Morer, A., Escalona, R. C., & Lázaro, L. (2025). Phenomenology of repetitive and restrictive behaviors and sensory phenomena in neurodevelopmental disorders: An exploratory study. *BMC Psychiatry, 25*(1), 163. DOI: 10.1186/s12888-025-06569-x PMID: 39994553

Zhou, X., Jia, F., Bambling, M., Edirippulige, S., Whiteford, H., & Lin, J. (2025). Associations between multiple neurodevelopmental disorders and mental health in children. *Journal of Affective Disorders, 120397*. Advance online publication. DOI: 10.1016/j.jad.2025.120397 PMID: 41072876

Chapter 3
The Clinical Continuum Hypothesis

ABSTRACT

This chapter examines how the clinical continuum hypothesis reframes neurodevelopmental conditions as graded variations rather than discrete disorders. Drawing from genetic research, cognitive studies, neuroimaging, and educational theory, it shows that traits associated with Autism, ADHD, Learning Disorders, and related profiles distribute along overlapping developmental pathways. These insights challenge rigid diagnostic categories and highlight the limits of label-based interventions. The chapter argues for assessment and instructional models that respond to functional profiles, emphasize transdiagnostic processes, and adapt across the lifespan. By treating education as a dynamic neurocognitive environment, the continuum perspective supports inclusive, flexible, and context-responsive practice that aligns with contemporary neurodevelopmental science.

FROM CATEGORICAL NOSOLOGY TO DIMENSIONAL CONTINUUM THINKING

For much of the twentieth century, neurodevelopmental disorders (NDDs) such as autism spectrum disorder (ASD), attention-deficit/hyperactivity disorder (ADHD), and specific learning disorders (SLDs) were conceptualized as discrete, categorical entities, each defined by specific diagnostic thresholds (Morris-Rosendahl & Crocq, 2020; Thapar et al., 2017). This categorical model was reinforced through successive iterations of diagnostic manuals, which aimed to improve reliability but often at the expense of acknowledging developmental and cognitive overlap among disorders. However, recent advances in genetics, neuroimaging, and behavioral sciences have

DOI: 10.4018/979-8-3693-6331-7.ch003

challenged these boundaries, proposing instead that NDDs exist along a continuum of neurodevelopmental variation (Dell'Osso et al., 2019; Louveau et al., 2023).

The neurodevelopmental continuum hypothesis, introduced by Dell'Osso and colleagues (2019), posits that disorders historically treated as distinct may represent variations along shared etiological and neurobiological dimensions. This framework has since evolved into what some scholars describe as a neurodevelopmental gradient hypothesis, recognizing gradations of cognitive, motor, and socio-emotional functioning rather than binary distinctions between disorder and typicality (Zoccante et al., 2021). Neurodevelopmental diversity, in this view, encompasses a spectrum that ranges from typical variation to clinically significant impairment, influenced by genetic, environmental, and experiential factors (Pallanti & Salerno, 2023).

From an educational standpoint, this theoretical shift has profound implications. Schools and learning environments have historically relied on diagnostic labeling to allocate resources and interventions. Yet, as Antolini and Colizzi (2023) argue, NDDs "do not stop at childhood"; they evolve dynamically across the lifespan, demanding an educational model responsive to ongoing developmental variation. Recognizing a continuum of functioning encourages educators to move from remediation toward accommodation, adapting environments, pedagogies, and assessment methods to meet diverse neurocognitive profiles rather than pathologizing difference. This mirrors Leifler's (2022) call for educational inclusion frameworks that transcend medicalized categories and instead cultivate "functional participation" across the continuum of ability.

TRANSDIAGNOSTIC AND MULTIDIMENSIONAL PERSPECTIVES

Recent research in developmental neuroscience and psychiatry has begun to erode rigid diagnostic boundaries, replacing them with transdiagnostic frameworks emphasizing shared mechanisms across conditions (Astle et al., 2022; Michelini et al., 2024). Astle et al. (2022) describe this as the "transdiagnostic revolution," noting that deficits in executive functioning, attention, or social cognition often traverse multiple diagnostic categories. Similarly, Hoza and Shoulberg (2024) underscore the necessity of examining these cross-cutting mechanisms in childhood to inform early, flexible interventions.

The continuum and transdiagnostic paradigms intersect in their critique of categorical diagnostic thinking, emphasizing that comorbidity once viewed as a confounding factor may in fact be intrinsic to the neurodevelopmental architecture (Bonti et al., 2024; Dewey, 2018; Hansen et al., 2018). High comorbidity rates among ADHD, ASD, and learning disorders (Eberhard et al., 2022; Francés et al.,

2023) reinforce the argument that distinct labels capture only fragments of broader developmental processes. In empirical studies, for example, Deserno et al. (2023) demonstrated through data-driven clustering that children with ASD and ADHD exhibit overlapping cognitive phenotypes rather than discrete clusters. Canu et al. (2022a, 2022b) further supported this notion through ocular-motor task performance, revealing gradations of impairment across disorders that support a shared developmental pathway.

In educational contexts, these findings call for transdiagnostic instructional design. Teachers and specialists must focus less on labels and more on the underlying cognitive and affective processes influencing learning such as attention regulation, working memory, sensory modulation, and executive control (Herreras, 2025). For instance, students diagnosed with ASD or SLD may both benefit from explicit executive function training, multisensory instruction, and structured routines. By aligning pedagogy with transdiagnostic constructs rather than categorical diagnoses, educators can address the core learning mechanisms implicated across the neurodevelopmental continuum.

The growing emphasis on dimensionality has also influenced assessment and intervention frameworks. Rodríguez et al. (2021) propose a Response to Intervention (RtI) model that replaces static diagnostic thresholds with data-driven, tiered support systems. This approach allows for dynamic monitoring of students' needs and outcomes, resonating with Dell'Osso et al.'s (2019) vision of developmental gradation. By continuously adapting instruction based on performance data rather than fixed categories, RtI embodies the educational application of the continuum hypothesis, transforming classrooms into living systems of neurodevelopmental monitoring and support.

BRIDGING CLINICAL AND EDUCATIONAL DIMENSIONS OF THE CONTINUUM

The integration of clinical and educational perspectives is essential to operationalizing the neurodevelopmental continuum in real-world settings. Aravamuthan et al. (2024) conceptualize NDDs as disorders of "neurodevelopmental trajectory," suggesting that the brain's plasticity and adaptive capacity render early intervention crucial but also lifelong in nature. Within educational systems, this aligns with the principle of lifespan inclusion, which emphasizes continuity of support from early childhood through adulthood (Antolini & Colizzi, 2023).

Recent findings in genetics and molecular neuroscience provide further evidence of shared developmental mechanisms that blur the distinction between NDDs and later psychiatric or neurodegenerative conditions. Iannotta et al. (2025) identified genetic overlap between schizophrenia and neurodevelopmental disorders, supporting the neurodevelopmental continuum hypothesis at a molecular level. Likewise, Dafsari et al. (2025) demonstrated that mutations in the autophagy-related EPG5 gene link NDDs with early-onset parkinsonism, while Ouellette and Lacoste (2021) proposed a vascular continuum spanning developmental to degenerative disorders. These discoveries reinforce that neurodevelopment is not a finite phase but an evolving biological process across the lifespan.

Educational systems must therefore move beyond the school-based intervention paradigm, acknowledging that neurodevelopmental challenges persist into adulthood and influence vocational, social, and civic participation (Antolini & Colizzi, 2023; Leifler, 2022). This requires a continuum-based educational model emphasizing longitudinal tracking, coordinated transition planning, and adaptive pedagogy that evolves with developmental changes. The recognition of such continuity can mitigate the "service cliff" often faced by students with NDDs upon exiting formal education (Morris et al., 2025). In this sense, the neurodevelopmental continuum reframes inclusive education as a dynamic, lifelong process rather than a static provision of accommodations. Educators become mediators of developmental trajectories, translating insights from neuroscience into adaptive teaching practices that honor each learner's evolving cognitive ecology. By viewing the brain and its environment as co-constructors of function, the educational space becomes not merely remedial, but regenerative, supporting growth across the gradient of neurodiversity (Pallanti & Salerno, 2023).

THE GENETIC, COGNITIVE, AND FUNCTIONAL DIMENSIONS OF THE NEURODEVELOPMENTAL CONTINUUM

The genetic evidence for a neurodevelopmental continuum has expanded dramatically over the past decade, reshaping the way both clinicians and educators conceptualize individual differences in cognitive and adaptive functioning. Molecular and genomic research increasingly reveals that neurodevelopmental disorders are not discrete categories but share a polygenic, overlapping risk structure (Iannotta et al., 2025; Wong et al., 2025). Studies on gene variants such as *SETBP1*,linked to transcriptional regulation and neuronal differentiation, demonstrate that small variations in genetic expression can yield a wide phenotypic range, from subtle cognitive differences to severe developmental impairment (Wong et al., 2025). This reinforces the notion that biological gradients, rather than strict dichotomies, underlie the expression of neurodevelopmental conditions. Iannotta et al. (2025) further emphasize the genetic intersection between schizophrenia and NDDs, suggesting a continuum of vulnerability in which developmental anomalies predispose individuals to later psychiatric manifestations. Similarly, Dafsari et al. (2025) identify mutations in *EPG5*, a key autophagy factor, as linking early neurodevelopmental disorders with later neurodegenerative pathologies such as Parkinsonism. Ouellette and Lacoste (2021) extend this framework through the vascular continuum hypothesis, positing shared mechanisms of inflammation, oxidative stress, and blood–brain barrier dysfunction across developmental and degenerative spectra.

Collectively, these findings support the neurodevelopmental gradient model (Dell'Osso et al., 2019; Zoccante et al., 2021), which posits a biologically continuous process of brain maturation and differentiation. For educators, this has profound implications: it positions learning differences not as isolated anomalies but as expressions of developmental variation within a complex genetic-environmental matrix. The biological continuum reinforces the need for tiered educational responses, where interventions are adjusted according to functional profiles rather than categorical labels (Rodríguez et al., 2021).

From a pedagogical standpoint, the genetic continuum also demands longitudinal educational planning. Antolini and Colizzi (2023) argue that since neurodevelopmental expression evolves dynamically from childhood into adulthood, educational systems must adopt a lifespan perspective to track learners' adaptive functioning. This is particularly relevant as genetic predispositions interact continuously with environmental and instructional factors, shaping trajectories of resilience or vulnerability. Thus, genetic understanding informs but not determines education, guiding differentiated instruction that supports neuroplasticity rather than reinforcing deterministic labels.

COGNITIVE PHENOTYPES AND SHARED FUNCTIONAL NETWORKS

Neurocognitive research corroborates the continuum perspective through evidence of shared functional impairments across NDDs. Herreras (2025) identifies consistent executive-functioning deficits, such as difficulties in inhibition, cognitive flexibility, and working memory, across autism, ADHD, and specific learning disorders, suggesting a common functional substrate. Similarly, Canu et al. (2022a, 2022b) demonstrate that visual search and ocular-motor control, often disrupted in ASD and ADHD, form a continuum of attentional and perceptual control rather than disorder-specific deficits. These findings challenge traditional remediation models that isolate conditions, instead advocating for transdiagnostic cognitive training that targets overlapping functional systems.

Neuroimaging evidence aligns with this convergence. Louveau et al. (2023) argue that neural networks associated with executive functioning, language, and social cognition exhibit graded variations across diagnostic categories, emphasizing the artificiality of categorical distinctions. Such evidence suggests that educational interventions should prioritize core cognitive domains such as executive control, attention regulation, social inference, whose optimization benefits learners across the neurodevelopmental spectrum.

Educational research mirrors these neurological findings. Leifler (2022) and Astle et al. (2022) highlight the relevance of transdiagnostic pedagogical models, where instructional differentiation is guided by functional assessment rather than diagnostic classification. For example, executive-function-based interventions, such as self-monitoring checklists, structured routines, and goal-management training, have been shown to improve learning outcomes across heterogeneous student groups. This reinforces the continuum hypothesis by demonstrating that cognitive support strategies can be universally beneficial while still adaptable to individual neurocognitive profiles.

The neurodevelopmental continuum also manifests in patterns of adaptive functioning, how individuals translate cognitive capacities into everyday skills. Isaksson et al. (2025) demonstrate, through a co-twin control design, that adaptive functioning varies continuously across neurodevelopmental conditions, independent of formal diagnosis. This aligns with the perspective of Pallanti and Salerno (2023), who propose that neurodevelopmental differences should be viewed through a dimensional lens of adaptation rather than pathology. Within educational practice, this translates into recognizing that functional outcomes, self-regulation, communication, social interaction, are influenced not only by neurological architecture but also by contextual affordances such as inclusive pedagogy and environmental scaffolding.

Empirical research supports this ecological interpretation. Hansen et al. (2018) and Francés et al. (2023) document extensive comorbidity and overlapping functional impairments among children referred to mental health services, suggesting that adaptive challenges are the cumulative expression of shared neurodevelopmental vulnerabilities. Educationally, this requires moving beyond categorical inclusion (e.g., autism-friendly classrooms) toward functionally inclusive learning environments that accommodate a broad range of sensory, cognitive, and behavioral needs.

Furthermore, educational interventions informed by the continuum hypothesis must be preventive as well as responsive. Early universal screening, flexible curriculum adaptation, and dynamic progress monitoring, hallmarks of the RtI framework (Rodríguez et al., 2021), can identify emerging developmental difficulties before they consolidate into fixed disabilities. These systemic measures align with the transdiagnostic revolution (Astle et al., 2022), embedding developmental monitoring within everyday pedagogy rather than deferring it to clinical diagnosis.

FROM NEUROSCIENCE TO EDUCATIONAL DESIGN

The challenge for modern education lies not in assimilating neurobiological data, but in translating it into meaningful pedagogical practice. Aravamuthan et al. (2024) emphasize that understanding NDDs requires a conceptual framework of developmental plasticity, where brain and behavior co-evolve through continuous learning experiences. In the classroom, this principle translates to adaptive teaching environments that mirror the brain's capacity for dynamic reorganization. Rather than teaching to fixed deficits, educators can design learning contexts that stimulate cognitive networks across modalities, visual, auditory, kinesthetic, thus engaging distributed systems implicated in the continuum (Louveau et al., 2023).

Neuroscientific and educational collaboration therefore becomes central to continuum-informed education. The work of Gachechiladze and Dougherty (2024) underscores that neurodevelopmental genetic associations extend "across the translational space–time continuum," highlighting the need for cross-disciplinary dialogue. This translational ethos calls for educators to act as mediators between scientific knowledge and lived classroom realities. Through evidence-based strategies, explicit executive function instruction, assistive technology integration, and environmental modifications, educators operationalize the continuum model within practical, inclusive pedagogy.

Finally, Morris et al. (2025) argue for moving away from psychopathological frameworks toward dimensional models of neurodiversity that recognize adaptive potential. Within such a paradigm, education becomes not a site of normalization but of empowerment: a developmental ecosystem that cultivates each learner's strengths while accommodating diversity in cognition and behavior. The neurodevelopmental continuum, therefore, provides the theoretical foundation for a neuro-inclusive educational philosophy, one that views diversity as an organizing principle rather than an exception to be managed.

EDUCATIONAL IMPLICATIONS AND THE PEDAGOGY OF THE NEURODEVELOPMENTAL CONTINUUM

The clinical-to-educational translation of the neurodevelopmental continuum demands a reorientation of how schools conceptualize difference. Historically, educational systems have operated within categorical diagnostic frameworks, mirroring medical taxonomies such as DSM and ICD classifications. While these models provided organizational clarity, they have also perpetuated fragmented intervention systems, where students are supported based on labels rather than dynamic profiles of functioning (Dell'Osso et al., 2019; Morris-Rosendahl & Crocq, 2020).

Under the continuum hypothesis, neurodevelopmental variation is understood as part of a continuous distribution of human cognitive, emotional, and adaptive traits (Antolini & Colizzi, 2023). This reframes the educational challenge from *categorizing deficits* to *understanding profiles of functioning*. Educators, psychologists, and clinicians thus collaborate to map students' strengths and vulnerabilities across multiple domains, executive, sensory, linguistic, socio-emotional, and adaptive functioning, rather than focusing solely on diagnosis-based accommodations (Herreras, 2025; Louveau et al., 2023).

Such an approach aligns with transdiagnostic, and dimensional models proposed in contemporary psychiatric and neurodevelopmental research (Astle et al., 2022; Hoza & Shoulberg, 2024; Michelini et al., 2024). In educational contexts, this means adopting flexible pedagogical strategies that address overlapping needs common across autism spectrum disorder (ASD), ADHD, and specific learning disorders (SLD), including attention regulation, working memory, and emotional self-regulation (Aravamuthan et al., 2024; Herreras, 2025). By doing so, schools can shift from "label-based support" to "function-based inclusion."

Emerging evidence indicates that school environments can serve as neurodevelopmental ecosystems, shaping neural plasticity, executive functioning, and social learning throughout development (Leifler, 2022; Rodríguez et al., 2021). Within the neurodevelopmental continuum framework, the educational setting is not merely a backdrop for instruction but an active neurocognitive and psychosocial milieu that can either reinforce or mitigate developmental disparities.

Canu et al. (2022a, 2022b) demonstrated that perceptual and attentional control functions, fundamental to classroom participation, exist along a continuum across NDDs, suggesting that universal design principles in learning environments may benefit all learners. This aligns with Louveau et al.'s (2023) call to move "beyond classifications" toward integrated interventions that transcend categorical diagnostic lines. Educational spaces, therefore, can be designed to accommodate neurodiversity gradients rather than isolated disabilities.

A practical implication of this approach is the Universal Design for Learning (UDL) framework, which provides multiple means of engagement, representation, and expression. Such frameworks operationalize the continuum model by recognizing that the same learning outcome may be achieved through different cognitive and sensory pathways (Antolini & Colizzi, 2023; Pallanti & Salerno, 2023). In this light, neurodiversity is not a challenge to be remediated but a variable to be designed for.

Antolini and Colizzi (2023) emphasize that neurodevelopmental trajectories persist and evolve across the lifespan, often revealing new challenges in adolescence and adulthood. The "childhood-bound" perception of NDDs limits the capacity of educational systems to prepare students for postsecondary education and employment. Educational inclusion, therefore, must adopt a lifespan developmental model that ensures continuity of care and support from childhood through adulthood (Aravamuthan et al., 2024).

The continuum perspective provides an epistemological bridge between pediatric and adult services (Iannotta et al., 2025), advocating for an integrated support system in which neurodevelopmental differences are viewed as lifelong profiles of functioning rather than time-limited disorders. This calls for a paradigm shift in teacher education, which traditionally separates special and general pedagogical training. Teachers trained within the continuum framework develop neuroeducational literacy, the capacity to translate neuroscientific and clinical insights into pedagogical strategies (Andréou et al., 2025; Leifler, 2022). Moreover, executive function interventions, including metacognitive scaffolding, working memory training, and emotional regulation supports, should be embedded within the school curriculum as preventive and developmental tools (Herreras, 2025). These approaches resonate with the Response to Intervention (RTI) model proposed by Rodríguez et al. (2021), which emphasizes data-driven tiered supports that can flexibly address varying degrees of learning and behavioral needs within the same classroom.

Education systems have a pivotal role as mediating institutions between clinical neurodevelopmental science and societal inclusion (Dell'Osso et al., 2019; Pallanti & Salerno, 2023). To operationalize the neurodevelopmental continuum, collaboration between educators, psychologists, neuroscientists, and policymakers is essential (Aravamuthan et al., 2024). The translational framework described by Gachechiladze and Dougherty (2024) supports this integration, showing that molecular and genetic findings have direct implications for understanding cognitive and adaptive variability in educational contexts.

Interdisciplinary teams can also enhance the ecological validity of educational assessment. Rather than relying solely on standardized testing, educators should incorporate context-sensitive evaluation, considering environmental, sensory, and social moderators of student functioning (Bonti et al., 2024; Morris-Rosendahl & Crocq, 2020). Similarly, the neurodevelopmental-to-neurodegenerative continuum described by Ouellette and Lacoste (2021) underscores the importance of lifelong educational engagement in maintaining cognitive reserve and mental flexibility. In practical terms, this means that schools function as health-promoting environments, offering cognitive and emotional enrichment that may buffer against later neurodegenerative risks. Educational inclusion, therefore, becomes both an equity practice and a public health intervention.

Policy frameworks must evolve alongside theoretical advances in neurodevelopmental science. Current policies often define eligibility for services based on binary diagnostic criteria, which exclude many individuals who fall within the "gray zones" of the continuum (Dewey, 2018; Hansen et al., 2018; Thapar et al., 2017). Moving toward dimensional legislation, that is, policy recognizing the spectrum of neurodevelopmental variation, would ensure access to accommodations, flexible assessment, and support structures for all students with functional needs, not only for those formally diagnosed. The work of Francés et al. (2023) and Eberhard et al. (2022) illustrates that the prevalence of comorbid and subthreshold profiles is much higher than previously estimated, reinforcing the necessity of flexible educational policy. Furthermore, inclusive education must be informed by transdiagnostic evidence that cognitive, emotional, and social impairments are shared across conditions (Deserno et al., 2023; Michelini et al., 2024; Morris et al., 2025).

At a systemic level, the educational environment becomes a diagnostic and therapeutic ecosystem, where the goal is not normalization but participation, learning, and self-regulation across the lifespan (Zoccante et al., 2021). This redefinition of inclusion, from access to adaptation, represents the pedagogical embodiment of the neurodevelopmental continuum hypothesis.

RETHINKING NEURODEVELOPMENT THROUGH AN EDUCATIONAL LENS

The clinical continuum hypothesis invites a profound reframing of how educational systems conceptualize and respond to neurodevelopmental diversity. Rather than perceiving NDDs as categorical deficits, contemporary transdiagnostic and gradient models (Astle et al., 2022; Dell'Osso et al., 2019; Michelini et al., 2024) suggest a continuum of neurocognitive functioning, distributed across all individuals. This paradigm resonates strongly with the pedagogical notion of neurodiversity, which posits that variations in cognitive and behavioral functioning represent natural forms of human heterogeneity rather than pathology (Morris et al., 2025).

Within this perspective, education becomes not merely compensatory but *transformative*: it offers neuroplastic contexts where cognitive profiles can evolve in interaction with supportive environments (Leifler, 2022; Rodríguez et al., 2021). The implication is that schools function as neurodevelopmental ecosystems, shaping trajectories that may either exacerbate or mitigate functional challenges over time (Aravamuthan et al., 2024). Critically, these findings demand a rethinking of the boundaries between clinical and educational practice. Traditional divisions, where medical disciplines diagnose and educators implement, no longer suffice in capturing the complexity of overlapping phenotypes (Bonti et al., 2024; Hansen et al., 2018). Instead, a model of *bidirectional collaboration* is emerging, in which educational research informs neurodevelopmental diagnostics and vice versa (Antolini & Colizzi, 2023; Pallanti & Salerno, 2023).

Furthermore, contemporary evidence from neuroimaging and behavioral genetics reveals that the boundaries between NDDs are porous and often artificial (Deserno et al., 2023; Louveau et al., 2023). Shared neurobiological pathways, particularly those involving executive function, attention, and social cognition, underpin conditions such as autism spectrum disorder (ASD), ADHD, and specific learning disorders (Herreras, 2025; Canu et al., 2022a, 2022b). Dell'Osso et al. (2019) described this as a *neurodevelopmental gradient*, emphasizing dimensional variation rather than discrete categorization.

Educationally, this translates into a need for transdiagnostic intervention frameworks that target core cognitive and emotional processes across conditions (Hoza & Shoulberg, 2024). For example, executive functioning and metacognitive skills are foundational not only for academic success but also for adaptive functioning across the NDD spectrum (Herreras, 2025; Rodríguez et al., 2021). Rather than offering condition-specific programs, schools might develop *function-specific learning environments*, where attention regulation, emotional control, and working memory are systematically scaffolded for all learners. Michelini et al. (2024) and Morris et al. (2025) advocate the inclusion of NDDs within broader transdiagnostic psychiatric frameworks, aligning with a dimensional model of neurodiversity. Such alignment bridges educational inclusion with clinical innovation, moving beyond remediation toward *participatory neuroeducation*. This model emphasizes students' agency, adaptive functioning (Isaksson et al., 2025), and social integration over mere symptom reduction.

NEURODEVELOPMENT ACROSS THE LIFESPAN

The transition from childhood to adulthood presents a critical phase in the neurodevelopmental continuum. Research demonstrates that while neuroplasticity decreases with age, adaptive learning remains possible across the lifespan (Antolini & Colizzi, 2023; Thapar et al., 2017). Educational institutions, therefore, must extend their reach beyond compulsory schooling to encompass lifelong learning ecosystems that support adults with NDDs in vocational, higher education, and community contexts.

Findings from longitudinal and genetic studies further support the lifespan persistence of neurodevelopmental traits (Iannotta et al., 2025; Wong et al., 2025). For instance, SETBP1 variants and other gene-environment interactions demonstrate overlapping pathways between childhood NDDs and adult psychiatric or neurodegenerative conditions (Dafsari et al., 2025; Ouellette & Lacoste, 2021). This overlap substantiates the *developmental continuity hypothesis*, reinforcing the necessity of consistent educational and psychosocial support across decades.

From an educational point of view this perspective challenges the tendency to "graduate" neurodivergent individuals out of support systems upon adulthood. Instead, it advocates for adaptive pedagogical design, integrating universal design for learning (UDL), vocational mentoring, and environmental modifications to sustain cognitive participation and self-efficacy (Leifler, 2022). In such frameworks, neurodevelopment is not an event bounded by childhood, but an evolving process shaped by social and institutional contexts.

Additionally, from a policy perspective, the neurodevelopmental continuum carries profound implications for equity, ethics, and governance. Traditional nosological conventions, based on categorical classifications (DSM-5; ICD-11), are increasingly misaligned with emerging scientific evidence (Morris-Rosendahl & Crocq, 2020; Stein et al., 2020). A continuum-based model requires policy systems that recognize individual variability rather than diagnostic thresholds as the primary determinant of educational support (Francés et al., 2023). Ethically, the movement toward dimensionality aligns with the principle of *developmental justice*, the right of every individual to environments that nurture cognitive and emotional growth (Pallanti & Salerno, 2023). This entails rethinking funding structures, diagnostic criteria, and training standards for educators and clinicians alike. As Dell'Osso et al. (2019) and Gachechiladze & Dougherty (2024) argue, integrating molecular genetics with psychosocial data can bridge the translational gap between laboratory and classroom, enabling data-driven yet person-centered educational policies. Moreover, as Louveau et al. (2023) emphasize, the future of NDD research lies "beyond classifications", toward interventions that flexibly respond to the individual's neurobiological and social profile. Educational policy must thus evolve toward neuroecological inclusion, where architecture, curriculum, and pedagogy coalesce to create cognitively ergonomic and socially supportive environments for all learners.

Future research should pursue three converging directions that bridge neuroscience, education, and clinical practice. First, longitudinal neuroeducational mapping is essential for integrating neuroscientific and educational datasets to monitor cognitive, emotional, and adaptive trajectories from early childhood through adulthood, thereby revealing how educational environments interact with neurodevelopmental processes across time (Isaksson et al., 2025; Zoccante et al., 2021). Second, interdisciplinary professional development must become a central objective, equipping educators with neurobiological literacy and clinicians with pedagogical understanding to promote authentic translational collaboration between schools and healthcare systems (Aravamuthan et al., 2024; Leifler, 2022). Third, ecological neuroeducation models should be designed and empirically tested to create adaptive learning environments aligned with the neurodevelopmental gradient, emphasizing flexibility, sensory regulation, and executive function support to enhance participation and learning outcomes for all students (Antolini & Colizzi, 2023; Canu et al., 2022a). Critically, this integration must remain ethically grounded, prioritizing the dignity, autonomy, and self-definition of neurodivergent individuals. As Dell'Osso et al. (2019) note, the gradient model resists both pathologization and idealization, affirming diversity while recognizing the need for structured, evidence-based supports.

The clinical continuum hypothesis, viewed through the educational lens, reveals a paradigm shift in understanding human diversity. Neurodevelopment is not a fixed biological trajectory but a lifelong dialogue between brain, behavior, and environment. Educational systems, as the most pervasive social institutions in human life, possess the transformative potential to translate neurodevelopmental science into inclusive practice. To realize this vision, interdisciplinary collaboration must deepen, linking molecular biology with pedagogy, policy with psychology, and ethics with neuroscience. A continuum approach reframes neurodevelopmental differences not as boundaries of normality but as gradients of potential, each requiring adaptive ecosystems to flourish. Such a shift holds the promise of dissolving the traditional divisions between medicine and education, enabling a society where every neurodevelopmental profile finds both understanding and opportunity.

Table 1. The clinical continuum of neurodevelopmental disorders across the lifespan

Developmental Stage	Dominant Neurodevelopmental Manifestations	Functional and Psychosocial Challenges	Educational and Support Priorities
Early Childhood	Early language delays, atypical social engagement, attentional regulation difficulties, emerging executive function vulnerabilities	Delayed communication, limited self-regulation, difficulties in social interaction, increased parental stress	Early identification, developmental screening, parent-mediated interventions, language and play-based support
Preschool and Early School Years	More defined neurodevelopmental profiles (e.g., SLD, ADHD, ASD), difficulties in attention, language, and emerging academic skills	Learning readiness difficulties, behavioral regulation challenges, peer interaction problems	Individualized educational planning, structured classroom environments, targeted academic and socioemotional interventions
Middle Childhood	Academic skill deficits, executive function difficulties, comorbidity across neurodevelopmental domains	Academic underachievement, reduced self-esteem, increased anxiety or behavioral difficulties	Multidimensional assessment, academic accommodations, executive function training, psychosocial support
Adolescence	Persistence or transformation of symptoms, emotional dysregulation, increased demands on autonomy and organization	Identity challenges, risk-taking behaviors, school disengagement, peer and emotional difficulties	Transition planning, mental health support, self-advocacy skill development, flexible educational pathways

continued on following page

Table 1. Continued

Developmental Stage	Dominant Neurodevelopmental Manifestations	Functional and Psychosocial Challenges	Educational and Support Priorities
Young Adulthood	Residual attentional, executive, or social communication difficulties, increased functional demands	Challenges in higher education, vocational training, and independent living	Disability services, academic and workplace accommodations, mentoring and transition support
Adulthood	Long-term neurodevelopmental imprint, adaptive coping strategies, variable symptom expression	Employment instability, social participation challenges, mental health vulnerability	Lifelong learning opportunities, inclusive employment practices, psychosocial and community-based support

REFERENCES

Andreou, G., Lymperopoulou, V., & Aslanoglou, V. (2025, September). *Structural language in neurodevelopmental disorders: Comparison between Autism Spectrum Disorder (ASD) and Developmental Language Disorder (DLD).* In *Frontiers in Education* (Vol. 10, p. 1641303). Frontiers. DOI: 10.3389/feduc.2025.1641303

Antolini, G., & Colizzi, M. (2023, April). *Where do neurodevelopmental disorders go? Casting the eye away from childhood towards adulthood.* In *Healthcare* (Vol. 11, No. 7, p. 1015). MDPI. DOI: 10.3390/healthcare11071015

Aravamuthan, B., Turnacioglu, S. O., & Gropman, A. L. (2024). *Neurodevelopmental Disabilities: Conceptual Framework.* In *Swaiman's Pediatric Neurology* (pp. 636–641). Elsevier. DOI: 10.1016/B978-0-443-10944-7.00068-8

Astle, D. E., Holmes, J., Kievit, R., & Gathercole, S. E. (2022). Annual Research Review: The transdiagnostic revolution in neurodevelopmental disorders. *Journal of Child Psychology and Psychiatry, and Allied Disciplines, 63*(4), 397–417. DOI: 10.1111/jcpp.13481 PMID: 34296774

Bonti, E., Zerva, I. K., Koundourou, C., & Sofologi, M. (2024). The high rates of comorbidity among neurodevelopmental disorders: Reconsidering the clinical utility of distinct diagnostic categories. *Journal of Personalized Medicine, 14*(3), 300. DOI: 10.3390/jpm14030300 PMID: 38541042

Canu, D., Ioannou, C., Müller, K., Martin, B., Fleischhaker, C., Biscaldi, M., Beauducel, A., Smyrnis, N., van Elst, L. T., & Klein, C. (2022a). Evidence towards a continuum of impairment across neurodevelopmental disorders from basic ocular-motor tasks. *Scientific Reports, 12*(1), 16521. DOI: 10.1038/s41598-022-19661-z PMID: 36192503

Canu, D., Ioannou, C., Müller, K., Martin, B., Fleischhaker, C., Biscaldi, M., Beauducel, A., Smyrnis, N., van Elst, L. T., & Klein, C. (2022b). Visual search in neurodevelopmental disorders: Evidence towards a continuum of impairment. *European Child & Adolescent Psychiatry, 31*(8), 1–18. DOI: 10.1007/s00787-021-01756-z PMID: 33751240

Dafsari, H. S., Deneubourg, C., Singh, K., Maroofian, R., Suprenant, Z., Kho, A. L., Ingham, N. J., Steel, K. P., Sheshadri, P., Baur, F., Hentrich, L., Gerisch, B., Zamani, M., Alves, C., Siddiqui, A., Dafsari, H. S., Salari, M., Lang, A. E., Harris, M., & Jungbluth, H. (2025). Mutations in the key autophagy tethering factor EPG5 link neurodevelopmental and neurodegenerative disorders including early-onset parkinsonism. *Annals of Neurology*, *98*(5), 932–950. Advance online publication. DOI: 10.1002/ana.78013 PMID: 41053928

Dell'Osso, L., Lorenzi, P., & Carpita, B. (2019). The neurodevelopmental continuum towards a neurodevelopmental gradient hypothesis. *Journal of Psychopathology*, *25*(4), 179–182. https://hdl.handle.net/11568/1172285

Deserno, M. K., Bathelt, J., Groenman, A. P., & Geurts, H. M. (2023). Probing the overarching continuum theory: Data-driven phenotypic clustering of children with ASD or ADHD. *European Child & Adolescent Psychiatry*, *32*(10), 1909–1923. DOI: 10.1007/s00787-022-01986-9 PMID: 35687205

Dewey, D. (2018). What is comorbidity and why does it matter in neurodevelopmental disorders? *Current Developmental Disorders Reports*, *5*(4), 235–242. DOI: 10.1007/s40474-018-0152-3

Eberhard, D., Billstedt, E., & Gillberg, C. (2022). Neurodevelopmental disorders and comorbidity in young adults attending a psychiatric outpatient clinic. *Psychiatry Research*, *313*, 114638. DOI: 10.1016/j.psychres.2022.114638 PMID: 35597136

Francés, L., Ruiz, A., Soler, C. V., Francés, J., Caules, J., Hervás, A., Carretero, C., Cardona, B., Quezada, E., Fernández, A., & Quintero, J. (2023). Prevalence, comorbidities, and profiles of neurodevelopmental disorders according to the DSM-5-TR in children aged 6 years old in a European region. *Frontiers in Psychiatry*, *14*, 1260747. DOI: 10.3389/fpsyt.2023.1260747 PMID: 38025459

Gachechiladze, M. A., & Dougherty, J. D. (2024). Neurodevelopmental genetic associations across the translational space-time continuum. *Biological Psychiatry*, *95*(9), 825–827. DOI: 10.1016/j.biopsych.2024.02.1010 PMID: 38599709

Hansen, B. H., Oerbeck, B., Skirbekk, B., Petrovski, B. É., & Kristensen, H. (2018). Neurodevelopmental disorders: Prevalence and comorbidity in children referred to mental health services. *Nordic Journal of Psychiatry*, *72*(4), 285–291. DOI: 10.1080/08039488.2018.1444087 PMID: 29488416

Herreras, E. B. (2025). Executive functioning profiles in children with neurodevelopmental disorders. *Behavioral Sciences (Basel, Switzerland)*, *15*(9), 1256. DOI: 10.3390/bs15091256 PMID: 41009286

Hoza, B., & Shoulberg, E. K. (2024). Transdiagnostic considerations are critical to understanding childhood neurodevelopmental disorders. *Frontiers in Human Neuroscience*, *18*, 1385873. DOI: 10.3389/fnhum.2024.1385873 PMID: 38774129

Iannotta, F., La Monica, I., Di Iorio, M. R., Freda, V., Sica, A., de Bartolomeis, A., Pastore, L., Iasevoli, F., & Lombardo, B. (2025). A paradigmatic case of genetic overlap between neurodevelopment disorders and schizophrenia aligning with the neurodevelopmental continuum hypothesis. *International Journal of Molecular Sciences*, *26*(9), 3970. DOI: 10.3390/ijms26093970 PMID: 40362210

Isaksson, J., Eklund, F., Remnélius, K. L., Black, M. H., & Bölte, S. (2025). Neurodevelopmental conditions and adaptive functioning–a co-twin control study. *Journal of Child Psychology and Psychiatry, and Allied Disciplines*, jcpp.70073. Advance online publication. DOI: 10.1111/jcpp.70073 PMID: 41188694

Leifler, E. (2022). *Educational inclusion for students with neurodevelopmental conditions.* Karolinska Institutet. https://hdl.handle.net/10616/48104

Louveau, C., Ellul, P., Iftimovici, A., Dubreucq, J., Laidi, C., Leyrolle, Q., Purper-Ouakil, D., Jacquemont, S., Lyonnet, S., Barthélémy, C., Krebs, M.-O., Bai, J., Olivier, P., & Chaumette, B. (2023). Neurodevelopmental disorders (NDD) without boundaries: Research and interventions beyond classifications. *Journal of Neural Transmission (Vienna, Austria)*, *130*(3), 473–479. DOI: 10.1007/s00702-023-02586-w PMID: 36719463

Michelini, G., Carlisi, C. O., Eaton, N. R., Elison, J. T., Haltigan, J. D., Kotov, R., Krueger, R. F., Latzman, R. D., Li, J. J., Levin-Aspenson, H. F., Salum, G. A., South, S. C., Stanton, K., Waldman, I. D., & Wilson, S. (2024). Where do neurodevelopmental conditions fit in transdiagnostic psychiatric frameworks? Incorporating a new neurodevelopmental spectrum. *World Psychiatry; Official Journal of the World Psychiatric Association (WPA)*, *23*(3), 333–357. DOI: 10.1002/wps.21225 PMID: 39279404

Morris, I., Michelini, G., & Wilson, S. (2025). Moving toward transdiagnostic dimensional models of neurodiversity and mental health (and away from models of psychopathology). *Journal of Psychopathology and Clinical Science*, *134*(5), 483–485. DOI: 10.1037/abn0001007 PMID: 40310186

Morris-Rosendahl, D. J., & Crocq, M. A. (2020). Neurodevelopmental disorders, The history and future of a diagnostic concept. *Dialogues in Clinical Neuroscience*, *22*(1), 65–72. DOI: 10.31887/DCNS.2020.22.1/macrocq PMID: 32699506

Ouellette, J., & Lacoste, B. (2021). From neurodevelopmental to neurodegenerative disorders: The vascular continuum. *Frontiers in Aging Neuroscience*, *13*, 749026. DOI: 10.3389/fnagi.2021.749026 PMID: 34744690

Pallanti, S., & Salerno, L. (2023). Neurodevelopmental disorders (NDDs): Beyond the clinical definition and translational approach. *Children (Basel, Switzerland)*, *10*(1), 99. DOI: 10.3390/children10010099 PMID: 36670649

Rodríguez, C., Areces, D., García, T., Cueli, M., & Gonzalez-Castro, P. (2021). Neurodevelopmental disorders: An innovative perspective via the response to intervention model. *World Journal of Psychiatry*, *11*(11), 1017–1026. DOI: 10.5498/wjp.v11.i11.1017 PMID: 34888170

Stein, D. J., Szatmari, P., Gaebel, W., Berk, M., Vieta, E., Maj, M., de Vries, Y. A., Roest, A. M., de Jonge, P., Maercker, A., Brewin, C. R., Pike, K. M., Grilo, C. M., Fineberg, N. A., Briken, P., Cohen-Kettenis, P. T., & Reed, G. M. (2020). Mental, behavioral and neurodevelopmental disorders in the ICD-11: An international perspective on key changes and controversies. *BMC Medicine*, *18*(1), 21. DOI: 10.1186/s12916-020-1495-2 PMID: 31983345

Thapar, A., Cooper, M., & Rutter, M. (2017). Neurodevelopmental disorders. *The Lancet. Psychiatry*, *4*(4), 339–346. DOI: 10.1016/S2215-0366(16)30376-5 PMID: 27979720

Wong, M. M., Kampen, R. A., Braden, R. O., Alagöz, G., Hildebrand, M. S., Dingemans, A. J., Corbally, J., den Hoed, J., Mendoza, E., Claassen, W. J. J., Barnett, C., Barnett, M., Brusco, A., Carli, D., de Vries, B. B. A., Elmslie, F., Ferrero, G. B., Jansen, N. A., van de Laar, I. M. B. H., & Fisher, S. E. (2025). SETBP1 variants outside the degron disrupt DNA-binding, transcription and neuronal differentiation capacity to cause a heterogeneous neurodevelopmental disorder. *Nature Communications*, *16*(1), 9021. DOI: 10.1038/s41467-025-64074-x PMID: 41073373

Zoccante, L., Ciceri, M. L., Chamitava, L., Di Gennaro, G., Cazzoletti, L., Zanolin, M. E., Darra, F., & Colizzi, M. (2021). Postural control in childhood: Investigating the neurodevelopmental gradient hypothesis. *International Journal of Environmental Research and Public Health*, *18*(4), 1693. DOI: 10.3390/ijerph18041693 PMID: 33578752

Chapter 4
Neurodevelopmental Disorders During Early Childhood

ABSTRACT

This chapter presents early childhood as a period where rapid neural growth makes development especially sensitive to both risk and enrichment. It describes how genetic factors, hormonal regulation, metabolic processes, and environmental exposures converge to influence early trajectories linked to NDDs. Early differences in motor control, sensory reactivity, communication, attention, and social engagement offer valuable clues, though they may shift as children mature. The chapter highlights the importance of family involvement, culturally grounded practices, and accessible early education, noting that intervention is most effective when flexible, sustained, and aligned with developmental timing. By framing early neurodevelopmental variation as part of a broader continuum, the chapter underscores the role of integrated systems in supporting resilience and long-term participation.

INTRODUCTION

Early childhood represents one of the most dynamic and sensitive phases of human neurodevelopment, characterized by rapid growth, intense synaptogenesis, and profound plasticity. During this period, the brain's architecture is molded by a complex interplay of genetic, biological, and environmental factors, laying the foundation for later cognitive, emotional, and behavioral functioning. This developmental window is critical for establishing neural circuits that underlie language, attention, social cognition, and executive control, functions that are often disrupted in neurodevelopmental disorders (NDDs) (Bale et al., 2010; Meredith, 2015). Because

DOI: 10.4018/979-8-3693-6331-7.ch004

early neural processes are both highly plastic and highly vulnerable, perturbations during this time can produce long-lasting alterations in brain structure and function that manifest as atypical developmental trajectories (Cioni et al., 2016).

Neurodevelopmental disorders arise from disruptions in the timing or integrity of key neurobiological processes, often beginning before birth and continuing throughout early childhood. Genetic susceptibility, prenatal environmental insults, and postnatal exposures interact to influence neurodevelopmental outcomes (Carlsson et al., 2021). Research has shown that early adversity, including toxicant exposure, hormonal dysregulation, or chronic stress, can alter neuronal connectivity and affect the maturation of brain regions involved in sensory processing and self-regulation (Bale et al., 2010; Makris et al., 2023). These findings underscore that early childhood is not merely a preparatory phase for later learning but a biologically sensitive period in which experience exerts a disproportionate influence on brain organization.

From a clinical perspective, the early years provide a unique opportunity for preventive and therapeutic intervention. Neuroplasticity during infancy and preschool years enables compensatory mechanisms and reorganization of neural networks when appropriate interventions are introduced (Inguaggiato et al., 2017). Early evidence reviewed by Micai et al. (2020) suggests that the first years of life contain meaningful behavioral signals that may indicate emerging neurodevelopmental difficulties. While no single marker is definitive, patterns involving early social engagement, communication, motor coordination, and sensory responsiveness collectively offer valuable insight into developmental risk, yet it is emphasized the need for more consistent methods and long-term studies, underscoring that early identification remains both possible and important for supporting children's developmental trajectories.

Conversely, missed opportunities for detection or treatment can entrench atypical developmental patterns, leading to cumulative functional impairments across social, academic, and emotional domains (Duncan & Matthews, 2018). Hadders-Algra (2021) emphasizes that early diagnostic efforts must consider the age-dependent expression of symptoms: the same underlying neurobiological disturbance may produce different behavioral phenotypes at 6 months versus 3 years. This developmental fluidity complicates early identification but also highlights why interventions need to be dynamic and responsive to changing neurodevelopmental demands.

Environmental context further shapes early trajectories. Factors such as prenatal exposure to endocrine disruptors, malnutrition, and psychosocial deprivation have been linked to neurodevelopmental risk (Scattolin et al., 2022; Zhang et al., 2025). At the same time, enriched caregiving environments, responsive parenting, and early educational engagement can buffer biological vulnerabilities and foster adaptive neural development (Bosch et al., 2022). Thus, risk and resilience in early NDD expression are the outcomes of an ongoing dialogue between biology and experience rather than fixed determinants.

A growing body of neuroimaging research suggests that the first five years of life are decisive for establishing cortical connectivity patterns associated with later executive and socio-emotional functioning (Cioni et al., 2016; Meredith, 2015). During this period, even subtle disruptions in sensory processing or attention regulation may presage later diagnoses of autism spectrum disorder, attention-deficit/hyperactivity disorder, or learning disabilities. The conceptual shift from categorical diagnoses to developmental continua has further illuminated that early difficulties often cut across diagnostic boundaries (Finlay-Jones et al., 2019). Consequently, researchers and practitioners increasingly emphasize early transdiagnostic screening and support rather than waiting for symptom consolidation in later childhood.

Recognizing the early developmental window as both a risk and opportunity period reframes how professionals across disciplines approach NDDs. Rather than viewing early symptoms as precursors of fixed conditions, this perspective situates them within a dynamic system of interacting biological and environmental forces. For educators, clinicians, and policymakers alike, this underscores the importance of coordinated surveillance, early support, and family-centered care during infancy and preschool years.

NEUROBIOLOGICAL AND ENVIRONMENTAL DETERMINANTS IN EARLY NEURODEVELOPMENT

Understanding neurodevelopmental disorders (NDDs) during early childhood requires examining how neurobiological processes unfold within their environmental contexts (King-Dowling et al., 2019). Early brain development is a progressive orchestration of cell proliferation, migration, differentiation, and synaptic refinement, processes guided by genetic instructions but profoundly shaped by experience and environmental exposures (Cioni et al., 2016). When the timing or sequence of these processes is disrupted, even subtly, the consequences can manifest as lasting cognitive, sensory, or emotional difficulties. The early years thus represent a developmental crucible in which both biological vulnerabilities and environmental stressors can converge, amplifying risk for atypical outcomes (Meredith, 2015).

Neural Plasticity, Sensitive Periods, and Early Vulnerability

Neuroplasticity, the brain's capacity to reorganize structure and function in response to experience, is greatest during early childhood. This flexibility enables rapid learning and adaptation but also increases susceptibility to perturbations (Meredith, 2015). Sensitive periods define windows of heightened responsiveness when specific neural circuits are particularly malleable. For example, early disruptions in sensory input or emotional regulation can alter the maturation of fronto-striatal and limbic pathways, which later support attention, social behavior, and executive control (Cioni et al., 2016). In typically developing infants, these systems integrate progressively; in contrast, in children at risk for NDDs, connectivity may be inefficient or delayed, limiting compensatory reorganization.

Neuroplasticity is not uniformly beneficial; it amplifies both positive and negative experiences. As Inguaggiato et al. (2017) highlight, the same mechanisms that allow early rehabilitation can also consolidate maladaptive patterns if early adversity or inadequate stimulation occurs. Early intervention capitalizes on plasticity to steer neural circuits toward more adaptive trajectories, but such interventions must align with developmental timing to be effective (Hadders-Algra, 2021). The interdependence of neural systems implies that perturbations in one domain, such as motor control or sensory integration, can cascade into cognitive or social impairments.

Genetic and Epigenetic Contributions

NDDs have complex genetic architectures involving multiple susceptibility loci and gene–environment interactions. Studies of twins and siblings reveal that heritability is significant but not deterministic; shared environmental influences also play meaningful roles (Carlsson et al., 2021). Genetic liability shapes the brain's baseline organization, influencing susceptibility to environmental insults. Recent advances suggest that epigenetic mechanisms, heritable modifications in gene expression without changes to DNA sequence, mediate how early experiences, including stress and caregiving quality, regulate neurodevelopmental risk (Bale et al., 2010). For instance, early-life stress can alter methylation patterns of genes associated with glucocorticoid regulation, affecting stress reactivity and emotional processing throughout life (Makris et al., 2023). Hannigan et al. (2023) further demonstrate that genetic liability interacts with developmental milestones: children with higher polygenic risk for NDDs exhibit subtle delays in early motor and language milestones even before formal diagnosis. These findings underscore that biological vulnerability is expressed developmentally, often preceding recognizable symptoms. Thus, identifying early deviations from typical developmental trajectories can signal underlying neurobiological divergence long before behavioral phenotypes stabilize.

Hormonal regulation is a crucial mediator of brain development during early life. Cortisol, oxytocin, and sex steroids influence neuronal differentiation, synaptogenesis, and myelination. When early life stress dysregulates these systems, it can shape the architecture of neural networks related to emotion and cognition (Makris et al., 2023). Chronic stress in infancy is associated with hyperactivation of the hypothalamic–pituitary–adrenal (HPA) axis, leading to structural alterations in the amygdala and hippocampus, regions essential for emotional regulation and memory (Bale et al., 2010).

Early adversity does not simply increase risk; it can shift developmental trajectories, altering how children respond to later experiences. Makris et al. (2023) emphasize that hormonal dysregulation during sensitive periods can have cascading effects on attention, social behavior, and executive functions, which are core features across multiple NDDs. Importantly, such stress-related pathways interact with genetic predispositions and environmental inputs, making early neurodevelopmental outcomes the product of cumulative and transactional influences.

Beyond genetics and stress, neurodevelopment depends on adequate cellular energy metabolism. The brain's high metabolic demand during early development means that even minor disruptions in mitochondrial or glucose pathways can influence neural maturation. Oyarzábal et al. (2021) identify metabolic dysfunction as a shared pathway in multiple NDDs, where impaired energy utilization limits synaptic activity and myelination. This metabolic fragility can be exacerbated by nutritional deficiencies, infections, or environmental toxins during early childhood. Consequently, addressing metabolic health early, through nutrition, early screening, and treatment of metabolic anomalies, may serve as a preventive strategy in vulnerable populations.

Environmental Exposures and Toxicants

While genetic and hormonal factors establish intrinsic vulnerability, environmental toxicants represent modifiable extrinsic risks. Exposure to endocrine disruptors such as bisphenol A (BPA) has been linked to altered neurodevelopmental outcomes in children under 12 years (Zhang et al., 2025). BPA interferes with estrogenic signaling critical for synaptic plasticity and cognitive function. Prenatal or early postnatal exposure has been associated with attention deficits, behavioral dysregulation, and reduced IQ, reflecting the brain's particular sensitivity to hormonal mimics during rapid development.

Scattolin et al. (2022) similarly emphasize the environmental dimension, noting that pollutants, psychosocial stressors, and socioeconomic disadvantage collectively impact neurodevelopment. Children in low-resource settings often experience cumulative exposure, such as poor nutrition, environmental toxins, and limited stimulation, that compounds neurobiological vulnerability. These findings highlight why early public health interventions must extend beyond the clinic to address environmental justice and social determinants of health.

Sociodemographic variables intersect with biological risk in shaping developmental outcomes. Bosch et al. (2022) report that prevalence and severity of NDDs among Spanish school-age children correlate with parental education and income, suggesting that environmental enrichment and access to early screening are protective. Early cognitive and socio-emotional development benefits from responsive caregiving and structured learning opportunities, which stimulate neuroplasticity in adaptive ways. Conversely, deprivation or chaotic environments can reinforce maladaptive neural pathways (Smythe et al., 2024).

Cultural context also determines how early symptoms are recognized and addressed. Families' interpretations of developmental differences influence help-seeking behaviors, affecting the timing of diagnosis and intervention (Magaya & Tafirenyika, 2025). Therefore, a multidisciplinary approach must consider not only biological and environmental mechanisms but also the sociocultural frameworks that mediate children's experiences and access to care.

The convergence of these factors, genetic, metabolic, hormonal, and environmental, illustrates the inherently multifactorial nature of early NDDs. Neurobiological development does not proceed in isolation but within an ecosystem of contextual influences. For instance, children exposed to environmental toxins may exhibit heightened vulnerability if genetic regulation of detoxification enzymes is less efficient (Zhang et al., 2025). Similarly, early stress may have amplified effects when hormonal systems are already dysregulated (Makris et al., 2023). Understanding NDDs in early childhood thus demands models that integrate molecular biology, neuroscience, and developmental ecology.

Importantly, neurodevelopmental outcomes are not predetermined by early adversity. The same plasticity that renders the brain vulnerable also enables recovery and adaptation. Protective environments that characterized by nurturing relationships, early learning experiences, and access to intervention, can mitigate biological risks and promote resilience (Hadders-Algra, 2021; Inguaggiato et al., 2017). This bidirectional dynamic reframes NDDs as developmental processes rather than static disorders, emphasizing the potential for positive change when timely supports are provided.

EARLY DEVELOPMENTAL MARKERS AND IDENTIFICATION PATHWAYS

Early identification of neurodevelopmental disorders (NDDs) is a cornerstone of effective intervention. Yet, detecting atypical trajectories during infancy and preschool years remains one of the field's most complex challenges. In early development, the distinction between variation and deviation is often blurred. Many features of NDDs, such as language delays, sensory sensitivities, or social disengagement, may emerge gradually or fluctuate across developmental phases. Therefore, understanding early markers requires a multidimensional framework that integrates neurobiological, behavioral, and contextual evidence.

Several studies increasingly reveal that NDDs do not arise abruptly but evolve from subtle differences observable during the first years of life (Finlay-Jones et al., 2019). For instance, atypicalities in gaze behavior, motor coordination, and emotional regulation during infancy may foreshadow later diagnoses of autism spectrum disorder (ASD) or attention-deficit/hyperactivity disorder (ADHD). These early deviations often reflect underlying alterations in neural connectivity rather than the behavioral manifestations themselves (Cioni et al., 2016).

Cioni and colleagues (2016) describe early neurobehavioral profiles in infants with cerebral palsy risk, highlighting how differences in general movements, such as reduced variability or monotonous sequences, signal atypical brain organization. Similarly, early atypicalities in sensory reactivity have been associated with later social-communication difficulties, suggesting that early sensory processing may be a transdiagnostic marker across NDDs (Hadders-Algra, 2021). The challenge for clinicians lies in distinguishing transient delays from persistent divergences indicative of enduring neurodevelopmental differences.

Cognitive and socio-emotional markers provide crucial windows into early neurodevelopment. Infants and toddlers who later meet criteria for NDDs often display atypical patterns in attention shifting, working memory, and social orienting (Meredith, 2015). For example, slower habituation to novel stimuli or difficulty disengaging attention can indicate atypical cortical processing, reflecting immature or inefficient neural synchronization. These neurocognitive patterns often precede overt behavioral symptoms, underscoring the need for early screening tools that capture underlying processing styles rather than surface behaviors.

Socio-emotional markers are particularly salient in early detection. Reduced shared attention, limited response to name, and diminished imitation behaviors during the first two years are among the earliest reliable indicators of atypical social development (Finlay-Jones et al., 2019). However, these behaviors are influenced by context: environmental stress, inconsistent caregiving, or cultural differences in parent–child interaction can mimic or mask early risk signs (Magaya & Tafirenyika, 2025). Thus, interpretation of socio-emotional markers must be grounded in a nuanced understanding of each child's ecological context.

PATHWAYS TO EARLY IDENTIFICATION

Motor development provides one of the most accessible entry points for early screening. Subtle deviations in postural control, reaching, and coordination often reflect underlying neural dysmaturation (Inguaggiato et al., 2017). The *General*

Movements Assessment (GMA), for example, has demonstrated strong predictive validity for cerebral palsy and related motor disorders in infants as young as three months (Cioni et al., 2016). Hadders-Algra (2021) notes that early motor patterns can also inform broader developmental predictions because they depend on integrated sensorimotor and cognitive processes.

Sensory processing is another early domain where atypical patterns emerge. Infants with heightened or diminished sensory reactivity may show altered responses to sound, touch, or visual stimuli, which can disrupt attention and learning. These differences often co-occur with later difficulties in self-regulation and social engagement (Bale et al., 2010). Importantly, sensory-based interventions during early childhood, such as structured tactile play or auditory integration therapy, can support adaptive regulation when delivered in developmentally appropriate ways (Inguaggiato et al., 2017).

Language and Communication Development

Language acquisition is among the most sensitive indicators of neurodevelopmental integrity. Early communication difficulties, ranging from delayed babbling to reduced joint attention, often serve as the first observable sign of divergence (Finlay-Jones et al., 2019). Research shows that even subtle delays in prelinguistic communication (gesture use, pointing, and vocal imitation) can signal atypical underlying neurocognitive processes. However, variability in early language milestones is common, making it essential to interpret language markers alongside motor, cognitive, and social cues.

Duncan and Matthews (2018) argue that language impairments often represent a shared manifestation of diverse neurobiological mechanisms, such as deficits in auditory processing or working memory. This overlap reinforces the idea that language-based markers are not disorder-specific but reflect convergent pathways of neural vulnerability. Early speech–language interventions, therefore, should adopt a transdiagnostic lens, targeting underlying cognitive and perceptual skills rather than focusing solely on surface linguistic performance.

Parental Observations and Early Screening Tools

Parents and caregivers are often the first to detect developmental anomalies. Their observations provide ecologically valid data that complement professional assessment. Tools such as the *Ages and Stages Questionnaires* (ASQ) and the *Modified Checklist for Autism in Toddlers* (M-CHAT) rely heavily on parental input and have been instrumental in early screening efforts (Bosch et al., 2022). However, the accuracy of parent-reported data is influenced by cultural expectations, health literacy, and parental stress (Magaya & Tafirenyika, 2025). Clinicians must therefore interpret screening results within the family's cultural and contextual frame.

In recent years, digital platforms and artificial intelligence have expanded early identification possibilities. Computer-vision systems analyzing infant movement or gaze patterns can detect subtle behavioral markers that escape human observation (Makris et al., 2023). While promising, these technologies raise ethical and practical questions concerning privacy, data validity, and equitable access. For multidisciplinary teams, integrating traditional observation with digital analytics may represent the next frontier in early developmental surveillance.

Despite advances in screening, diagnostic delays remain pervasive. Smythe et al. (2024) highlight how systemic inequities, including socioeconomic disadvantage, healthcare access, and professional training, significantly affect early identification rates. In many contexts, children with early signs of NDDs are not referred for formal assessment until school age, missing the optimal window for intervention. This gap is particularly evident in low- and middle-income settings, where screening infrastructure and specialist availability are limited (Bosch et al., 2022).

Moreover, diagnostic frameworks themselves can impede early detection. Categorical systems such as the DSM-5 or ICD-11 rely on behavioral manifestations that may not be fully expressed in early childhood (Hadders-Algra, 2021). As a result, clinicians often hesitate to label very young children, preferring to adopt "watchful waiting" approaches. However, this caution can inadvertently delay support, reinforcing developmental disparities. A shift toward dimensional assessment, evaluating strengths and vulnerabilities across domains, may improve sensitivity to early risk while reducing stigma.

Effective early detection of NDDs requires collaboration across health, education, and social sectors. Pediatricians, psychologists, speech–language pathologists, occupational therapists, and educators each contribute unique perspectives. The multidisciplinary model has evolved into a more integrated transdisciplinary framework in which professionals share conceptual understandings and intervention goals (Carlsson et al., 2021). This approach not only enhances diagnostic accuracy but also supports family-centered care by reducing fragmentation across services. Carlsson et al. (2021) emphasize that early identification should be conceived as an iterative, developmental process rather than a one-time diagnostic act. Continuous monitoring and information exchange between professionals allow for adjustment as the child's profile evolves. For example, early sensory–motor anomalies may initially suggest a motor disorder but later reveal broader attentional or language implications. Such developmental fluidity necessitates flexible, ongoing assessment frameworks that can capture the child's evolving strengths and challenges.

Ethical practice in early identification extends beyond accurate diagnosis. It requires sensitivity to the potential psychosocial impact of early labeling, particularly when diagnostic certainty is low. Families often experience anxiety or guilt when informed of possible developmental differences; thus, communication must emphasize developmental support rather than deficit (Magaya & Tafirenyika, 2025). Culturally responsive assessment is also essential: tools standardized in one population may misclassify children from diverse linguistic or cultural backgrounds. Engaging community voices in screening design enhances validity and trust.

Additionally, the ethical principle of "do no harm" mandates that early screening should always link to tangible support. Identifying risk without accessible follow-up services can exacerbate parental distress and reinforce inequities. Consequently, early identification systems must be embedded within broader service networks that ensure prompt access to developmental support, parental guidance, and educational resources (Smythe et al., 2024).

Early Intervention and Educational Implications

The early years represent a critical window not only for identifying neurodevelopmental vulnerabilities but also for shaping adaptive developmental outcomes through targeted intervention. During this period of heightened neural plasticity, experiences, both enriching and adverse, leave enduring imprints on brain organization and function (Cioni et al., 2016; Meredith, 2015). Effective early intervention (EI) leverages this neuroplastic potential by creating supportive environments that promote compensatory neural mechanisms and minimize cascading effects of early dysfunction (Inguaggiato et al., 2017). Rather than narrowly remediating symptoms, early intervention frameworks now emphasize developmental optimization, supporting the child's capacity to engage, learn, and participate across contexts (Hadders-Algra, 2021).

Neuroscientific research has established that the first five years of life are characterized by extraordinary synaptic density and dynamic reorganization of cortical networks (Meredith, 2015). During this phase, sensory, motor, and associative regions of the brain are particularly responsive to environmental input, providing a biological rationale for early intervention. Disruptions in synaptic pruning, myelination, or neurotransmitter regulation during this sensitive period can lead to enduring alterations in brain architecture that underlie NDDs (Bale et al., 2010; Cioni et al., 2016). However, this same plasticity allows for reorganization and compensatory growth when children receive enriched, structured, and emotionally secure experiences.

Studies in infants at elevated neurodevelopmental risk demonstrate that early behavioral and sensory-motor interventions can normalize activation patterns in brain regions associated with social cognition and attention (Inguaggiato et al., 2017). This evidence reinforces that intervention should begin at the earliest detectable sign of atypical development, ideally before behavioral symptoms consolidate into fixed patterns. Delaying support until formal diagnosis, as often happens in conventional systems, misses this period of maximal malleability (Duncan & Matthews, 2018; Hadders-Algra, 2021).

Contemporary early intervention models emphasize family-centered, relationship-based practice as a foundation for developmental progress. Parents are not passive recipients of expert guidance but active co-therapists in their child's learning process (Wong et al., 2025). When caregivers are empowered with knowledge and strategies to interpret and respond to their child's cues, everyday interactions become potent opportunities for developmental growth.

Family-implemented programs such as responsive parenting interventions and naturalistic developmental behavioral models have demonstrated efficacy in enhancing social communication and adaptive behavior among children with early signs of ASD and related conditions (Wong et al., 2025). These programs not only support children's skill acquisition but also reduce parental stress and improve family cohesion, outcomes particularly critical in multidisciplinary care (Magaya & Tafirenyika, 2025).

Importantly, family-centered approaches must account for sociocultural variation in caregiving beliefs and practices. Magaya and Tafirenyika (2025) found that early childhood educators and parents in Zimbabwe's mainstream settings often employed adaptive coping strategies to manage learners with NDDs despite limited formal resources. This underscores that cultural sensitivity and contextual adaptation are vital for intervention sustainability, especially in under-resourced regions.

Early Educational Engagement

Educational settings play a pivotal role in identifying and supporting children with early NDDs. High-quality early childhood education programs provide structured social interaction, language exposure, and cognitive stimulation, factors known to enhance developmental outcomes (Gutiérrez-Ruiz & Santoya Montes, 2025). Early educational inclusion also fosters a sense of belonging and continuity that mitigates later academic and social difficulties.

Research demonstrates that children with NDDs benefit most from early education when educators receive targeted training in neurodevelopmental principles and differentiated instruction (Magaya & Tafirenyika, 2025). For instance, embedding sensory regulation activities into classroom routines supports attention and participation for children with sensory processing challenges (Boulton et al., 2023). Similarly, scaffolding executive function skills through play-based tasks can strengthen cognitive flexibility and self-regulation in children with attentional or language-related vulnerabilities (Lee et al., 2025).

Nevertheless, inclusion without adequate teacher preparation can exacerbate stress for both educators and learners. Teacher training programs should therefore integrate neurodevelopmental literacy, understanding how brain–behavior relationships shape learning and behavior, to create responsive educational environments (Cook et al., 2025). Such environments blend structure with flexibility, recognizing that development unfolds along diverse trajectories rather than through uniform milestones.

The success of early intervention is closely linked to its timing, intensity, and individualization. Evidence from systematic reviews and meta-analyses underscores that earlier initiation and greater dosage are associated with better cognitive, language, and adaptive outcomes across NDD populations (Aldharman et al., 2023; Smythe et al., 2024). However, the most effective interventions are not merely intensive but developmentally attuned, responsive to each child's current level of functioning and evolving needs. Inguaggiato et al. (2017) emphasize that early programs must dynamically adjust to developmental change, integrating ongoing assessment and flexible goal setting. This approach contrasts with rigid programmatic models that treat intervention as a static protocol. By continuously aligning goals with the child's emerging skills and family context, practitioners can maximize gains and prevent regression. Moreover, early intervention should address the full range of functioning, including emotional well-being and family adaptation, rather than narrowly targeting symptom reduction (Purpura et al., 2025). Purpura and colleagues (2025) illustrate how pre-emptive early intervention can enhance social engagement and motor control even before formal diagnosis, demonstrating that proactive support fosters resilience rather than pathologizing risk. Their findings reflect a broader paradigm shift from a deficit-based to a strengths-based orientation, viewing early intervention as a process of nurturing potential.

Because NDDs share overlapping etiological and behavioral pathways, early intervention must adopt transdiagnostic frameworks that target common developmental mechanisms such as attention regulation, sensory integration, and social motivation (Cortese et al., 2023; Finlay-Jones et al., 2019). Interventions grounded in these cross-cutting domains can benefit children with diverse diagnostic profiles, promoting efficiency and inclusivity in service delivery.

Interdisciplinary collaboration enhances both diagnostic precision and intervention quality. Pediatricians, psychologists, educators, speech-language pathologists, and occupational therapists contribute distinct expertise to a unified developmental plan (Carlsson et al., 2021). For example, coordinated sensory–motor and communication therapies can address the interplay between perceptual and linguistic development. Hannigan et al. (2023) demonstrate that developmental milestones, particularly motor and language acquisition, are genetically intertwined across NDDs, supporting an integrated approach rather than isolated treatment silos. Effective collaboration depends on shared conceptual frameworks and communication across disciplines.

Interprofessional training programs and case-based learning models foster mutual understanding and streamline care coordination (Cook et al., 2025). When educators, clinicians, and families operate from common developmental principles, intervention becomes more coherent, reducing fragmentation and improving long-term outcomes.

ADDRESSING ENVIRONMENTAL AND SOCIOECONOMIC BARRIERS

Despite compelling evidence for early intervention efficacy, access remains uneven globally. Socioeconomic disparities, healthcare infrastructure, and cultural stigma often hinder early identification and service utilization (Smythe et al., 2024). In many low-resource settings, parents report long delays between initial concerns and professional support (Bosch et al., 2022). Overcoming these barriers requires systemic strategies that combine public health, education, and social policy initiatives.

Community-based intervention models offer promising solutions. Smythe et al. (2024) advocate for task-sharing approaches in which non-specialist health or education workers receive structured training to deliver evidence-based developmental support. These scalable models have demonstrated success in expanding early services to underserved populations. Similarly, parental empowerment programs that teach everyday stimulation techniques can extend developmental benefits beyond formal therapy sessions (Magaya & Tafirenyika, 2025). At the macro level, policy alignment between health and education sectors is crucial. Integrated service delivery ensures that early screening results translate into intervention pathways rather than bureaucratic dead ends. National early childhood policies emphasizing inclusive education and developmental surveillance, such as the frameworks emerging in parts of Europe and Asia, offer replicable models for global adaptation (Cha et al., 2025).

Outcome evaluation in early intervention has evolved beyond measuring isolated skill gains to assessing broader developmental competencies and quality of life. Sustained improvement depends on continuity between early programs and later educational environments (Hadders-Algra, 2021). Long-term follow-ups reveal that early benefits can fade without consistent reinforcement, underscoring the need for coordinated transition planning from preschool to primary education (Duncan & Matthews, 2018).

Moreover, outcome measures should reflect family perspectives and cultural priorities. Standardized tests may not capture meaningful changes in daily participation, emotional regulation, or parent–child relationships (Wong et al., 2025). Combining quantitative and qualitative indicators yields a more comprehensive understanding of impact. As early intervention systems mature, data-driven evaluation can inform policy decisions, resource allocation, and professional training priorities.

Families, Systems, and Futures

The success of early identification and intervention for neurodevelopmental disorders (NDDs) does not rest solely on clinical precision or pedagogical skill; it depends equally on the systems that sustain and amplify developmental change. Families, schools, and policy environments collectively shape whether early gains translate into lasting inclusion and well-being. A systems perspective recognizes that children with NDDs develop within interconnected contexts where biological, psychological, and social processes interact continuously. The family thus serves as the primary developmental environment, while education systems and social policy provide the scaffolds that sustain early growth trajectories (Smythe et al., 2024; Wong et al., 2025).

Families play an indispensable role in supporting children with NDDs. Early caregiving environments influence brain maturation through daily routines, emotional attunement, and interactive play (Makris et al., 2023). Secure, responsive relationships regulate stress physiology and enhance synaptic integration, providing a biological buffer against neurodevelopmental risk (Bale et al., 2010). Conversely, chronic family stress, low socioeconomic resources, or inconsistent caregiving can magnify developmental difficulties (Scattolin et al., 2022).

Parenting a child with early-diagnosed NDDs can be emotionally and practically demanding. Feelings of uncertainty, grief, or stigma may emerge following diagnosis, while navigating complex service systems often adds further strain (Magaya & Tafirenyika, 2025). Yet research consistently highlights the adaptive capacity of families when given appropriate support. Parent-implemented interventions that teach communication, regulation, and play strategies not only enhance child outcomes but also strengthen parental efficacy and mental health (Wong et al., 2025). This dual benefit underscores that effective early support must view parents as co-agents of change, not merely caregivers.

In many cultural contexts, extended family networks provide essential caregiving and emotional resources. However, beliefs about disability and development can shape whether families seek or accept early support. In collectivist settings, for example, NDDs may be interpreted through moral, spiritual, or social lenses rather than medical models (Magaya & Tafirenyika, 2025). Therefore, culturally grounded approaches are crucial for ensuring that early identification and intervention efforts align with families' values and lived experiences.

Strong family–professional partnerships are central to sustaining early intervention gains. When families and practitioners collaborate as equals, they co-construct intervention goals that reflect both developmental priorities and daily realities. Research on family-centered practice demonstrates that shared decision-making and mutual respect enhance engagement and adherence, leading to better outcomes

(Wong et al., 2025). However, such partnerships are not automatically achieved. Families frequently encounter fragmented systems, separate health, education, and social services that operate in silos (Smythe et al., 2024). Parents may have to repeat their child's history across multiple professionals or navigate inconsistent eligibility criteria for services. Establishing interdisciplinary coordination structures, such as case conferences, shared electronic records, and keyworker models, reduces these burdens and ensures coherent care (Carlsson et al., 2021).

The tone and language professionals use when communicating developmental concerns also shape families' long-term engagement. Framing differences as developmental diversity rather than deficit fosters acceptance and optimism. Purpura et al. (2025) emphasize that pre-emptive interventions framed around potential rather than pathology enhance family motivation and reduce stigma. This strengths-based language aligns with contemporary paradigms of neurodiversity, emphasizing empowerment and inclusion rather than remediation.

Schools as Developmental Continuums

Educational systems serve as the bridge between early intervention and lifelong learning. Schools are often where the enduring effects of early developmental differences become most visible. For children with NDDs, transitions from early childhood programs into formal schooling can be challenging if continuity of support is not maintained (Duncan & Matthews, 2018). Sustaining developmental progress requires communication between early intervention teams and school personnel to ensure consistent strategies and expectations.

Teachers play a crucial role in identifying residual or emerging needs as the child's learning demands increase. Studies in mainstream classrooms reveal that educators' neurodevelopmental literacy, their understanding of brain–behavior relationships, directly influences classroom adaptation and inclusion (Magaya & Tafirenyika, 2025). Professional development in NDD-informed pedagogy helps teachers balance structure and flexibility, enabling individualized approaches that accommodate attention, sensory, or language differences (Cook et al., 2025).

Schools that adopt inclusive frameworks from the outset, integrating universal design for learning, social–emotional learning curricula, and sensory-friendly environments, create continuity with early intervention principles. Such practices benefit not only children with NDDs but also their typically developing peers, fostering empathy and diversity awareness (Gutiérrez-Ruiz & Santoya Montes, 2025). Inclusion becomes a shared educational philosophy rather than a reactive accommodation.

Despite growing awareness of early developmental support, systemic inequities persist across regions and populations. Access to early diagnosis and intervention is strongly correlated with socioeconomic status, parental education, and geographic location (Bosch et al., 2022; Marschik et al., 2017; Smythe et al., 2024). In many low-resource settings, children are identified late, and families face long waiting lists or prohibitive costs for specialized services (Aldharman et al., 2023). Addressing these disparities requires coordinated policy and cross-sector investment. Smythe et al. (2024) highlight that achieving equitable early developmental outcomes by 2030 necessitates integrated frameworks linking health, education, and social protection. Task-shifting approaches, training community health workers or educators to deliver basic developmental support, have shown promise in scaling early intervention. Similarly, public awareness campaigns can normalize early screening, reduce stigma, and promote help-seeking behavior.

National policy reforms should also embed early identification and intervention within universal early childhood systems rather than treating them as specialized add-ons. When developmental monitoring becomes a standard component of pediatric care or preschool enrollment, disparities in access narrow significantly (Cha et al., 2025). Policies emphasizing early inclusion and multi-agency collaboration can create systemic continuity that sustains developmental progress across transitions.

Recent studies reveal that early developmental interventions have enduring effects on cognitive, emotional, and adaptive functioning when followed by sustained support (Hadders-Algra, 2021). Early motor, sensory, and language gains predict later academic achievement and psychosocial well-being, but the stability of these gains depends on environmental continuity. Without reinforcement, early improvements may attenuate as new developmental challenges emerge (Duncan & Matthews, 2018). Neurodevelopmental pathways remain dynamic throughout childhood and adolescence. Genetic and environmental influences continue to interact, shaping the trajectory of strengths and difficulties (Hannigan et al., 2023). Thus, developmental surveillance should extend beyond early childhood, transitioning into adolescent and educational monitoring systems that provide adaptive scaffolding at each stage. The continuum of care perspective recognizes that NDDs are lifelong conditions requiring evolving, flexible support rather than time-limited interventions.

A future-oriented perspective on NDDs in early childhood calls for a cultural shift from a model of deficit correction to one of developmental diversity. Recognizing neurodiversity means valuing individual cognitive and behavioral differences as natural variations within the human spectrum. Educational and social systems grounded in neuroinclusion not only improve participation for children with NDDs but enrich collective understanding of learning and adaptation (Cortese et al., 2023; Finlay-Jones et al., 2019).

In this context, early childhood becomes the launch point for inclusive citizenship. Children who experience acceptance and meaningful participation early are more likely to develop positive self-identity, resilience, and community belonging. Parents and educators, as agents of inclusion, can model empathy and flexibility that extend beyond the classroom. Systemically, investment in early childhood inclusion yields long-term social and economic benefits by reducing the need for later remediation and promoting workforce participation among individuals with NDDs (Smythe et al., 2024). Families and systems form the connective tissue that binds biological potential to social opportunity. The sustainability of early intervention outcomes depends on the alignment of family empowerment, inclusive education, and equitable policy. When these elements converge, children with NDDs can thrive not despite their differences but through them, contributing uniquely to their communities.

INTEGRATION AND FUTURE DIRECTIONS

The preceding chapters have traced the complex landscape of neurodevelopmental disorders (NDDs) in early childhood, spanning neurobiological mechanisms, environmental determinants, early markers, intervention strategies, and systemic considerations. A coherent understanding of early NDDs requires synthesizing these domains, recognizing that development is an emergent, dynamic process shaped by the interplay of biology, experience, and social context (Cioni et al., 2016; Meredith, 2015). Integration across these levels offers a roadmap for optimizing early identification, intervention, and lifelong inclusion.

a. *Neurobiological and Environmental Interdependence.* Early brain development is characterized by sensitive periods of heightened plasticity, during which neural circuits governing attention, language, motor control, and social cognition are particularly malleable (Inguaggiato et al., 2017; Meredith, 2015). Genetic predispositions interact with environmental exposures, including stress, toxins, and socioeconomic disadvantage, to shape developmental trajectories (Bale et al., 2010; Scattolin et al., 2022; Zhang et al., 2025). Hormonal regulation and energy metabolism further modulate neurodevelopment, highlighting the multiplicity of pathways that converge to influence early risk and resilience (Makris et al., 2023; Oyarzábal et al., 2021).
b. *Early Markers and Multidimensional Identification.* Early deviations in motor skills, sensory responsiveness, language acquisition, attention, and socio-emotional engagement offer measurable windows into atypical neurodevelopment (Finlay-Jones et al., 2019; Hadders-Algra, 2021). Identifying these markers requires multidimensional assessment, integrating behavioral observation, parental input, and, increasingly, digital and AI-based analytic tools (Boulton et al., 2023; Makris et al., 2023). Sensory-motor, language, and socio-emotional domains serve as transdiagnostic markers, allowing early intervention before

categorical diagnoses are feasible, while minimizing risks of over- or misdiagnosis (Duncan & Matthews, 2018; Finlay-Jones et al., 2019).

c. *Early Intervention and Neurodevelopmental Optimization.* Interventions are most effective when they begin at the earliest detectable signs, are developmentally attuned, and are embedded within family and educational contexts (Hadders-Algra, 2021; Inguaggiato et al., 2017; Purpura et al., 2025). Family-centered approaches empower caregivers as co-agents of change, strengthening both child outcomes and family resilience (Wong et al., 2025). Early educational environments that incorporate inclusive practices, scaffolding strategies, and neurodevelopmental literacy among teachers reinforce gains achieved in clinical or home-based interventions (Gutiérrez-Ruiz & Santoya Montes, 2025; Magaya & Tafirenyika, 2025).
d. *Systems and Equity Considerations.* Developmental gains are contingent on coherent systems that bridge early intervention, schooling, and community support. Fragmented services, socioeconomic disparities, and cultural misalignments threaten the sustainability of progress (Bosch et al., 2022; Smythe et al., 2024). Conversely, integrated policy frameworks, task-shared interventions, and culturally responsive programs extend early support to underserved populations, reducing inequities and enhancing long-term developmental outcomes (Aldharman et al., 2023; Cha et al., 2025).
e. *Family and Community Dynamics.* Families act as both the primary developmental environment and as the interface with broader social systems. Strengthening family–professional partnerships, fostering parental self-efficacy, and attending to socio-cultural beliefs about NDDs enhance engagement and sustainability (Magaya & Tafirenyika, 2025; Wong et al., 2025). Community acceptance, inclusive schooling, and policy support amplify the effects of early intervention, allowing neurodevelopmental potential to manifest within socially meaningful contexts.

IMPLICATIONS FOR MULTIDISCIPLINARY PRACTICE

The synthesis of evidence across neurobiology, behavioral markers, and systemic supports underscores the critical need for transdisciplinary collaboration. Pediatricians, psychologists, occupational and speech therapists, educators, and social service providers each contribute domain-specific expertise, yet their effectiveness is maximized when they operate within shared conceptual frameworks that prioritize developmental trajectories over static diagnoses (Carlsson et al., 2021; Cortese et al., 2023). Effective multidisciplinary practice begins with early and proactive engagement, screening for transdiagnostic markers in infancy and toddlerhood to initiate timely support and capitalize on critical periods of neuroplasticity (Finlay-Jones et al., 2019). Interventions must remain flexible and individualized, adjusting intensity and content based on ongoing assessments rather than adhering to rigid protocols (Inguaggiato et al., 2017; Purpura et al., 2025). Central to this approach is the empowerment of families, positioning parents as co-therapists and integrating their cultural and contextual knowledge into intervention planning, which enhances both engagement and developmental outcomes (Magaya & Tafirenyika, 2025; Wong et al., 2025). Continuity across environments is also essential; aligning early intervention, preschool programs, and elementary education ensures that gains achieved in early childhood are maintained and reinforced within social and academic contexts (Duncan & Matthews, 2018; Gutiérrez-Ruiz & Santoya Montes, 2025).

Despite substantial progress, significant gaps remain in understanding early NDDs and optimizing interventions. Longitudinal research is particularly critical to map developmental trajectories and identify which early markers reliably predict outcomes across diverse contexts (Hannigan et al., 2023). Future studies should explore the interplay between genetic susceptibility and environmental enrichment or adversity, clarifying how these factors jointly shape neurodevelopmental outcomes (Bale et al., 2010; Carlsson et al., 2021). There is also a pressing need to evaluate transdiagnostic interventions that target core neurocognitive and socio-emotional mechanisms rather than focusing exclusively on disorder-specific symptoms (Cortese et al., 2023; Finlay-Jones et al., 2019). The integration of technology-mediated monitoring offers promising opportunities for early detection and ongoing developmental surveillance, but these approaches must be rigorously assessed for validity, accessibility, and ethical implications (Makris et al., 2023). Additionally, research on cultural and equity-sensitive implementation is essential to ensure that interventions are effective and sustainable in diverse sociocultural and resource-constrained contexts, including low- and middle-income countries (Magaya & Tafirenyika, 2025; Smythe et al., 2024).

Translating these insights into practice requires alignment across health, education, and social policy. Universal developmental surveillance should be embedded into routine pediatric care and early childhood programs to facilitate timely identification of at-risk children (Cha et al., 2025). Integrated service delivery models that promote cross-sector collaboration help to reduce fragmentation and ensure seamless transitions from home to school and beyond (Carlsson et al., 2021; Smythe et al., 2024). Equitable allocation of resources is essential to target under-resourced populations and reduce disparities in access, timing, and intensity of early intervention (Aldharman et al., 2023; Bosch et al., 2022). Strengthening professional training in neurodevelopmental literacy among educators, clinicians, and allied health professionals further enables responsive, evidence-informed practice and supports the implementation of effective, individualized interventions (Cook et al., 2025; Magaya & Tafirenyika, 2025).

Ethical practice in early childhood NDD support requires balancing early identification with sensitivity to parental and community concerns. Interventions should prioritize developmental support over premature labeling, emphasizing strengths, potential, and functional engagement (Purpura et al., 2025; Wong et al., 2025). Culturally adapted practices that respect family beliefs and local caregiving norms enhance participation, sustain engagement, and increase the likelihood of long-term developmental benefits (Magaya & Tafirenyika, 2025). Furthermore, ethical oversight is paramount when employing emerging technologies for early detection, ensuring privacy, informed consent, and equitable access (Makris et al., 2023). A neuroinclusive approach emphasizes that interventions are not simply corrective but developmental in nature, fostering autonomy, participation, and well-being.

Integrating neurobiological, behavioral, familial, educational, and systemic perspectives highlights that early childhood is both a window of vulnerability and an opportunity for resilience. When interventions are timely, individualized, and embedded in supportive systems, children with NDDs can achieve developmental gains that extend into school-age years and beyond (Duncan & Matthews, 2018; Hadders-Algra, 2021). A lifespan perspective encourages the design of developmental continuums linking early intervention with ongoing educational and community supports. Sustaining these gains across childhood, adolescence, and early adulthood requires continuous monitoring, adaptation, and coordination across multiple ecological layers, recognizing that neurodevelopmental care is dynamic, longitudinal, and participatory, engaging families, professionals, and communities as co-constructors of developmental opportunity.

As a conclusion, early childhood represents a critical nexus of risk and opportunity for neurodevelopment. NDDs emerge from complex, interacting genetic, neurobiological, and environmental factors. Early markers across motor, sensory, cognitive, and socio-emotional domains enable timely detection, while family-centered, developmentally attuned interventions leverage neuroplasticity to optimize trajectories. Sustaining progress demands multidisciplinary collaboration, integrated educational and social systems, and culturally responsive practice. Future directions emphasize transdiagnostic approaches, equitable access, longitudinal monitoring, and ethical deployment of emerging technologies. By reconceptualizing neurodevelopmental differences as expressions of developmental diversity rather than pathology, society can promote inclusion, participation, and well-being from the earliest years onward. In the context of schooling and community life, early intervention serves not merely as a remedial measure but as a foundation for equitable developmental opportunity, social participation, and lifelong potential.

REFERENCES

Aldharman, S. S., Al-Jabr, K. H., Alharbi, Y. S., Alnajar, N. K., Alkhanani, J. J., Alghamdi, A., & Abdellatif, R. (2023). Implications of early diagnosis and intervention in the management of neurodevelopmental delay (NDD) in children: A systematic review and meta-analysis. *Cureus*, *15*(5), e38745. DOI: 10.7759/cureus.38745 PMID: 37303321

Bale, T. L., Baram, T. Z., Brown, A. S., Goldstein, J. M., Insel, T. R., McCarthy, M. M., Nemeroff, C. B., Reyes, T. M., Simerly, R. B., Susser, E. S., & Nestler, E. J. (2010). Early life programming and neurodevelopmental disorders. *Biological Psychiatry*, *68*(4), 314–319. DOI: 10.1016/j.biopsych.2010.05.028 PMID: 20674602

Bosch, R., Pagerols, M., Rivas, C., Sixto, L., Bricollé, L., Español-Martín, G., Prat, R., Ramos-Quiroga, J. A., & Casas, M. (2022). Neurodevelopmental disorders among Spanish school-age children: Prevalence and sociodemographic correlates. *Psychological Medicine*, *52*(14), 3062–3072. DOI: 10.1017/S0033291720005115 PMID: 33436129

Boulton, K. A., Hodge, M. A., Jewell, A., Ong, N., Silove, N., & Guastella, A. J. (2023). Diagnostic delay in children with neurodevelopmental conditions attending a publicly funded developmental assessment service: Findings from the Sydney Child Neurodevelopment Research Registry. *BMJ Open*, *13*(2), e069500. DOI: 10.1136/bmjopen-2022-069500 PMID: 36725093

Carlsson, T., Molander, F., Taylor, M. J., Jonsson, U., & Bölte, S. (2021). Early environmental risk factors for neurodevelopmental disorders: A systematic review of twin and sibling studies. *Development and Psychopathology*, *33*(4), 1448–1495. DOI: 10.1017/S0954579420000620 PMID: 32703331

Cha, J. H., Ryu, S., Park, M., Lim, B. C., Kim, Y. J., & Moon, J. H. (2025). Developmental screening of neurodevelopmental disorders before age 6: A nationwide health screening program. *Pediatric Research*, *97*(5), 1566–1574. DOI: 10.1038/s41390-024-03516-6 PMID: 39261660

Cioni, G., Inguaggiato, E., & Sgandurra, G. (2016). Early intervention in neurodevelopmental disorders: Underlying neural mechanisms. *Developmental Medicine and Child Neurology*, *58*(S4), 61–66. DOI: 10.1111/dmcn.13050 PMID: 27027609

Cook, B., Van Bockstaele, S., Crow, S. B., Sainsbury, D., Butterworth, S., & Filson, S. (2025). Neurodevelopmental disorders in children with cleft lip and palate: A systematic review. *European Child & Adolescent Psychiatry*, *34*(6), 1731–1738. DOI: 10.1007/s00787-024-02636-y PMID: 39724178

Cortese, S., Solmi, M., Michelini, G., Bellato, A., Blanner, C., Canozzi, A., Eudave, L., Farhat, L. C., Højlund, M., Köhler-Forsberg, O., Leffa, D. T., Rohde, C., de Pablo, G. S., Vita, G., Wesselhoeft, R., Martin, J., Baumeister, S., Bozhilova, N. S., Carlisi, C. O., & Correll, C. U. (2023). Candidate diagnostic biomarkers for neurodevelopmental disorders in children and adolescents: A systematic review. *World Psychiatry; Official Journal of the World Psychiatric Association (WPA)*, *22*(1), 129–149. DOI: 10.1002/wps.21037 PMID: 36640395

Duncan, A. F., & Matthews, M. A. (2018). Neurodevelopmental outcomes in early childhood. *Clinics in Perinatology*, *45*(3), 377–392. DOI: 10.1016/j.clp.2018.05.001 PMID: 30144844

Finlay-Jones, A., Varcin, K., Leonard, H., Bosco, A., Alvares, G., & Downs, J. (2019). Very early identification and intervention for infants at risk of neurodevelopmental disorders: A transdiagnostic approach. *Child Development Perspectives*, *13*(2), 97–103. DOI: 10.1111/cdep.12319

Gutiérrez-Ruiz, K., & Santoya Montes, Y. (2025). Early detection of neurodevelopmental disorders as a strategy for educational inclusion in early childhood education. *International Journal of Disability Development and Education*, *72*(3), 523–539. DOI: 10.1080/1034912X.2024.2370799

Hadders-Algra, M. (2021). Early diagnostics and early intervention in neurodevelopmental disorders: Age-dependent challenges and opportunities. *Journal of Clinical Medicine*, *10*(4), 861. DOI: 10.3390/jcm10040861 PMID: 33669727

Hannigan, L. J., Askeland, R. B., Ask, H., Tesli, M., Corfield, E., Ayorech, Z., Magnus, P., Njølstad, P. R., Øyen, A.-S., Stoltenberg, C., Andreassen, O. A., Ronald, A., Smith, G. D., Reichborn-Kjennerud, T., & Havdahl, A. (2023). Developmental milestones in early childhood and genetic liability to neurodevelopmental disorders. *Psychological Medicine*, *53*(5), 1750–1758. DOI: 10.1017/S0033291721003330 PMID: 37310338

Inguaggiato, E., Sgandurra, G., & Cioni, G. (2017). Brain plasticity and early development: Implications for early intervention in neurodevelopmental disorders. *Neuropsychiatrie de l'Enfance et de l'Adolescence*, *65*(5), 299–306. DOI: 10.1016/j.neurenf.2017.03.009

King-Dowling, S., Proudfoot, N. A., & Obeid, J. (2019). Comorbidity among chronic physical health conditions and neurodevelopmental disorders in childhood. *Current Developmental Disorders Reports*, *6*(4), 248–258. DOI: 10.1007/s40474-019-00173-x

Lee, D., Boulton, K. A., Sun, C., Phillips, N. L., Munro, M., Kumfor, F., Demetriou, E. A., & Guastella, A. J. (2025). Attention and executive delays in early childhood: A meta-analysis of neurodevelopmental conditions. *Molecular Psychiatry*, *30*(5), 1906–1914. DOI: 10.1038/s41380-024-02802-3 PMID: 39489868

Magaya, M., & Tafirenyika, J. (2025). Early childhood development teachers' coping strategies for managing learners with neurodevelopmental disorders in Zimbabwe's mainstream classes. *Early Child Development and Care*, *195*(5-6), 1–13. DOI: 10.1080/03004430.2025.2499455

Makris, G., Eleftheriades, A., & Pervanidou, P. (2023). Early life stress, hormones, and neurodevelopmental disorders. *Hormone Research in Paediatrics*, *96*(1), 17–24. DOI: 10.1159/000523942 PMID: 35259742

Marschik, P. B., Pokorny, F. B., Peharz, R., Zhang, D., O'Muircheartaigh, J., Roeyers, H., Bölte, S., Spittle, A. J., Urlesberger, B., Schuller, B., Poustka, L., Ozonoff, S., Pernkopf, F., Pock, T., Tammimies, K., Enzinger, C., Krieber, M., Tomantschger, I., Bartl-Pokorny, K. D., & Kaufmann, W. E.BEE-PRI Study Group. (2017). A novel way to measure and predict development: A heuristic approach to facilitate the early detection of neurodevelopmental disorders. *Current Neurology and Neuroscience Reports*, *17*(5), 43. DOI: 10.1007/s11910-017-0748-8 PMID: 28390033

Meredith, R. M. (2015). Sensitive and critical periods during neurotypical and aberrant neurodevelopment: A framework for neurodevelopmental disorders. *Neuroscience and Biobehavioral Reviews*, *50*, 180–188. DOI: 10.1016/j.neubiorev.2014.12.001 PMID: 25496903

Micai, M., Fulceri, F., Caruso, A., Guzzetta, A., Gila, L., & Scattoni, M. L. (2020). Early behavioral markers for neurodevelopmental disorders in the first 3 years of life: An overview of systematic reviews. *Neuroscience and Biobehavioral Reviews*, *116*, 183–201. DOI: 10.1016/j.neubiorev.2020.06.027 PMID: 32610179

Oyarzábal, A., Musokhranova, U., Barros, L. F., & García-Cazorla, A. (2021). Energy metabolism in childhood neurodevelopmental disorders. *EBioMedicine*, *69*, 103474. DOI: 10.1016/j.ebiom.2021.103474 PMID: 34256347

Purpura, G., Annunziata, S., Biancardi, S., Brivio, M., Caporali, C., Mantegazza, G., Piazza, E., Restelli, A., & Cavallini, A. (2025). Clinical and research insights from pre-emptive early intervention for neurodevelopmental disorders: A case series. *Children (Basel, Switzerland)*, *12*(11), 1489. DOI: 10.3390/children12111489 PMID: 41300607

Scattolin, M. A. D. A., Resegue, R. M., & Rosário, M. C. D. (2022). The impact of the environment on neurodevelopmental disorders in early childhood. *Jornal de Pediatria*, *98*, 66–72. DOI: 10.1016/j.jped.2021.11.002 PMID: 34914896

Smythe, T., Scherer, N., Nanyunja, C., Tann, C. J., & Olusanya, B. O. (2024). Strategies for addressing the needs of children with or at risk of developmental disabilities in early childhood by 2030: A systematic umbrella review. *BMC Medicine*, *22*(1), 51. DOI: 10.1186/s12916-024-03265-7 PMID: 38302917

Wong, C. H., Chan, C. K., & Wong, P. W. (2025). Family outcomes of a parent-implemented early intervention for neurodevelopmental disorders: An exploratory study of family dynamics. *Frontiers in Public Health*, *13*, 1614089. DOI: 10.3389/fpubh.2025.1614089 PMID: 40697837

Zhang, J., Yuan, M., Liu, Y., Zhong, X., Wu, J., & Chen, W. (2025). Bisphenol A exposure and neurodevelopmental disorders and problems in children under 12 years of age: A systematic review and meta-analysis. *Journal of Hazardous Materials*, *137731*, 137731. Advance online publication. DOI: 10.1016/j.jhazmat.2025.137731 PMID: 40054188

Chapter 5

The School Experience of Students With Neurodevelopmental Disorders:

Symptomatology and Support Mechanisms

ABSTRACT

This chapter considers how neurodevelopmental differences influence students' daily lives in school, focusing on the interplay between symptom expression, instructional demands, and social dynamics. It draws on educational-psychology research to show that teacher understanding, emotional attunement, and access to collaborative support strongly affect engagement and adjustment. The analysis highlights early identification, relational practices, and coordinated interventions as key to building inclusive learning environments. Ultimately, the chapter argues that genuine inclusion arises from school cultures that recognize neurodiversity as part of human variation, not as a deviation to be contained.

CONCEPTUALIZING NEURODEVELOPMENTAL DISORDERS IN EDUCATIONAL CONTEXTS

The education of students with neurodevelopmental disorders (NDDs) stands at the intersection of neuroscience, pedagogy, and psychology. Schools increasingly encounter diverse learners whose developmental trajectories and cognitive profiles diverge from normative expectations, requiring teachers, leaders, and support pro-

DOI: 10.4018/979-8-3693-6331-7.ch005

fessionals to adapt practices that are both inclusive and effective. Educational psychology, grounded in the understanding of learning processes, social development, and emotional well-being, provides a powerful lens through which to interpret and enhance the school experiences of these students. The objective of this chapter is to examine critically how neurodevelopmental symptomatology shapes students' educational experiences and to evaluate the multi-layered support mechanisms that promote inclusion, well-being, and learning outcomes. Drawing from recent empirical and conceptual contributions (e.g., Bölte et al., 2021; Ferreira et al., 2025; Leifler, 2022; Nadeau et al., 2020; Van der Merwe et al., 2020), this discussion integrates contemporary perspectives on inclusive education, teacher competence, and cross-professional collaboration within the context of neurodevelopmental diversity.

Globally, inclusion has evolved from a moral imperative to a policy requirement. Yet inclusion for students with NDDs often remains conditional, fragmented, or unevenly implemented (Edin, 2025; Leifler et al., 2024). Schools are frequently expected to balance competing demands such as academic standards, behavioral management, and psychosocial care, within limited resources (Van Herwege et al., 2019). Moreover, teacher preparedness, attitudes toward neurodivergence, and systemic support play crucial roles in shaping both educational outcomes and teachers' sense of efficacy (Bettcher Erickson, 2025; Obiweluozo et al., 2021). Justus et al. (2025) highlight a notable disconnect between what teachers perceive and what they feel equipped to handle in their daily practice. Although many educators believe that ASD and ADHD are increasingly common in their classrooms, they also report lacking the preparation needed to teach students with these profiles confidently and effectively. The study suggests that this is less a matter of reluctance and more a reflection of limited training opportunities that focus on neurodevelopmental diversity. By drawing attention to this gap, it is emphasized the importance of strengthening teacher education so that educators gain both the knowledge and the practical tools required to support learners with complex developmental needs. Their findings reinforce a broader message within the field: when teachers feel more competent and supported, students with ASD and ADHD are more likely to experience inclusive, responsive, and meaningful learning environments.

Thus, the "school experience" of NDD students cannot be understood solely in terms of individual deficits; it reflects a complex ecology of relationships, structures, and pedagogical practices that interact dynamically with neurocognitive profiles.

Neurodevelopmental disorders encompass a range of conditions characterized by early-onset impairments in cognition, behavior, and motor or social functioning (Vargo, 2015). These include autism spectrum disorder (ASD), attention-deficit/hyperactivity disorder (ADHD), specific learning disorders, developmental coordination disorder, and intellectual disabilities. Educationally, these disorders influence how students perceive, process, and respond to instruction. As Nadeau et al. (2020)

argue, each condition manifests along a spectrum, and their overlapping features, such as executive dysfunction, attention difficulties, or challenges in emotional regulation, create varied classroom implications.

Understanding symptomatology through an educational psychology lens emphasizes not pathology but functioning and adaptation. Tajik-Parvinchi et al. (2021) highlight the role of cognitive control and emotion regulation in predicting mental health outcomes among children with NDDs, suggesting that deficits in these domains can exacerbate learning stressors. Similarly, Beckman et al. (2016) found significant associations between NDDs, school well-being, and social interaction difficulties, indicating that academic struggles often coexist with relational and emotional challenges. The emphasis, therefore, shifts from "fixing" deficits toward supporting developmental trajectories that optimize engagement and learning. Hadders-Algra (2022) underscores that neurodevelopment occurs within dynamic brain-environment interactions, and early interventions, especially in low- and middle-income contexts, are crucial to promoting school readiness. The recognition that neurodevelopmental variations represent both risk and potential necessitates a strength-based educational framework. Rather than focusing exclusively on impairments, schools must cultivate adaptive environments that align with diverse neurocognitive and emotional needs (Leifler, 2022). Educational psychology contributes here by linking cognitive theory with classroom practice, illuminating how motivation, self-efficacy, and social belonging interact with neurological development.

From the perspective of educational inclusion, Leifler (2022) and Bölte et al. (2021) demonstrate that policy commitments to inclusion in Sweden have yielded variable implementation, with persistent gaps between ideals and practice. Teachers' relational competence, as examined by Ewe and Aspelin (2025), becomes central to bridging these gaps. Their "microscopic relational analysis" reveals that teachers who demonstrate empathy, attunement, and reflective understanding foster more positive learning climates for students with NDDs. This echoes the educational psychology principle that learning is socially mediated, cognitive engagement is inseparable from relational trust.

Contemporary educational systems increasingly frame NDDs within the neurodiversity paradigm, which views neurological differences as part of human variation rather than pathology. Jack et al. (2025) explore this paradigm through the voices of neurodivergent student teachers, who emphasize the tension between institutional expectations of conformity and personal identity negotiation. Their findings suggest that even within teacher education, systemic biases and structural rigidity persist, reinforcing the need for institutional self-reflection and cultural change. Such insights resonate with Angelides et al. (2012) earlier argument that inclusive education requires leadership grounded in moral purpose and collaboration, not merely procedural compliance.

In summary, educational conceptualizations of neurodevelopmental disorders must transcend categorical labels and embrace a developmental-ecological view, one that considers biological, psychological, and contextual dimensions. The next section builds on this foundation by critically exploring the symptomatology of NDDs and its direct implications for learning processes, emotional adjustment, and social participation in schools.

SYMPTOMATOLOGY AND EDUCATIONAL IMPLICATIONS

Understanding symptomatology is central to interpreting how neurodevelopmental disorders (NDDs) shape students' learning and participation. From an educational-psychology perspective, the "symptom" is not merely a diagnostic marker but an observable pattern of cognitive, behavioral, or socio-emotional functioning that interacts with the school environment. Vargo (2015) notes that attention deficits, executive-function weaknesses, and sensory sensitivities affect not only information processing but also self-regulation and motivation, domains crucial for academic success. When unaddressed, these manifestations can create cumulative barriers to achievement and self-esteem.

Tajik-Parvinchi et al. (2021) demonstrate that impairments in cognitive control and emotion regulation predict later mental-health difficulties in children with NDDs, suggesting that the ability to manage emotions and focus attention mediates the relationship between neurological variation and school adaptation. Educational psychology translates these findings into practical guidance: teachers must scaffold metacognitive and emotional-regulation skills, rather than interpret behavioral dysregulation as defiance. Ferreira et al. (2025) further emphasize that working-memory capacity, often reduced in NDD populations, can be strengthened through adaptive training programs that combine cognitive exercises with motivational feedback. Such interventions link neuroscience and classroom pedagogy, embodying the principle that neurodevelopmental limitations are modifiable through contextually rich learning experiences.

Beckman et al. (2016) provide population-based evidence from Sweden showing that NDD symptomatology correlates with poorer self-rated health, greater peer exclusion, and reduced school satisfaction. These findings underline the psychosocial cost of untreated cognitive and behavioral difficulties. Within an educational-psychology framework, symptom expression thus has dual significance: it reveals underlying neural differences while simultaneously reflecting the school's capacity or incapacity to accommodate them.

The behavioral manifestations of NDDs frequently disrupt traditional classroom management systems. Hyperactivity, impulsivity, and repetitive behaviors can challenge teachers' authority and peers' tolerance, increasing the risk of exclusionary practices (Van der Merwe et al., 2020). Teachers newly entering the profession often report uncertainty in differentiating instruction and managing diverse needs effectively. Van der Merwe et al. (2020) found that novice teachers employ intuitive "learning-support strategies," yet many still operate within deficit-based frames. In contrast, Leifler (2022) argues for shifting focus toward participation and belonging, positioning behavior as communication within relational contexts.

Ewe and Aspelin (2025) provide microscopic analyses of teacher–student interactions, revealing that relational competence, teachers' sensitivity to micro-signals of engagement or distress, predicts more positive outcomes than procedural interventions alone. Their findings resonate with the socio-constructivist roots of educational psychology: learning emerges through guided participation and attuned interaction. When teachers interpret symptomatic behaviors through relational rather than disciplinary lenses, students' agency and self-concept improve markedly.

Social participation is equally shaped by peer dynamics. Beckman et al. (2016) and Bölte et al. (2021) report persistent friendship barriers and stigma associated with visible symptomatology. Inclusive education, therefore, must extend beyond placement to cultivate peer cultures of acceptance. Sorkos et al. (2025) contribute here through the notion of critical interculturalism, which frames inclusion as shared responsibility and informal leadership among students. Encouraging such agency aligns with psychological theories of prosocial development and belongingness as buffers against marginalization.

Early Identification And Educational Readiness

Gutiérrez-Ruiz and Santoya Montes (2025) highlight early detection as a linchpin of educational inclusion. Their study shows that systematic screening in early-childhood education enables timely interventions that enhance later academic trajectories. Hadders-Algra (2022) adds that promoting early brain development, particularly in under-resourced settings, can prevent learning disparities before formal schooling begins. From an educational-psychology standpoint, these insights reaffirm the importance of developmental timing: cognitive plasticity and socio-emotional learning capacities are highest in early years, making responsive environments crucial.

Symptom expression also interacts with systemic variables such as teacher beliefs and school climate. Bettcher Erickson (2025) demonstrates that educators' perceptions of neurodivergence directly affect their self-efficacy in teaching students with ADHD and autism. Teachers who conceptualize neurodivergence through a strengths-based lens report greater confidence and persistence, whereas deficit-oriented attitudes correspond with frustration and avoidance. This attitudinal dimension echoes findings by Magnin et al. (2021), who observed ambivalence among medical educators toward supporting students with NDDs in higher education. Both studies illustrate how implicit biases can amplify the impact of symptomatology by shaping expectations and opportunities.

Similarly, Obiweluozo et al. (2021) show that teachers of children with NDDs experience elevated stress levels, particularly when institutional support is inadequate. Their intervention using blended rational-emotive behavioral therapy (REBT) significantly reduced stress and improved coping, reinforcing the reciprocal relationship between teacher well-being and student support quality. Educational psychology situates this reciprocity within social-cognitive theory: teachers' self-efficacy and emotional regulation influence classroom climate, which in turn affects student engagement and symptom expression.

Technological And Pedagogical Innovation

Frei-Landau et al. (2023) explore how digital neuroscience tools integrated into special-needs-teacher education can enhance understanding of neural mechanisms underlying behavior. Their findings indicate that experiential learning with neuroimaging simulations fosters empathy and more precise pedagogical decision-making. In the same vein, Eguia et al. (2025) review motor-learning strategies that leverage technology and embodied practice to improve coordination and executive-function outcomes in children with NDDs. Together, these studies represent an emerging neuro-pedagogical paradigm, aligning neuroscientific insights with applied educational interventions.

Technology also facilitates adaptive feedback and self-paced learning approaches particularly beneficial for students whose symptomatology includes attention or processing-speed limitations (Ferreira et al., 2025). Yet, as Berglund Melendez et al. (2020) caution from a school-health perspective, digital tools must complement, not replace, human relationships. School nurses in their study stressed that successful technological interventions depend on empathetic dialogue and consistent follow-up, reiterating that psychosocial support remains central even in technologically enriched settings.

Intersectionality Of Symptoms And Context

The heterogeneity of NDD symptomatology challenges one-size-fits-all interventions. Edin (2025) notes that special-educational solutions often segregate rather than integrate, partly because school leaders interpret diagnostic labels as mandates for separate provision. This institutional reflex to isolate rather than accommodate reinforces stigma and undermines inclusive goals. Conversely, Leifler et al.

(2024) show that perceived inclusiveness improves when schools implement multi-perspective approaches combining teacher, student, and parent input. Their research demonstrates that acknowledging subjective experiences of inclusion yields more accurate assessments of support effectiveness.

The intersection of symptomatology with socioeconomic status and cultural norms also demands attention. Van Herwegen et al. (2019) found that professionals' understanding of NDDs varies with cultural expectations about normal behavior, influencing resource allocation. Educational psychology must therefore contextualize symptoms within ecological systems theory, recognizing that learning behaviors are co-constructed by biological predispositions and environmental affordances.

Taken together, contemporary evidence portrays neurodevelopmental symptomatology not as static pathology but as a dynamic interaction between neurological functions, psychological processes, and educational environments. Effective teaching for students with NDDs requires deep diagnostic insight combined with relational and systemic awareness. When schools interpret symptoms through ecological and developmental frameworks, they move from remediation toward empowerment.

THE SCHOOL EXPERIENCE, INCLUSION, AND SUPPORT SYSTEMS

Inclusion for students with neurodevelopmental disorders (NDDs) is not a static condition but a dynamic, context-dependent process involving curriculum access, social belonging, and emotional safety. Educational psychologists have long emphasized that belonging and perceived competence are as vital as instructional differentiation for sustained engagement. Leifler (2022) conceptualizes inclusion for students with neurodevelopmental conditions as an interplay between participation, recognition, and agency. Her research across Swedish schools revealed that formal inclusion policies often coexist with subtle exclusion through low expectations or social isolation. This duality reflects what Edin (2025) calls *segregated inclusion*, a paradox in which pupils remain physically present in mainstream schools yet pedagogically and socially peripheral.

Bölte et al. (2021) similarly found that inclusive practice for students with NDDs in Sweden varied markedly between schools despite common policy frameworks. The determining factors were teachers' attitudes, leadership vision, and access to multidisciplinary support. Their findings align with Angelides et al. (2012), who argued more than a decade earlier that sustainable inclusion requires leadership grounded in empathy and collaboration rather than bureaucratic compliance. When inclusion is treated as a moral and relational practice, rather than a procedural obligation, teachers and students alike experience higher levels of trust and engagement.

Ewe and Aspelin (2025) deepen understanding of inclusion by examining teachers' *relational competence*, their ability to perceive and respond to students' subtle emotional cues. Their micro-analytic observations reveal that attuned relational exchanges create psychological safety, enabling students with NDDs to risk participation and manage anxiety. In contrast, misattuned interactions often intensify behavioral symptoms. These findings substantiate social-cognitive theories within educational psychology, which view relationships as mediators between internal regulation and academic performance.

The importance of relational sensitivity extends beyond teachers. Beckman et al. (2016) demonstrated that students with NDDs report lower peer acceptance and increased loneliness, factors strongly linked to diminished academic motivation. Addressing these relational inequities requires intentional cultivation of peer empathy. Sorkos et al. (2025) show that empowering students as *informal leaders for intercultural responsibility* can foster solidarity and dismantle stigma. Translated to neurodiversity contexts, such peer-led agency represents a promising model for transforming classroom culture from tolerance to mutual respect.

Teachers' interpretations of neurodiversity decisively shape the lived school experience of students. Bettcher Erickson (2025) found that educators' perceptions of neurodivergence directly affect their self-efficacy in teaching learners with ADHD and autism. Teachers who adopt a competence-oriented understanding of NDDs express greater confidence, adaptability, and persistence, whereas deficit-based beliefs correspond with anxiety and withdrawal. Similarly, Van der Merwe et al. (2020) reported that recently qualified teachers in South Africa rely heavily on informal problem-solving and peer consultation because pre-service training rarely equips them with explicit learning-support strategies. These results underscore the critical role of initial teacher education in integrating neuroscientific and psychological knowledge with inclusive pedagogy. Jack et al. (2025) extend this discussion by exploring the experiences of *neurodivergent student teachers* themselves. Their qualitative analysis highlights systemic rigidity within teacher-education institutions, where neurodivergent trainees often must "change themselves" to fit established norms rather than prompting programmatic adaptation. This reversal of responsibility reveals how ableist assumptions remain embedded within educational culture. Educational psychology advocates reciprocal adaptation, schools and individuals co-regulating to achieve inclusion, which suggests the need for reflective institutional reform. Magnin et al. (2021) provide complementary insights from higher education: medical teachers, though generally sympathetic, expressed uncertainty about how to manage students with NDDs effectively. This uncertainty mirrors patterns observed in school contexts and reinforces the necessity of systemic professional development across educational levels.

Support Systems and Multidisciplinary Collaboration

Effective inclusion depends not only on teacher competence but also on *collaborative infrastructure*. Salminen et al. (2025) synthesized evidence from Europe indicating that professional collaboration, among teachers, psychologists, therapists, and health professionals, substantially improves outcomes for students with NDDs. Their scoping review emphasizes that collaboration must be intentional, structured, and reciprocal; ad-hoc cooperation seldom yields sustainable results. Similarly, Berglund Melendez et al. (2020) illustrate how school nurses play pivotal roles in identifying difficulties early and coordinating care, yet their effectiveness is contingent on shared communication channels with educators. This underscores an ecological principle central to educational psychology: support functions best within coherent systems, not isolated interventions.

Nadeau et al. (2020) propose a taxonomy of *educational adjustments*, curricular, behavioral, and environmental, that should operate synergistically. Their framework advocates for individualized supports (e.g., sensory accommodations, structured routines) embedded within universally designed learning environments. By contrast, Edin (2025) warns that when adjustments occur only through special-class placement, they risk reinforcing segregation. A balanced model would combine flexible classroom adaptation with specialist consultation, maintaining both belonging and targeted expertise.

The necessity of early, continuous coordination resonates with Gutiérrez-Ruiz and Santoya Montes (2025), who argue that detection and support mechanisms must function as a continuum from early childhood through adolescence. Continuity prevents the "support cliff" often experienced during educational transitions. Educational psychology research confirms that stability in adult relationships and intervention methods enhances self-efficacy and resilience among students with NDDs.

Teachers' psychological resources directly affect their capacity to sustain inclusive environments. Obiweluozo et al. (2021) found that teachers of children with NDDs exhibit high stress levels, largely due to workload and emotional labor. Their application of blended rational-emotive behavioral therapy (REBT) produced significant reductions in burnout indicators, suggesting that emotional-regulation training should accompany pedagogical training. Educational psychology situates such findings within self-determination theory: teachers, like students, require autonomy, competence, and relatedness to thrive. Institutions that neglect staff well-being inadvertently compromise student inclusion. Ranieri et al. (2024) expand the scope to informal learning environments e.g. museums, community centers, after-school programs, arguing that inclusive support should transcend classroom walls. Their review identifies informal education as fertile ground for social participation, especially for students whose symptomatology complicates traditional schooling.

Integrating formal and informal supports aligns with ecological models of learning that recognize development across multiple contexts.

Rapid advances in neuroscience and digital pedagogy are reshaping how schools conceptualize support. Frei-Landau et al. (2023) document that training special-needs-teacher candidates in digital neuroscience enhances both theoretical understanding and empathetic responsiveness. When teachers visualize brain processes underlying attention or sensory regulation, they adopt more patient and strategic approaches. Likewise, Ferreira et al. (2025) show that computerized working-memory training yields cognitive gains that translate into improved classroom behavior when accompanied by reflective discussions, a clear example of bridging neuroscience and humanistic pedagogy. Eguia et al. (2025) focus on *motor-learning strategies* as inclusive tools, demonstrating that task-oriented practice improves executive function and emotional regulation, especially when framed within supportive social contexts. These innovations illustrate a paradigm shift toward embodied, experiential learning that values movement, feedback, and multisensory engagement, all consistent with educational psychology's holistic understanding of development.

Professional Collaboration as a Cornerstone of Inclusion

Educational inclusion for students with neurodevelopmental disorders (NDDs) is sustained not merely through individual teacher competence but through systemic collaboration among educational, clinical, and social professionals. Salminen et al. (2025) demonstrated that cross-disciplinary teamwork significantly enhances identification accuracy, consistency in intervention, and family-school alignment. Collaboration functions as a protective factor against fragmentation, ensuring that cognitive, behavioral, and emotional supports complement one another. This systemic approach reflects Bronfenbrenner's ecological theory of human development, which underpins educational psychology's commitment to the interconnectedness of school systems.

Van Herwegen et al. (2019) found that professionals frequently recognize the educational needs of children with NDDs, such as structure, predictability, and emotional regulation, but face systemic constraints, including time limitations and inconsistent access to expertise. Consequently, the *intention* to collaborate does not always translate into *capacity* to do so. When multidisciplinary collaboration is institutionalized through shared planning sessions and co-reflection, as Berglund Melendez et al. (2020) advocate, schools transform into inclusive ecosystems rather than isolated service providers.

At the same time, Nadeau et al. (2020) emphasize that collaboration must preserve the integrity of each professional domain. Teachers, psychologists, and therapists contribute distinct but complementary perspectives. The danger arises when roles blur without accountability creating confusion rather than synergy. Therefore, professional development should not only teach inclusion as a concept but also model it as a collaborative practice.

Leadership profoundly determines whether collaboration flourishes or fades. Angelides et al. (2012) introduced the notion of *special teachers as leaders*, demonstrating that when teachers with expertise in inclusion are empowered to mentor colleagues, entire school cultures shift. Their model integrates relational leadership with distributed responsibility, principles reaffirmed by Leifler et al. (2024), who found that teachers' and parents' perceptions of inclusion improve markedly when leadership communicates a coherent vision of neurodiversity.

Edin (2025) cautions, however, that leadership discourse can sometimes mask structural inequalities. School leaders may interpret inclusion as a logistical task, allocating resources, rather than an ethical stance requiring transformation of underlying values. To sustain inclusion, leaders must navigate tensions between efficiency and empathy, accountability and care. From an educational psychology standpoint, leadership that fosters psychological safety among staff enables the relational and reflective practices essential for supporting students with NDDs.

EARLY DETECTION AND PREVENTIVE INTERVENTION

Preventive strategies are equally critical to shaping the educational trajectories of students with NDDs. Gutiérrez-Ruiz and Santoya Montes (2025) argue that early detection of developmental differences, especially in early childhood education, lays the foundation for later inclusion. When educators receive training to recognize atypical developmental markers, interventions can target self-regulation, language, and executive functioning before academic demands amplify difficulties. Hadders-Algra (2022) reinforces this view, asserting that promoting school readiness in children at risk of NDDs requires attention to both neurobiological and socio-environmental factors, particularly in low- and middle-income contexts. Early intervention embodies educational psychology's preventive ethos: supporting growth rather than remediating failure.

At the classroom level, ongoing research highlights the importance of cognitive and emotional scaffolding. Ferreira et al. (2025) demonstrated that targeted working-memory training enhances learning outcomes for students with NDDs, but only when integrated with reflective metacognitive guidance. Similarly, Tajik-Parvinchi et al. (2021) found that cognitive control and emotion regulation are predictive of mental health and academic adjustment. These findings converge on a key principle: cognitive development and emotional well-being are mutually reinforcing, necessitating instruction that combines structure with empathy.

Eguia et al. (2025) broaden the scope by demonstrating how motor learning strategies, task-oriented and feedback-based, improve coordination, attention, and self-esteem. Movement-based learning aligns with embodied cognition theories in educational psychology, which posit that learning emerges through sensory-motor engagement rather than abstract cognition alone. This body of evidence calls for inclusive curricula that integrate physical, cognitive, and emotional domains rather than treating them as separate silos.

Teachers' Emotional Labor and Resilience

Inclusive education also demands recognition of teachers' psychological realities. Obiweluozo et al. (2021) showed that emotional exhaustion among teachers of children with NDDs stems not only from workload but from the empathic strain of navigating behavioral crises and systemic under-support. Their intervention blended rational-emotive behavioral therapy, enhanced teachers' emotional resilience, demonstrating that mental-health literacy is essential for sustainable inclusion. From a humanistic educational psychology perspective, teacher well-being is both a moral imperative and a prerequisite for effective pedagogy. Irungu et al. (2025) show that teachers working with students who have neurodevelopmental disorders experience heightened stress, but that structured emotional-regulation support can make a meaningful difference. Their blended REBT intervention significantly reduced teachers' stress levels, underscoring the importance of supporting educators' psychological well-being as part of effective inclusive education.

Jack et al. (2025) remind us that the conversation about inclusion must also encompass *neurodivergent educators*. Their qualitative study revealed how institutional expectations often impose conformity rather than valuing diversity within teaching staff. Recognizing neurodivergent teachers as assets rather than anomalies advances the very inclusion schools seek to model for students.

The inclusion of students with NDDs also intersects with ethical and cultural dimensions of education. Güneş Özler and Yıldız (2025) explored privacy education for young children with neurodevelopmental disabilities and found that collaboration between teachers and parents enhances children's understanding of boundaries, autonomy, and self-protection. This highlights the importance of ethical literacy teaching respect for personal and cognitive diversity as part of the curriculum.

Sorkos et al. (2025) offer a complementary perspective through their study on student agency and critical interculturalism, emphasizing the value of peer leadership in promoting respect for difference. Transposed to neurodiversity, such frameworks advocate moving beyond accommodation toward *critical inclusion*, a pedagogy that challenges societal hierarchies of normalcy and competence. Educational psychology, with its dual concern for individual development and social context, is well positioned to lead this epistemic shift.

Recent advances suggest that digital tools can extend inclusivity when thoughtfully implemented. Frei-Landau et al. (2023) demonstrated that integrating digital neuroscience into teacher education cultivates multifaceted learning outcomes, including empathy, critical reflection, and scientific literacy. Technology becomes not merely a tool for remediation but a bridge between cognitive understanding and relational awareness. This aligns with Vargo's (2015) earlier proposition that educators' knowledge of neurodevelopmental processes empowers them to design instruction that respects both neurological diversity and individual dignity.

However, technology is not a panacea. Its ethical use demands ongoing reflection on accessibility, privacy, and the potential for over-mechanization. Educational psychology contributes here by evaluating how digital interventions influence motivation, attention, and identity formation, ensuring that technological inclusion remains human-centered.

The broader ecology of inclusion extends beyond formal schooling. Ranieri et al. (2024) argue that informal educational settings e.g. libraries, after-school programs, museums, serve as inclusive laboratories where social participation can flourish free from academic pressure. Integrating these settings into educational planning fosters a sense of community belonging crucial to students' psychosocial development. This resonates with ecological systems theory, affirming that learning occurs across nested contexts of family, school, and community. Vargo (2015) underscores that educators must see themselves as developmental guides rather than mere instructors, recognizing that each student's trajectory is shaped by intersecting biological, psychological, and social variables. Inclusion thus becomes an ongoing negotiation of meaning rather than a destination, a stance that defines the ethical heart of educational psychology.

Looking ahead, research must deepen its focus on *relational and contextual* variables. Quantitative measures of inclusion such as test scores, attendance, diagnostic prevalence, must be complemented by qualitative insights into belonging, identity, and voice. As Leifler et al. (2024) demonstrate, inclusion is most meaningful when assessed through multiple perspectives: teachers, parents, and students. Expanding such participatory methodologies can democratize research and align it with the lived realities of neurodivergent learners.

Furthermore, integration between neuroscience and pedagogy, as exemplified by Frei-Landau et al. (2023) and Ferreira et al. (2025), should evolve toward ethically grounded practice frameworks. Digital and cognitive innovations must be guided by principles of dignity, autonomy, and equity. Collaboration between researchers, practitioners, and policymakers will remain central to translating evidence into sustainable action.

The school experience of students with neurodevelopmental disorders encapsulates both the promise and the challenge of inclusive education. Across the studies reviewed, a coherent pattern emerges inclusion succeeds when schools cultivate empathy, relational attunement, professional collaboration, and reflective leadership. Educational psychology offers the conceptual and ethical scaffolding to realize this vision bridging neuroscience, pedagogy, and human development.

Ultimately, the measure of an inclusive school lies not in policy compliance but in everyday interactions that affirm every learner's right to belonging, competence, and joy. As Angelides et al. (2012) remind us, leadership for inclusion begins not with systems but with people, teachers who listen, students who lead, and communities that believe diversity is not an obstacle to learning but its greatest resource.

Table 1. School-age manifestations of neurodevelopmental disorders and support mechanisms

Neurodevelopmental Profile	Typical School-Based Challenges	Impact on Learning and Participation	Evidence-Based Support Mechanisms
Specific Learning Disorders (SLD)	Difficulties in reading, writing, and/or mathematics; slow academic progress despite effort	Academic underachievement, frustration, reduced academic self-concept	Structured literacy and numeracy instruction, individualized accommodations, assistive technology, executive function support
Attention-Deficit/ Hyperactivity Disorder (ADHD)	Inattention, impulsivity, hyperactivity, difficulty following routines and classroom rules	Disrupted learning, incomplete tasks, strained teacher–student relationships	Classroom structure, behavioral interventions, self-regulation strategies, collaboration with mental health services
Autism Spectrum Disorder (ASD)	Difficulties in social communication, sensory sensitivities, rigidity in routines	Social isolation, challenges in group work, uneven academic performance	Visual supports, predictable routines, inclusive practices, individualized educational planning
Communication Disorders	Difficulties in understanding or expressing language, pragmatic challenges	Reduced classroom participation, literacy difficulties, peer interaction problems	Speech and language therapy, language-rich instruction, pragmatic communication support
Motor and Tic Disorders	Poor fine and/or gross motor coordination, involuntary movements	Difficulties with handwriting, physical tasks, increased performance-related stress	Occupational therapy input, task adaptation, reduced motor demands
Co-occurring / Comorbid Profiles	Overlapping academic, attentional, and socioemotional difficulties	Increased risk for school disengagement and emotional distress	Multidisciplinary assessment, integrated academic and psychosocial interventions

REFERENCES

Angelides, P., Savva, K., & Hajisoteriou, C. (2012). Leading inclusion: Special teachers as leaders in the development of inclusive education. *International Studies in Educational Administration (Commonwealth Council for Educational Administration & Management (CCEAM)), 40*(1). https://search.informit.org/doi/10.3316/aeipt.190896

Beckman, L., Janson, S., & von Kobyletzki, L. (2016). Associations between neurodevelopmental disorders and factors related to school, health, and social interaction in schoolchildren: Results from a Swedish population-based survey. *Disability and Health Journal, 9*(4), 663–672. DOI: 10.1016/j.dhjo.2016.05.002

Berglund Melendez, A., Malmsten, M., Einberg, E. L., Clausson, E. K., & Garmy, P. (2020). Supporting students with neurodevelopmental disorders in school health care—School nurses' experiences. *International Journal of Environmental Research and Public Health, 17*(16), 5752. DOI: 10.3390/ijerph17165752

Bettcher Erickson, N. D. (2025). *The impact of educators' perceptions of neurodivergence on their self-efficacy in teaching middle and high school students with ADHD and autism in rural Minnesota* (Publication No. 1022) [Master's thesis, Minnesota State University Moorhead]. *Dissertations, Theses, and Projects*. https://red.mnstate.edu/thesis/1022

Bölte, S., Leifler, E., Berggren, S., & Borg, A. (2021). Inclusive practice for students with neurodevelopmental disorders in Sweden. *Scandinavian Journal of Child and Adolescent Psychiatry and Psychology, 9*(9), 15. DOI: 10.21307/sjcapp-2021-002

Edin, J. (2025). Segregated special educational solutions designated for pupils diagnosed with neurodevelopmental disorders: School leaders' understandings. *Education Inquiry, 16*(2), 139–156. DOI: 10.1080/20004508.2023.2173120

Eguia, K. F., Ng, S. S., Capio, C. M., & Wong, T. W. (2025). Applications of motor learning strategies in children with neurodevelopmental disorders: A scoping review. *Review Journal of Autism and Developmental Disorders*, •••, 1–16. DOI: 10.1007/s40489-025-00500-6

Ewe, L. P., & Aspelin, J. (2025). Teachers' relational competence towards students with neurodevelopmental symptoms: A microscopic relational analysis. *Journal of Research in Special Educational Needs, 25*(2), 225–236. DOI: 10.1111/1471-3802.12721

Ferreira, P., David, C., Costescu, C., Vera, L., Herrera, G., Lopes, S., Stilwell, D., Ferreira, A., Domingues, D., Brito, J., Campos, J., Paiva, A. M., Simão, A. M. V., Heldal, I., Stefanut, T., Rosan, A., & Trindade, F. (2025). Neurodevelopmental disorders: Assessing and training working memory. *BMC Psychology*, *13*(1), 1163. DOI: 10.1186/s40359-025-02912-9

Frei-Landau, R., Grobgeld, E., & Guberman, R. (2023). Implementing digital neuroscience in special-needs-teacher education: Exploring student-teachers' multifaceted learning outcomes related to teaching children with neurodevelopmental disorders. *Frontiers in Psychology*, *14*, 1232315. DOI: 10.3389/fpsyg.2023.1232315

Güneş Özler, N., & Yıldız, G. (2025). Privacy education for young children with neurodevelopmental disabilities: Teacher and parent experiences. *Sex Education*, **•••**, 1–16. DOI: 10.1080/14681811.2025.2476147

Hadders-Algra, M. (2022). The developing brain: Challenges and opportunities to promote school readiness in young children at risk of neurodevelopmental disorders in low- and middle-income countries. *Frontiers in Pediatrics*, *10*, 989518. DOI: 10.3389/fped.2022.989518

Irungu, , S. KAli, , N. AMbugua, , G. (2025). Self-reported teachers 'self-efficacy in supporting children with neurodevelopmental disorders in early-year education within Kiambu county, Kenya. *African Journal of Emerging Issues, 7*(8), 1-14. https://ajoeijournal.org/sys/index.php/ajoei/article/view/819

Jack, C., Crane, L., Kenny, A., Blaisdell, C., & Davis, R. (2025). "There's only so much the school can change about itself… Before you need to change something about yourself"—A qualitative analysis of the experiences of neurodivergent student teachers. *Autism in Adulthood: Challenges and Management*, *7*(4), 435–446. DOI: 10.1089/aut.2024.0047

Justus, S. A., Pogue, E. M., & Simanovich, V. (2025). Pre-K–12 teachers' views on ASD + ADHD: Prevalence estimates and teaching preparedness. *Children (Basel, Switzerland)*, *12*(3), 342. DOI: 10.3390/children12030342

Leifler, E. (2022). *Educational inclusion for students with neurodevelopmental conditions* [Doctoral dissertation, Karolinska Institutet]. Karolinska Institutet University Press. ISBN 978-91-8016-640-9

Leifler, E., Borg, A., & Bölte, S. (2024). A multi-perspective study of perceived inclusive education for students with neurodevelopmental disorders. *Journal of Autism and Developmental Disorders*, *54*(4), 1611–1617. DOI: 10.1007/s10803-022-05643-7

Magnin, E., Ryff, I., & Moulin, T. (2021). Medical teachers' opinions about students with neurodevelopmental disorders and their management. *BMC Medical Education*, *21*(1), 16. DOI: 10.1186/s12909-020-02413-w

Nadeau, M. F., Massé, L., Argumedes, M., & Verret, C. (2020). Education for students with neurodevelopmental disabilities—Resources and educational adjustments. In *Handbook of Clinical Neurology* (Vol. 174, pp. 369–378). Elsevier. DOI: 10.1016/B978-0-444-64148-9.00027-2

Obiweluozo, P. E., Dike, I. C., Ogba, F. N., Elom, C. O., Orabueze, F. O., Okoye-Ugwu, S., Ani, C. K. C., Onu, A. O., Ukaogo, V., Obayi, L. N., Abonyi, S. E., Onu, J., Omenma, Z. O., Okoro, I. D., Eze, A., Igu, N. C. N., Onuigbo, L. N., Umeano, E. C., & Onyishi, C. N. (2021). Stress in teachers of children with neurodevelopmental disorders: Effect of blended rational emotive behavioral therapy. *Science Progress*, *104*(4), 00368504211050278. DOI: 10.1177/00368504211050278

Ranieri, J. M., Neil, N., Sadowski, M., & Azzam, M. (2024). Supporting inclusion in informal education settings for children with neurodevelopmental disorders: A scoping review. *Journal of Developmental and Physical Disabilities*, *36*(6), 955–993. DOI: 10.1007/s10882-024-09970-8

Salminen, J., Laasanen, M., Leinonen, L., Karukivi, M., Vornanen, R., Alin, M., & Kyttälä, M. (2025). Professional collaboration for children and adolescents with neurodevelopmental disorders: A scoping review. *European Journal of Special Needs Education*, *40*(1), 16–34. DOI: 10.1080/08856257.2024.2314422

Sorkos, G., Hajisoteriou, C., & Angelides, P. (2025). Students' agency for critical interculturalism: From intercultural responsibility to informal leadership. *Oxford Review of Education*, *51*(1), 109–128. DOI: 10.1080/03054985.2023.2296091

Tajik-Parvinchi, D., Farmus, L., Tablon Modica, P., Cribbie, R. A., & Weiss, J. A. (2021). The role of cognitive control and emotion regulation in predicting mental health problems in children with neurodevelopmental disorders. *Child: Care, Health and Development*, *47*(5), 608–617. DOI: 10.1111/cch.12868

Van Herwegen, J., Ashworth, M., & Palikara, O. (2019). Views of professionals about the educational needs of children with neurodevelopmental disorders. *Research in Developmental Disabilities*, *91*, 103422. DOI: 10.1016/j.ridd.2019.05.001

Vargo, F. E. (2015). *Neurodevelopmental disorders: A definitive guide for educators*. W. W. Norton & Company.

Chapter 6
Comprehending Neurodevelopmental Disorders and Challenges Among School-Aged Children

ABSTRACT

This chapter considers how NDDs unfold within bilingual and twice-exceptional populations, where language experience and cognitive strengths can distort typical pathways of identification and support. Bilingualism may mask or imitate symptoms, while high ability can hide significant learning needs, placing these learners at risk of delayed intervention. The chapter underscores the importance of culturally grounded assessment, careful interpretation of linguistic performance, and educational plans that balance strengths with targeted assistance. It argues that meaningful support emerges from integrated frameworks that address cognitive, linguistic, and emotional development together.

THE CASE OF BILINGUAL INDIVIDUALS

Bilingual children with neurodevelopmental disorders (NDDs) present unique challenges in both educational assessment and intervention. The intersection of language diversity and developmental difficulties complicates diagnosis, often masking or mimicking disorders such as developmental language disorder (DLD), dyslexia, or other specific learning disorders (SLDs) (Beauchamp et al., 2022; Choo et al., 2023). As multilingual student populations increase globally, schools

DOI: 10.4018/979-8-3693-6331-7.ch006

and clinicians face the pressing need for culturally and linguistically sensitive diagnostic tools that accurately differentiate typical second-language acquisition from genuine neurodevelopmental difficulties (Leung & Molnar, 2025; Li'el et al., 2019). Misdiagnosis, either over- or under-identification, can have lasting effects on educational attainment, self-esteem, and social integration (Norbury & Sparks, 2013; Vender et al., 2021).

Underestimation is a common concern, where challenges are attributed solely to bilingualism. In these cases, children may be overlooked for early intervention, widening achievement gaps (Beauchamp et al., 2022; Bonuck et al., 2022). Minority-language speakers are especially vulnerable due to limited access to assessment resources, language-appropriate tools, and professionals experienced in bilingual development (Vender et al., 2021; Yuen et al., 2025). Conversely, overestimation occurs when normal bilingual language development is misclassified as pathological, often due to cultural biases or inadequate norms derived from monolingual populations (Norbury & Sparks, 2013). These diagnostic challenges underline the necessity of nuanced assessment protocols incorporating both home and societal languages, longitudinal monitoring, and parental insights (Abutbul-Oz & Armon-Lotem, 2022; Li'el et al., 2019).

The cognitive profile of bilingual children can be both protective and complicating in the presence of NDDs. Research demonstrates that bilingual children often exhibit superior executive functions, including working memory and attentional control, compared to monolingual peers (Morales et al., 2013; TM et al., 2023). Yet, for bilingual children with neurodevelopmental difficulties, this cognitive advantage may obscure linguistic deficits, complicating identification (Choo et al., 2023; Yuen et al., 2025). Studies highlight the interaction between bilingualism and NDDs as multifaceted: while bilingual exposure can enhance domain-general cognition, it may simultaneously complicate the detection of language-specific disorders (Li'el et al., 2019).

Assessment biases remain a major barrier. Clinicians' implicit assumptions about language proficiency and cultural norms can lead to both over- and underestimation of abilities, underscoring the importance of training and standardized protocols for bilingual populations (Leung & Molnar, 2025; TM et al., 2023). Recommended practices include assessing children in both languages, considering socio-economic background, age of bilingual exposure, and language dominance (Abutbul-Oz & Armon-Lotem, 2022; Li'el et al., 2019). Incorporating parental questionnaires enriches diagnostic accuracy and offers insights into functional language use across contexts, particularly where standardized tools are insufficient (Abutbul-Oz & Armon-Lotem, 2022; Bonuck et al., 2022).

Developmental Language Disorder illustrates these complexities vividly. Bilingual children with DLD often show deficits in vocabulary acquisition, syntactic processing, and reading, which can be obscured by their bilingual environment (Peristeri & Tsimpli, 2023; Li'el et al., 2019). Standardized tests developed for monolingual children may misclassify these difficulties, highlighting the importance of adapted assessment tools such as sentence repetition tasks or receptive/expressive word class subtests (Li'el et al., 2019). Furthermore, factors like exposure frequency, literacy support, and cognitive engagement influence assessment outcomes, emphasizing individualized evaluation (Li'el et al., 2019).

Educational interventions for bilingual children with NDDs remain underdeveloped. Digital platforms, especially post-COVID-19, offer new opportunities for remote assessment and intervention (Andreou & Argatzopoulou, 2025; Borasio et al., 2023). Evidence supports integrating technology-based approaches with culturally responsive pedagogy, individualized language scaffolding, and executive function support to optimize academic outcomes (Beauchamp et al., 2022; Yuen et al., 2025). Systemic barriers, such as limited professional training and inequitable resource allocation, continue to restrict access, particularly for minority-language populations (Borasio et al., 2023; Vender et al., 2021).

Case studies highlight systemic disparities in assessment. For example, in Iceland, migrant children are more frequently referred for neurodevelopmental evaluation, often resulting in higher diagnosis rates of speech and language disorders (Gudmundsdottir et al., 2025). These findings stress the importance of early identification, policy support, and targeted interventions to ensure equitable educational opportunities. Overall, bilingual children with NDDs require a multidimensional approach that integrates linguistic, cognitive, and socio-emotional factors, emphasizing accurate diagnosis and supportive interventions to foster positive academic and social outcomes (Abutbul-Oz & Armon-Lotem, 2022; Borasio et al., 2023; Li'el et al., 2019,).

NEURODEVELOPMENTAL DISORDERS AND TWICE-EXCEPTIONAL INDIVIDUALS IN SCHOOLS

Twice-exceptional learners, students who demonstrate high cognitive or creative ability alongside neurodevelopmental disorders, present an additional layer of educational complexity. Their cognitive strengths often conceal underlying learning difficulties, delaying diagnosis and complicating intervention (Amend & Peters, 2021; Cheek et al., 2023). Effective support requires assessment frameworks that account for both strengths and weaknesses, enabling educators to design individualized learning plans that maximize potential while addressing needs (Baldwin et al., 2015; Stephens & Karnes, 2000).

Definitional variability has historically complicated research and practice. Terms like "gifted," "high potential," and "twice exceptional" have been used inconsistently, affecting identification and service provision (Desvaux et al., 2024; Johnsen, 2021; Popp et al., 2025). Twice-exceptional learners are heterogeneous, with diverse profiles of ability and disability, requiring tailored assessment methods and differentiated educational strategies (Cornoldi et al., 2023; Ronksley-Pavia, 2015; Reider Lewis, 2021).

High cognitive ability can mask learning disabilities such as ADHD, dyslexia, or DLD, leading to underidentification and unmet needs (Kranz et al., 2024; Lee & Olenchak, 2015; McCallum et al., 2013). Multidimensional assessment, encompassing cognitive, behavioral, and socio-emotional domains, is essential to detect both exceptional talent and developmental difficulties (Siegle et al., 2025; Trail, 2022; Weber et al., 2025). Interventions should support executive function, self-regulation, and resilience while leveraging cognitive strengths to facilitate academic and social engagement (Küry & Fischer, 2025; Stoeger & Zeidner, 2019).

The intersection of bilingualism and twice-exceptionality presents compounded challenges. Linguistic diversity can obscure cognitive profiles, necessitating culturally responsive assessment practices that consider home language, societal language, and executive function (Choo et al., 2023; Leung & Molnar, 2025). For instance, bilingual twice-exceptional learners may display high reasoning skills yet struggle in reading or language comprehension, emphasizing the need for individualized support and differentiated instruction.

Educational policy and systemic factors play a crucial role. Programs that explicitly acknowledge twice-exceptional profiles, provide teacher training, and offer transition support for minoritized students improve both identification and outcomes (Foley-Nicpon & Teriba, 2022; Hunter et al., 2024). Instructional strategies should include scaffolded learning, adaptive assessment, and enrichment opportunities, ensuring that students' exceptional abilities are nurtured while their developmental challenges are accommodated (Açar & Gül, 2025; Amend & Peters, 2021; Trail, 2022).

Socio-emotional support is equally critical. Twice-exceptional learners are at higher risk of anxiety, low self-esteem, and social difficulties when their exceptional talents or developmental difficulties are unrecognized (Rinn, 2024; Saß et al., 2025; Sharkey & Nickl-Jockschat, 2023). Targeted interventions focusing on self-regulation, emotional resilience, and peer interaction contribute significantly to academic persistence and well-being (Küry & Fischer, 2025; Stoeger & Zeidner, 2019).

Finally, research emphasizes the need for longitudinal studies to clarify the developmental trajectories of twice-exceptional learners. While existing evidence highlights challenges in identification and intervention, emerging frameworks suggest that comprehensive, multidisciplinary approaches such as integrating cognitive, linguistic, and socio-emotional dimensions, enhance outcomes for this population (Baldwin et al., 2015; Cheek et al., 2023; Romano et al., 2024; Ozturk

& Tan, 2025). Schools, clinicians, and policymakers must collaborate to ensure that twice-exceptional learners achieve both academic and personal growth.

In conclusion, twice-exceptional learners represent a complex intersection of high ability and neurodevelopmental challenges. Accurate identification, culturally responsive assessment, and tailored interventions are essential to promote equitable educational opportunities. Integrating cognitive, linguistic, and socio-emotional considerations enables schools to support the potential of these students while mitigating developmental barriers.

Table 1. Distinguishing neurodevelopmental disorders, bilingualism, and twice exceptionality in educational assessment

Profile / Population	Key Characteristics	Risk of Misinterpretation	Assessment Considerations	Educational Implications
Neurodevelopmental Disorder (e.g., SLD, ADHD, ASD)	Persistent difficulties across contexts, developmental history of early-emerging challenges	Under-identification or late diagnosis due to compensatory strategies	Multidimensional assessment, developmental history, cross-setting evidence	Individualized educational planning, targeted academic and psychosocial support
Bilingual / Multilingual Learners	Variability in language proficiency across languages, uneven academic performance	Misdiagnosis of language difference as disorder	Assessment in both languages, consideration of exposure, cultural and linguistic context	Language-sensitive instruction, avoidance of deficit-based labeling
Twice-Exceptional Learners (2e)	Coexistence of high ability and neurodevelopmental difficulties, uneven performance profiles	Masking of difficulties by strengths or under-recognition of giftedness	Strengths-based assessment, dynamic and qualitative evaluation	Dual support: enrichment alongside targeted intervention
Bilingual Learners with NDDs	Overlapping language-related and neurodevelopmental difficulties	Attribution of disorder-related difficulties solely to bilingualism	Differentiation between language acquisition processes and neurodevelopmental vulnerabilities	Integrated linguistic and neurodevelopmental support
Twice-Exceptional Bilingual Learners	Complex interaction of giftedness, bilingualism, and neurodevelopmental risk	High risk of misidentification or exclusion from support services	Highly individualized assessment, interdisciplinary collaboration	Flexible, inclusive educational pathways addressing both strengths and needs

REFERENCES

Abutbul-Oz, H., & Armon-Lotem, S. (2022). Parental questionnaires and assessment of bilingual children with developmental language disorder. *Journal of Speech, Language, and Hearing Research: JSLHR*, *65*(4), 1023–1038. DOI: 10.1044/2022_JSLHR-21-00567

Açar, D., & Gül, M. D. (2025). Academic and socio-emotional experiences of a twice-exceptional student. *Behavioral Sciences (Basel, Switzerland)*, *15*(10), 1349. DOI: 10.3390/bs15101349 PMID: 41153139

Amend, E. R., & Peters, D. B. (2021). The importance of accurate assessment of gifted students: Issues with misdiagnosis, missed diagnoses, and twice-exceptionality. In *Handbook for counselors serving students with gifts and talents* (pp. 713–731). Routledge., DOI: 10.4324/9781003235415-44

Andreou, G., & Argatzopoulou, S. (2025). Digital assessment tools for bilingual students with neurodevelopmental disorders. *International Journal of Learning Disabilities*, *58*(2), 145–160. DOI: 10.1002/ijl.441

Baldwin, L., Baum, S., Pereles, D., & Hughes, C. (2015). Twice-exceptional learners: The journey toward a shared vision. *Gifted Child Today*, *38*(4), 206–214. DOI: 10.1177/1076217515597277

Beauchamp, J., Choo, T. M., Nayeb, L., & Vender, E. (2022). Linguistically responsive assessment of bilingual students with neurodevelopmental disorders. *Frontiers in Education*, *7*, 908231. DOI: 10.3389/feduc.2022.908231

Beauchamp, M. L., Amorim, K., Wunderlich, S. N., Lai, J., Scorah, J., & Elsabbagh, M. (2022). Barriers to access and utilization of healthcare services for minority-language speakers with neurodevelopmental disorders: A scoping review. *Frontiers in Psychiatry*, *13*, 915999. DOI: 10.3389/fpsyt.2022.915999 PMID: 36090362

Bonuck, K., Peristeri, E., & Tsimpli, I. M. (2022). Developmental language disorder and literacy acquisition in bilingual children. *Journal of Child Language*, *49*(6), 1137–1156. DOI: 10.1017/S0305000921000651

Bonuck, K., Shafer, V., Battino, R., Valicenti-McDermott, R. M., Sussman, E. S., & McGrath, K. (2022). Language disorders research on bilingualism, school-age, and related difficulties: A scoping review of descriptive studies. *Academic Pediatrics*, *22*(4), 518–525. DOI: 10.1016/j.acap.2021.12.002 PMID: 34896271

Borasio, F., Mascheretti, S., & Lorusso, M. L. (2023). New digital technologies for diagnosis and rehabilitation of neurodevelopmental disorders. *Applied Sciences (Basel, Switzerland)*, *13*(22), 12165. DOI: 10.3390/app132212165

Borasio, G., & Morales, J., & TM, S. (2023). Remote interventions and bilingual neurodevelopmental support post-COVID-19. *International Journal of School & Educational Psychology*, *11*(3), 211–229.

Cheek, C. L., Garcia, J. L., Mehta, P. D., Francis, D. J., & Grigorenko, E. L. (2023). The exceptionality of twice-exceptionality: Examining combined prevalence of giftedness and disability using multivariate statistical simulation. *Exceptional Children*, *90*(1), 43–56. DOI: 10.1177/00144029221150929

Choo, A. L., Smith, S., Pratt, A., & Leon Guerrero, S. (2023). Multilingualism and neurodevelopmental disorders. *Frontiers in Psychology*, *14*, 1267023. DOI: 10.3389/fpsyg.2023.1267023 PMID: 37655193

Choo, T. M., Yuen, M., & Leung, K. (2023). Multilingualism and neurodevelopmental disorders: Implications for assessment and intervention. *Journal of Child Psychology and Psychiatry, and Allied Disciplines*, *64*(8), 953–968. DOI: 10.1111/jcpp.13745

Cornoldi, C., Giofrè, D., & Toffalini, E. (2023). Cognitive characteristics of intellectually gifted children with a diagnosis of ADHD. *Intelligence*, *97*, 101736. DOI: 10.1016/j.intell.2023.101736

Desvaux, T., Danna, J., Velay, J. L., & Frey, A. (2024). From gifted to high potential and twice-exceptional: A state-of-the-art meta-review. *Applied Neuropsychology. Child*, *13*(2), 165–179. DOI: 10.1080/21622965.2023.2252950 PMID: 37665678

Foley-Nicpon, M., & Teriba, A. (2022). Policy considerations for twice-exceptional students. *Gifted Child Today*, *45*(4), 212–219. DOI: 10.1177/10762175221110943

Gudmundsdottir, E., Frigge, H. M., Saemundsen, E., Jensinudottir, K. B., & Njardvik, U. (2025). Rate differences in referrals and diagnostic outcomes of neurodevelopmental disorders between children with native and migrant backgrounds: A retrospective cohort study. *European Child & Adolescent Psychiatry*, *34*(10), 1–13. DOI: 10.1007/s00787-025-02744-3 PMID: 40423709

Hunter, W., Rath, S., Barnes, K., Hilliard, L., White, C. L., & McGiffert-Sandoval, D. (2024). "Take me to the bridge": Transitional support for minoritized twice-exceptional learners. *Teaching Exceptional Children*, *57*(6), 448–457. DOI: 10.1177/00400599241239332

Johnsen, S. K. (2021). Definitions, models, and characteristics of gifted students. In *Identifying gifted students* (pp. 1–32). Routledge., DOI: 10.4324/9781003235682-1

Kranz, A. E., Serry, T. A., & Snow, P. C. (2024). Twice-exceptionality unmasked: A systematic narrative review of the literature on identifying dyslexia in the gifted child. *Dyslexia (Chichester, England)*, *30*(1), e1763. DOI: 10.1002/dys.1763 PMID: 38232949

Küry, L., & Fischer, C. (2025). The self-perceptions of twice-exceptional children: A systematic review. *Education Sciences*, *15*(1), 44. DOI: 10.3390/educsci15010044

Lee, K. M., & Olenchak, F. R. (2015). Individuals with a gifted/ADHD diagnosis: Identification, performance, outcomes, and interventions. *Gifted Education International*, *31*(3), 185–199. DOI: 10.1177/0261429414530712

Leung, K., & Molnar, M. (2025). Clinician biases in evaluating bilingual students with neurodevelopmental disorders. *Journal of Learning Disabilities*, *58*(6), 451–468.

Li'el, , NWilliams, , CKane, , R. (2019). Identifying developmental language disorder in bilingual children from diverse linguistic backgrounds. *International Journal of Speech-Language Pathology, 21*(6), 613–622. https://doi.org/10.1080/17549507.2018.1513073

McCallum, R. S., Bell, S. M., Coles, J. T., Miller, K. C., Hopkins, M. B., & Hilton-Prillhart, A. (2013). A model for screening twice-exceptional students (gifted with learning disabilities) within a response to intervention paradigm. *Gifted Child Quarterly*, *57*(4), 209–222. DOI: 10.1177/0016986213500070

Morales, J., Calvo, A., & Bialystok, E. (2013). Working memory development in monolingual and bilingual children. *Journal of Experimental Child Psychology*, *114*(2), 187–202. DOI: 10.1016/j.jecp.2012.09.002 PMID: 23059128

Norbury, C. F., & Sparks, A. (2013). Difference or disorder? Understanding variations in bilingual children. *Child Language Teaching and Therapy*, *29*(2), 131–144. DOI: 10.1177/0265659012450432

Ozturk, B., & Tan, S. (2025). Identifying twice-exceptional students: A systematic research review. *Journal of Advanced Academics*, 1932202X251387416. Advance online publication. DOI: 10.1177/1932202X251387416

Peristeri, E., & Tsimpli, I. M. (2023). Spelling and syntactic difficulties in bilingual children with developmental language disorder. *Applied Psycholinguistics*, *44*(1), 87–110. DOI: 10.1017/S0142716422000427

Popp, C., Stoeger, H., & Ziegler, A. (2025). Robust and semi-robust findings from giftedness and talent research: Results of a secondary-order scoping review on meta-analyses. *Journal for the Education of the Gifted*, *48*(4), 334–363. Advance online publication. DOI: 10.1177/01623532251372590

Reider Lewis, L. (2021). Twice-exceptionality: Maximizing academic & psychosocial success in youth. *Journal of Health Service Psychology*, *47*(4), 191–196. DOI: 10.1007/s42843-021-00046-1

Rinn, A. N. (2024). A critique on the current state of research on the social and emotional experiences of gifted individuals and a framework for moving the field forward. *Gifted Child Quarterly*, *68*(1), 34–48. DOI: 10.1177/00169862231197780

Romano, S., Esposito, D., Aricò, M., Arigliani, E., Cavalli, G., Vigliante, M., Penge, R., Sogos, C., Pisani, F., & Romani, M. (2024). Giftedness and twice-exceptionality in children suspected of ADHD or specific learning disorders: A retrospective study. *Sci*, *6*(2), 23. DOI: 10.3390/sci6020023

Ronksley-Pavia, M. (2015). A model of twice-exceptionality: Explaining and defining the apparent paradoxical combination of disability and giftedness in childhood. *Journal for the Education of the Gifted*, *38*(3), 318–340. DOI: 10.1177/0162353215592499

Saß, S., Köller, O., & Zimmermann, F. (2025). Smart but maladapted? Differences in the psychological functioning of intellectually gifted students compared with average-ability students. *Gifted Child Quarterly*, *69*(3), 219–236. Advance online publication. DOI: 10.1177/00169862241310871

Sharkey, R. J., & Nickl-Jockschat, T. (2023). The neurobiology of autism spectrum disorder as it relates to twice-exceptionality. *Neurobiology of Learning and Memory*, *200*, 107740. DOI: 10.1016/j.nlm.2023.107740 PMID: 36894126

Siegle, D., McCoach, D. B., Boldt, G. T., Hamilton, R., Gubbins, E. J., & Callahan, C. M. (2025). What really happens in gifted and talented education? A portrait of three states. *Gifted Child Quarterly*, 00169862251370673. Advance online publication. DOI: 10.1177/00169862251370673

Stephens, K. R., & Karnes, F. A. (2000). State definitions for the gifted and talented revisited. *Exceptional Children*, *66*(2), 219–238. DOI: 10.1177/001440290006600206

Stoeger, H., & Zeidner, M. (2019). Self-regulated learning in gifted, talented, and high-achieving learners. *High Ability Studies*, *30*(1–2), 1–8. DOI: 10.1080/13598139.2019.1601326

TM, S., Morales, J., & Borasio, G. (2023). Research design considerations for bilingual neurodevelopmental studies. *Frontiers in Psychology*, *14*, 1223456.

Trail, B. A. (2022). *Twice-exceptional gifted children: Understanding, teaching, and counseling gifted students*. Routledge. DOI: 10.4324/9781003261216

Vender, E., Beauchamp, J., & Nayeb, L. (2021). Equity and access in neurodevelopmental assessment of bilingual students. *International Journal of Bilingual Education and Bilingualism, 24*(7), 1014–1031.

Weber, C. L., Boswell, C., & Behrens, W. A. (2025). Recognizing learning and developmental differences. In *Exploring critical issues in gifted education* (pp. 33–62). Routledge. DOI: 10.4324/9781003380634-4

Yuen, H. K., Ge, H., Zhang, C., Wong, Y. T., Chan, E. Y., Tsang, W. W., & Capio, C. M. (2025). Effects of bilingualism on executive function of children with neurodevelopmental disorders: A scoping review. *Children (Basel, Switzerland), 12*(9), 1247. DOI: 10.3390/children12091247 PMID: 41007112

Yuen, M., Choo, T. M., & Leung, K. (2025). Cognitive and linguistic interplay in bilingual children with neurodevelopmental disorders. *Journal of Cognition and Development, 26*(3), 411–430.

Chapter 7
The Relationship Among Neurodevelopmental Disorders and Executive Functions

ABSTRACT

This chapter explores executive functions (EFs) as a dynamic bridge between neurodevelopment and learning, examining how they shape behavior, emotion, and academic performance in individuals with NDDs. It describes EF as a network-based system influenced by neural maturation, environmental structure, and emotional context. The text reviews evidence on developmental pathways, real-world assessment, and interventions that leverage neuroplasticity through cognitive, physical, and socio-emotional methods. By integrating scientific and educational perspectives, the chapter positions EF at the center of inclusive practice, highlighting its relevance for designing environments that support self-regulation and adaptive growth.

INTRODUCTION

Understanding the complex relationship between neurodevelopmental disorders (NDDs) and executive functions (EFs) is central to contemporary neuroscience and educational psychology. Executive functions, a broad construct encompassing inhibition, working memory, planning, cognitive flexibility, and emotional regulation, form the foundation of purposeful behavior, self-regulation, and adaptive learning throughout life (Cortés Pascual et al., 2019; De Felice et al., 2015; Zelazo & Carlson, 2020). When these mechanisms are disrupted, the effects extend beyond cognitive efficiency, influencing emotional adjustment, social participation, and educational

DOI: 10.4018/979-8-3693-6331-7.ch007

achievement. In individuals with NDDs such as attention-deficit/hyperactivity disorder (ADHD), autism spectrum disorder (ASD), and specific learning disorders (SLDs), executive dysfunction represents one of the most persistent and disabling characteristics (Crisci et al., 2021; Otterman et al., 2019).

Over the past two decades, advances in developmental neuroscience and longitudinal research have shown that executive function deficits appear early in childhood and evolve dynamically through adolescence and adulthood (Otterman et al., 2019; Sadozai et al., 2024). These developmental patterns mirror the gradual maturation of the prefrontal cortex and its integration with neural networks responsible for attention, motivation, and socio-emotional regulation. Consequently, EF should not be viewed as a fixed skill set but as a multidimensional construct shaped by biological, environmental, and educational influences. For educators and clinicians, this means executive functioning must be approached simultaneously as a neurocognitive capacity and a pedagogical objective, a dual perspective that requires collaboration across disciplines (Barnes, 2023; Southon, 2022).

FROM NEUROLOGY TO PEDAGOGY

Historically, the study of executive functions (EFs) has been dominated by neuropsychological and clinical perspectives that emphasized deficits in inhibition, planning, and working memory as diagnostic markers of neurodevelopmental disorders (NDDs) (Cortés Pascual et al., 2019; De Felice et al., 2015). However, recent research increasingly highlights the critical role of educational environments in shaping executive growth and compensatory strategies. Southon (2022), for instance, demonstrated that variability in EF performance among university students with subclinical neurodevelopmental traits predicted differences in academic achievement, even when intelligence and motivation were controlled for. These findings underscore the sensitivity of executive skills to contextual and instructional variables, such as teacher expectations, task structure, and feedback systems, revealing that schools are not merely sites for cognitive assessment but also environments for developing EF capacities.

Recent studies of early childhood further confirm that EF deficits represent a transdiagnostic vulnerability across NDDs. Otterman et al. (2019) found that children who displayed attentional and emotional regulation difficulties between ages 4 and 6 were significantly more likely to meet diagnostic criteria for ADHD, ASD, or language disorders by age 8. This convergence supports a dimensional model in which executive control functions as a common pathway linking diverse developmental outcomes. Educational research has translated these neurocognitive findings into pedagogical strategies that include structured routines, working-memory scaffolding, and metacognitive prompts embedded within classroom activities, all of which foster both academic achievement and socio-emotional learning (Barnes, 2023).

Within school settings, executive functions constitute a "hidden curriculum" underlying success in literacy, numeracy, and social adaptation. As Theodoratou et al. (2023) note, EF skills strongly predict quality of life and functional independence across the neurodevelopmental spectrum. Students with robust EF capacities can plan, self-monitor, and adapt flexibly to changing academic demands, whereas those with deficits often face cascading challenges such as poor organization, reduced persistence, and emotional dysregulation, difficulties that impede both learning and interpersonal relationships.

For educators, executive dysfunction should therefore be understood not merely as a behavioral issue but as a cognitive-regulatory challenge requiring instructional adaptation. Barnes (2023) emphasizes that interventions to strengthen EF are most effective when embedded in authentic classroom contexts rather than treated as isolated cognitive exercises. Techniques such as goal setting, verbal mediation, and feedback-driven reflection promote self-regulatory growth and metacognitive awareness, processes especially vital for learners with NDDs, who often rely on external scaffolding before internalizing control.

The interaction between EF and NDDs is multidirectional. Executive dysfunction can both contribute to and result from core neurodevelopmental symptoms. For instance, deficits in inhibition and cognitive shifting exacerbate repetitive behaviors in ASD (Perry et al., 2022), while limitations in working memory and attentional control intensify academic difficulties in SLDs (Filipe, 2021). Conversely, experiences of chronic academic failure or social rejection may further erode EF performance by increasing stress and emotional reactivity (Frazier et al., 2022). This circular relationship underscores the need for ecological models that integrate neurocognitive, emotional, and contextual determinants of functioning.

Furthermore, comorbidity across NDDs, such as the overlap between ADHD and SLDs, illustrates how shared executive mechanisms underpin distinct diagnostic categories (Crisci et al., 2021). Consequently, contemporary frameworks advocate a transdiagnostic approach that focuses on underlying processes rather than surface-level symptom clusters (Mareva et al., 2024). From an educational standpoint, this paradigm enables practitioners to move away from categorical labeling and toward individualized support grounded in cognitive-functional assessment. Executive functions thus operate as a bridge between neurobiological maturation and educational achievement. Zelazo and Carlson (2020) describe EF as the central mechanism through which children learn to regulate attention, emotion, and behavior in accordance with classroom norms and expectations. Inadequate EF development contributes to widening academic gaps, particularly for learners with neurodevelopmental vulnerabilities. For this reason, EF assessment and training have become integral to inclusive and special-needs education programs. Intervention studies show that executive functions are modifiable through targeted, experiential learning: physical activity, cognitive training, and emotion-regulation curricula have all been shown to enhance EF and, in turn, academic outcomes (Sung et al., 2022). These findings reinforce the view that education functions as a form of neurorehabilitation, a context in which cognitive flexibility, self-monitoring, and problem-solving are cultivated through structured practice and reflective feedback.

Integrating insights from neuroscience, psychology, and pedagogy, contemporary scholarship calls for a neuroeducational framework that situates executive functions at the core of learning processes. In this framework, educators are not merely transmitters of knowledge but facilitators of cognitive regulation, capable of designing environments that align with learners' neurodevelopmental profiles. As Southon (2022) and Barnes (2023) argue, such an approach bridges the traditional divide between clinical diagnosis and classroom practice, advancing educational equity through scientifically informed teaching.

DEFINING EXECUTIVE FUNCTIONS WITHIN NEURODEVELOPMENT

Executive functions (EFs) represent a constellation of higher-order cognitive processes that enable goal-directed behavior, adaptive problem-solving, and emotional regulation. Within neurodevelopment, these processes emerge gradually, reflecting the prolonged maturation of the prefrontal cortex and its reciprocal connections with parietal, limbic, and subcortical structures. According to Zelazo and Carlson (2020), EF serves as a "control system" that organizes thought and action in pursuit of internal goals, distinguishing human cognition by its flexibility and capacity for self-regulation.

Core components of EF (e.g. working memory, inhibitory control, and cognitive flexibility) develop along partially independent yet interrelated trajectories. Working memory supports the temporary storage and manipulation of information; inhibition enables the suppression of prepotent responses; and cognitive flexibility allows individuals to shift perspectives and strategies in response to changing demands (Otterman et al., 2019; Zelazo & Carlson, 2020). Together, these processes form the foundation for complex skills such as planning, decision-making, and emotional self-management.

In neurodevelopmental disorders (NDDs), impairments in one or more EF domains are common and often explain the heterogeneity of clinical presentations. For example, children with ADHD frequently show deficits in inhibitory control and working memory that undermine attentional persistence and task organization

(Crisci et al., 2021). Similarly, individuals with autism spectrum disorder (ASD) tend to experience difficulties in cognitive flexibility, leading to rigidity in thought and behavior (Perry et al., 2022). In specific learning disorders (SLDs), EF weaknesses in monitoring and planning often coexist with deficits in academic skills such as reading comprehension or problem-solving (Filipe, 2021; Frazier et al., 2022). These patterns suggest that EF operates as a transdiagnostic construct, one that cuts across categorical diagnoses and influences both cognitive and socio-emotional outcomes (Mareva et al., 2024).

Neuroimaging and longitudinal studies further clarify that EF is not a unitary capacity but a networked system that depends on the synchronization of distributed brain regions. Functional MRI findings demonstrate that efficient EF performance relies on the dynamic integration of prefrontal and parietal areas with the basal ganglia and cerebellum (Otterman et al., 2019). Disruptions in these circuits, whether due to genetic, environmental, or developmental factors, can produce a spectrum of executive impairments that vary in severity and domain specificity. Sadozai et al. (2024) emphasize that developmental trajectories of EF are influenced by both biological maturation and environmental scaffolding, underscoring the importance of early, supportive educational experiences.

From an educational perspective, defining EF within neurodevelopmental contexts extends beyond cognitive measurement. It involves recognizing how these functions interact with motivation, emotion, and context to shape learning outcomes. Barnes (2023) and Southon (2022) argue that EF should be considered both a neurocognitive foundation and an educational target: a set of skills that can be cultivated through intentional instructional design. Theodoratou et al. (2023) similarly propose that executive abilities mediate the relationship between neurocognitive health and quality of life, bridging the domains of neurology, psychology, and pedagogy.

Summarizing, executive functions represent an evolving, multidimensional construct whose development is integral to both neurological maturation and educational achievement. Within neurodevelopmental disorders, EF serves as a key explanatory framework linking neural variability to behavioral outcomes. Recognizing this interdependence allows researchers and educators to move toward a more integrated understanding of how brain, cognition, and learning co-develop, informing both diagnostic practices and intervention strategies across the lifespan (Sadozai et al., 2024; Zelazo & Carlson, 2020).

EXECUTIVE FUNCTIONS AND NEUROCOGNITIVE ARCHITECTURE

Executive functions are deeply embedded within the neurocognitive architecture of the developing brain (Christofani et al., 2020; McFall et al., 2017; Zelazo, 2015). They emerge from the coordinated activity of distributed neural networks, primarily involving the prefrontal cortex (PFC), anterior cingulate cortex, parietal lobes, and subcortical structures such as the basal ganglia and cerebellum. This network supports both the regulation of behavior and the integration of sensory, emotional, and cognitive information necessary for goal-oriented activity (Marini et al., 2020; Ruffini et al., 2024).

During development, the maturation of these neural circuits follows a hierarchical pattern, progressing from basic sensory-motor coordination toward increasingly abstract and self-regulatory forms of cognition. The prefrontal cortex plays a central role, acting as an integrative hub that modulates the activity of other cortical and subcortical areas. This top-down regulation enables individuals to inhibit automatic responses, maintain information over time, and flexibly adapt to new contexts (Crisci et al., 2021; Zelazo & Carlson, 2020).

In neurodevelopmental disorders, atypical structural and functional connectivity within these networks often underlies EF impairments. Neuroimaging evidence shows altered white matter integrity and atypical activation patterns in prefrontal and frontoparietal regions among individuals with ADHD, ASD, and SLDs (Perry et al., 2022; Southon, 2022). Such disruptions can lead to deficits in inhibitory control, working memory, and cognitive flexibility, which are core components that sustain everyday cognitive functioning. Moreover, the interaction between these neural differences and environmental experiences can exacerbate or mitigate functional outcomes, emphasizing the dynamic and plastic nature of the developing brain (Bombonato et al., 2024; Sung et al., 2022).

Recent studies have proposed that EF deficits in NDDs may not stem solely from localized dysfunctions but rather from broader inefficiencies in network integration and neural communication. Functional MRI studies suggest that the default mode network (DMN) and the central executive network (CEN) exhibit atypical synchronization in children with autism and ADHD, leading to difficulties in attentional switching and goal maintenance (Harkness et al., 2025; Mareva et al., 2024). These findings point toward the importance of examining EF within a systems-based perspective that recognizes the interdependence of neural and environmental factors.

Emotion and motivation also contribute significantly to EF architecture. Tajik-Parvinchi et al. (2021) demonstrate that emotional regulation and cognitive control share overlapping neural substrates, particularly within the orbitofrontal and anterior cingulate cortices. Dysregulation within these regions may explain the heightened emotional reactivity often observed in individuals with ADHD and ASD. This neurocognitive overlap highlights that EF is not purely cognitive, it also encompasses the integration of affective and motivational systems essential for adaptive behavior (Costescu et al., 2023; Frazier et al., 2022).

Overall, the neurocognitive architecture of executive functions reflects a complex interplay between structure, function, and experience. Within neurodevelopmental contexts, disruptions in these systems yield a spectrum of cognitive and behavioral outcomes, demonstrating that EF is both biologically grounded and environmentally modulated. Understanding this dual influence provides a foundation for developing targeted interventions that strengthen executive processes by leveraging neuroplasticity and contextual support (Filipe, 2021; Mareva et al., 2024; Perry et al., 2022).

TRANSLATING NEUROSCIENCE TO EDUCATIONAL PRACTICE

The translation of neuroscientific findings on executive functions (EFs) into educational contexts has reshaped contemporary views of learning, development, and intervention. Traditionally, schooling has focused primarily on the acquisition of academic content such as literacy, numeracy, and factual knowledge, while the underlying cognitive and regulatory processes supporting these skills received comparatively less attention (Bowers, 2016; Busso & Pollack, 2015). However, the growing body of evidence linking EF to academic achievement, emotional adjustment, and social competence has prompted a paradigm shift: educators now recognize EF not merely as a by-product of instruction, but as a critical foundation for learning itself (Barnes, 2023; Zelazo & Carlson, 2020).

Empirical studies have demonstrated robust associations between EF components, such as working memory, inhibition, and cognitive flexibility and success across domains of reading, writing, and mathematics (Bukva & Memisevic, 2025; Ruffini et al., 2024). These functions enable children to plan, monitor progress, and adapt strategies in response to challenges. Deficits in EF, conversely, are associated with difficulties in task organization, time management, and emotional regulation, factors that significantly influence classroom behavior and learning outcomes (Bowers, 2016; Carrick & Hamilton, 2023; Southon, 2022).

The pedagogical relevance of EF research extends beyond cognitive performance to encompass emotional and social learning (Busso & Pollack, 2015). Emotionally grounded executive skills, often described as "hot" executive functions, are particularly implicated in self-regulation and decision-making in social contexts (Lane et al., 2025; Tajik-Parvinchi et al., 2021). For students with neurodevelopmental disorders, these capacities are frequently compromised, leading to impulsivity, frustration intolerance, or disengagement (Frazier et al., 2022). Integrating emotional regulation training and metacognitive reflection within classroom instruction can therefore enhance both academic engagement and psychosocial adaptation (Cristofani et al., 2020; Costescu et al., 2023).

Moreover, understanding EF development underscores the need for individualized pedagogical frameworks. Since EF maturation occurs along extended trajectories, teaching methods that align with developmental readiness can optimize cognitive growth (Busso & Pollack, 2015). Adaptive instruction, scaffolded learning, and structured feedback foster self-monitoring and goal-directed persistence (Filipe, 2021; Marini et al., 2020). In this context, teachers function not merely as knowledge transmitters but as facilitators of cognitive regulation, helping students develop strategies to control attention, manage working memory load, and shift between mental sets (Barnes, 2023).

The intersection of neurology and pedagogy also supports the design of early interventions for children at risk of EF deficits. Targeted cognitive training, mindfulness-based approaches, and movement-based programs such as physical activity and play therapy have shown positive effects on EF development in children with neurodevelopmental conditions (Sung et al., 2022; Tao et al., 2025). These interventions harness neuroplastic mechanisms to enhance executive control and promote functional generalization beyond the training context (Bombonato et al., 2024). Importantly, the integration of neuroscientific insights into educational practice must be accompanied by sensitivity to contextual factors, including socio-economic background, cultural expectations, and classroom environment, which can either amplify or constrain EF development (Filipe, 2021; Perry et al., 2022).

In summary, the connection between neurology and pedagogy highlights a transformative framework for education, one that views cognitive control, emotional regulation, and metacognition as teachable competencies. Embedding EF-informed strategies within curricula not only benefits learners with neurodevelopmental disorders but enriches educational quality for all students. This neuroeducational approach bridges scientific understanding and classroom practice, ensuring that developmental diversity is recognized as a foundation for adaptive teaching and learning.

EXECUTIVE FUNCTION INTERVENTIONS AND NEUROPLASTICITY

The growing recognition of executive functions (EFs) as a cornerstone of adaptive behavior has led to a surge of intervention research aimed at enhancing these skills, particularly in children and adolescents with neurodevelopmental disorders (NDDs). The effectiveness of such intervention's rests on the principle of neuroplasticity, the brain's capacity to reorganize itself through experience and targeted stimulation (Khan et al., 2017). By designing interventions that engage the neural circuits underlying EF, researchers and clinicians aim to promote both structural and functional changes that translate into improved real-world performance (Bombonato et al., 2024; Capobianco et al., 2025; Khan et al., 2017). Neuroimaging evidence supports the notion that EF training can induce measurable changes in brain connectivity and activation patterns. For example, task-based and resting-state fMRI studies have shown increased activity in the dorsolateral prefrontal cortex and improved synchronization within frontoparietal networks following structured cognitive training (Lee et al., 2025; Mareva et al., 2024). These findings indicate that EF is not a static construct but a trainable system responsive to environmental input and intentional practice. The degree of generalization and specifically how far training effects extend to untrained tasks or broader domains of functioning remains an ongoing research challenge, with current evidence suggesting moderate but significant transfer when interventions are contextually embedded (Bombonato et al., 2024).

Intervention models can be broadly divided into direct cognitive training and ecologically oriented programs. Direct cognitive training often involves repetitive practice of specific EF components, such as working memory or inhibitory control, using computerized or game-based formats (Frazier et al., 2022). While these approaches may enhance targeted cognitive performance, their transfer to everyday skills is variable. In contrast, ecologically oriented interventions, such as classroom-based regulation programs, social-cognitive curricula, or mindfulness practices, seek to integrate EF enhancement within authentic learning and social environments, thereby increasing the likelihood of sustained behavioral change (Filipe, 2021; Perry et al., 2022).

Physical activity interventions have also emerged as effective tools for supporting EF development. Meta-analyses demonstrate that aerobic exercise, coordinated movement, and motor-skill activities improve attention, working memory, and inhibitory control in children with NDDs (Sung et al., 2022; Tao et al., 2025). These effects are thought to arise from enhanced dopaminergic modulation and increased prefrontal activation, illustrating the reciprocal relationship between motor and cognitive systems. Similarly, virtual reality (VR) and technology-assisted programs are being increasingly applied in rehabilitation contexts to strengthen EF through immersive, adaptive tasks that promote engagement and feedback-based learning (Capobianco et al., 2025).

The most effective EF interventions appear to share several key characteristics: they are multimodal, combining cognitive, emotional, and physical dimensions; developmentally tailored, matching the learner's neurocognitive readiness; and contextually grounded, ensuring that acquired skills are relevant to real-world demands (Barnes, 2023; Khan et al., 2017; Southon, 2022). Moreover, they emphasize metacognitive awareness, encouraging learners to reflect on their strategies and self-regulatory processes, an essential mechanism for sustaining long-term gains.

Ultimately, the neuroplastic basis of EF underscores the importance of early, targeted, and sustained interventions. By capitalizing on sensitive developmental periods and integrating scientific understanding into educational and therapeutic practices, it becomes possible to foster meaningful improvements in both cognitive control and adaptive functioning. This integrative perspective not only benefits individuals with neurodevelopmental conditions but also advances a broader vision of education and rehabilitation grounded in brain-based learning principles (Bombonato et al., 2024; Khan et al., 2017; Sadozai et al., 2024; Tao et al., 2025).

SOCIO-EMOTIONAL DIMENSIONS OF EXECUTIVE FUNCTION IN NEURODEVELOPMENTAL DISORDERS

Executive functions (EFs) extend beyond cognitive control to encompass essential socio-emotional processes that regulate behavior in interpersonal and adaptive contexts. Emotional regulation, empathy, and motivation are integral to EF functioning and are closely linked to the neural systems that underlie self-monitoring and goal-directed behavior (Cristofani et al., 2020; McDermott et al., 2013; Tajik-Parvinchi et al., 2021). In children and adolescents with neurodevelopmental disorders (NDDs), impairments in these domains often manifest as emotional dysregulation, impulsivity, or diminished social reciprocity, reflecting difficulties in coordinating cognitive and affective processes (Costescu et al., 2023; Frazier et al., 2022; McDermott et al., 2013).

The overlap between cognitive control and emotional processing is supported by neuroanatomical evidence. Regions such as the orbitofrontal cortex, anterior cingulate cortex, and amygdala are jointly implicated in both emotion regulation and decision-making, highlighting their functional interdependence (Tajik-Parvinchi et al., 2021). Disruptions within these neural circuits have been associated with difficulties in sustaining attention to social cues, inhibiting inappropriate emotional responses, and adapting behavior to situational demands (Harkness et al., 2025). For individuals with ADHD and ASD, such disruptions often result in heightened emotional reactivity and reduced tolerance to frustration, features that significantly impact peer relationships and academic engagement (Carrick & Hamilton, 2023; Frazier et al., 2022).

Empathy, another socio-emotional dimension linked to EF, relies on the integration of cognitive perspective-taking and affective resonance. Cristofani et al. (2020) demonstrated that the development of empathy in children with NDDs is strongly associated with executive control abilities, particularly working memory and inhibitory regulation. This association suggests that understanding and responding to others' emotions require cognitive flexibility and the suppression of self-centered reactions. Consequently, interventions that target EF may simultaneously enhance social cognition and emotional understanding, producing cascading benefits across multiple domains of functioning (Costescu et al., 2023; McDermott et al., 2013; Rosenblum et al., 2025).

Motivational systems also play a pivotal role in EF performance. Dopaminergic pathways connecting the prefrontal cortex and striatal regions modulate reward sensitivity and effort allocation, mechanisms essential for maintaining engagement and persistence in goal-oriented activities. Dysfunctions within these pathways, as observed in ADHD and related disorders, contribute to difficulties in sustaining effort and responding adaptively to feedback (Lane et al., 2025; Perry et al., 2022). Educational and therapeutic programs that incorporate motivational scaffolds, such as reward-based learning or self-efficacy enhancement, can thus support both cognitive regulation and socio-emotional resilience (Barnes, 2023; Filipe, 2021; McDermott et al., 2013).

Integrating socio-emotional components into EF frameworks reinforces the view that cognition and emotion are inseparable elements of adaptive human behavior. Effective self-regulation arises from the interplay of neural, cognitive, and affective systems, each influencing the other across development (McDermott et al., 2013). For learners with neurodevelopmental disorders, supporting this integration is essential for promoting autonomy, social participation, and mental well-being. As current research indicates, interventions that simultaneously address executive and emotional functions yield stronger and more sustainable outcomes than those focusing on cognition alone (Costescu et al., 2023; Frazier et al., 2022; Tajik-Parvinchi et al., 2021). Fogel (2022) further demonstrated that executive function difficulties were moderately associated with poorer daily functioning, and that the frequency of cognitive strategy use moderated this relationship. Specifically, higher strategy use amplified the link between EF challenges and daily organization, with those using strategies most frequently showing the strongest EF–functioning association. Common strategies included to-do lists, electronic reminders, and limiting distractions. These results suggest that cognitive strategies do not simply compensate for EF deficits but help individuals translate their executive capacities more effectively into everyday tasks.

DEVELOPMENTAL TRAJECTORIES AND ASSESSMENT OF EXECUTIVE FUNCTIONS IN NEURODEVELOPMENTAL DISORDERS

The development of executive functions (EFs) follows a protracted and nonlinear trajectory that mirrors the maturation of the prefrontal cortex and its extensive neural connections. During early childhood, foundational skills such as inhibitory control and attentional regulation begin to emerge, forming the basis for later-developing abilities like planning, problem-solving, and self-monitoring (Otterman et al., 2019; Zelazo & Carlson, 2020). Rapid growth in EF capacity is typically observed between ages three and five, coinciding with significant synaptic reorganization and the establishment of goal-directed behavior. These processes continue into adolescence and early adulthood as myelination and synaptic pruning enhance the efficiency of prefrontal networks (Sadozai et al., 2024).

In neurodevelopmental disorders (NDDs), however, this maturation is often delayed, atypical, or fragmented. Children with ADHD or ASD, for example, display early difficulties in working memory and cognitive flexibility that persist into later childhood and adolescence (Andersen et al., 2025; Perry et al., 2022). Such deviations reflect both structural and functional differences in prefrontal activation, as well as altered connectivity with limbic and parietal regions that mediate motivation and emotion regulation (Lee et al., 2025). The heterogeneity of these developmental patterns underscores the importance of viewing EF not as a singular construct but as a set of interdependent processes that follow individualized trajectories across the lifespan (Mareva et al., 2024).

Longitudinal evidence further supports the notion that early EF proficiency predicts a wide range of adaptive and academic outcomes. Preschoolers who demonstrate stronger inhibitory control and attentional regulation are more likely to exhibit higher literacy, numeracy, and social competence during later schooling (Bukva & Memisevic, 2025; Morosini et al., 2025). Conversely, early deficits in EF are associated with increased risk for learning disorders, behavioral difficulties, and emotional dysregulation (Tajik-Parvinchi et al., 2021). These findings highlight the value of early EF assessment as both a diagnostic and preventive measure, allowing for timely interventions that may alter developmental pathways before deficits become entrenched.

Assessing EF in NDDs presents unique challenges. Traditional psychometric measures often fail to capture the complexity and context-dependent nature of executive functioning. Laboratory-based tasks such as the Stroop or Wisconsin Card Sorting Test provide valuable insights into specific processes like inhibition or shifting but may not reflect real-world functioning in educational or social contexts (Barnes, 2023). Contemporary approaches emphasize ecological validity through performance-based assessments, teacher and parent rating scales, and dynamic testing methods that measure strategy use and learning potential (Crisci et al., 2021; Southon, 2022). Instruments like the Behavior Rating Inventory of Executive Function (BRIEF) and the Delis–Kaplan Executive Function System (D-KEFS) have become standard tools in clinical and educational evaluations, enabling a more nuanced understanding of EF profiles across different NDDs.

Moreover, digital technologies have introduced innovative approaches to EF assessment. Computerized platforms and virtual reality (VR) environments now allow for real-time monitoring of attention, planning, and self-regulation within simulated classroom or social scenarios (Bombonato et al., 2024; Capobianco et al., 2025). Such tools offer higher ecological validity and can dynamically adapt to the individual's performance, providing personalized data on strengths and weaknesses. Nevertheless, as Andersen et al. (2025) and Hamilton et al. (2025) caution, the integration of digital assessments requires critical validation to ensure that measures of executive performance align with neuropsychological theory and developmental norms.

Ultimately, understanding the developmental course of EF in neurodiverse learners demands a multidimensional assessment strategy, one that integrates cognitive testing, behavioral observation, and contextual analysis. This holistic perspective enables educators, psychologists, and clinicians to move beyond deficit identification toward the design of responsive, developmentally attuned interventions. By situating EF assessment within ecological and educational frameworks, practitioners can better support adaptive growth and lifelong self-regulation for learners across the neurodevelopmental spectrum.

EDUCATIONAL DESIGN AND PEDAGOGICAL IMPLICATIONS OF EXECUTIVE FUNCTION RESEARCH

The integration of executive function (EF) research into educational design has transformed the understanding of how cognitive regulation, emotional control, and learning processes interact within classroom environments. Executive functions serve not only as predictors of academic achievement but also as mechanisms through which students adapt to the dynamic demands of schooling (Barnes, 2023; Zelazo & Carlson, 2020). When educators recognize EF as a developmental and teachable capacity, rather than a fixed trait, instruction can evolve from behavior management toward fostering autonomy, reflection, and self-regulated learning.

Research has demonstrated that classroom structure and instructional quality substantially influence EF development. Predictable routines, clear expectations, and emotionally supportive relationships create conditions that enhance attention, working memory, and self-control (Cristofani et al., 2020; Southon, 2022). Conversely, chaotic or punitive environments increase cognitive load and stress reactivity, impairing students' capacity for regulation and problem-solving. These findings underscore the necessity of aligning pedagogical practices with students' neurocognitive profiles, particularly for those with neurodevelopmental disorders (NDDs) who may experience heightened sensitivity to environmental stressors (Frazier et al., 2022; Tajik-Parvinchi et al., 2021).

Instructional models grounded in executive-inclusive design emphasize explicit teaching of planning, goal setting, and metacognitive reflection. Barnes (2023) highlights that embedding EF instruction within authentic academic tasks, rather than as isolated drills, promotes greater transferability of skills. For example, project-based learning encourages strategic organization, sustained attention, and collaborative regulation, while feedback cycles promote self-monitoring and cognitive flexibility. Similarly, Southon (2022) found that university students with subclinical neurodevelopmental traits benefited significantly from structured reflection activities, suggesting that EF supports remain relevant across educational levels.

A critical pedagogical insight emerging from EF research is the importance of scaffolding cognitive load. Learners with EF difficulties often struggle when presented with unstructured, open-ended tasks that require simultaneous planning, monitoring, and execution (Filipe, 2021). Instructional scaffolds such as visual aids, stepwise task sequencing, and verbal mediation help externalize these processes, enabling gradual internalization of control (Barnes, 2023). Moreover, incorporating movement, mindfulness, and emotion-regulation exercises within the school day strengthens neural systems responsible for inhibition and attentional switching (Sung et al., 2022; Tao et al., 2025). These embodied practices transform learning from a purely cognitive endeavor into an integrative neuroeducational experience.

Technology offers additional pathways for individualized EF support. Adaptive digital platforms can monitor attention, pacing, and error patterns in real time, providing learners with immediate feedback and teachers with actionable data for differentiation (Bombonato et al., 2024). Virtual reality (VR) environments further extend this potential by situating EF training within immersive, contextually rich settings that mirror real-world challenges (Capobianco et al., 2025). However, as Hamilton et al. (2025) note, technological innovation must be accompanied by ethical reflection and teacher mediation to ensure that digital tools reinforce rather than replace self-regulatory growth.

From a policy standpoint, embedding EF-sensitive principles within curriculum frameworks can enhance educational equity. Teacher education programs should include training on the neurocognitive foundations of learning, equipping educators to interpret behaviors such as distractibility or disorganization through a developmental lens (Aravamuthan et al., 2024; Leifler, 2024). School systems that incorporate executive-inclusive assessment, focusing on planning, persistence, and reflection, shift the evaluative focus from outcomes to processes, aligning pedagogy with the neuroscience of learning (Barnes, 2023).

The translation of EF research into pedagogy reframes education as an applied neurodevelopmental science. Classrooms become environments for cultivating cognitive regulation, where instruction not only imparts knowledge but also builds the neural architecture for lifelong adaptability. This perspective situates teachers as cognitive designers, professionals who intentionally orchestrate tasks, emotions, and contexts to optimize executive growth for every learner.

TOWARD AN INTEGRATED NEUROEDUCATIONAL FRAMEWORK FOR EXECUTIVE FUNCTION

The convergence of neuroscience, psychology, and education has given rise to a new understanding of executive function (EF) as both a neurobiological and pedagogical construct. Rather than viewing EF deficits merely as symptoms of neurodevelopmental disorders (NDDs), contemporary perspectives conceptualize them as dynamic interactional phenomena shaped by neural maturation, emotional context, and environmental affordances (Aravamuthan et al., 2024; Bombonato et al., 2024). This integrated approach repositions EF at the center of a neuroeducational framework that bridges cognitive science with classroom practice.

At its core, this framework emphasizes reciprocity between brain and environment. The brain develops through experience, while learning environments can be designed to stimulate neuroplastic processes that strengthen executive networks (Capobianco et al., 2025; Zelazo & Carlson, 2020). Such a bidirectional model rejects the dichotomy between biological determinism and educational constructivism, proposing instead that cognition and context evolve together through cycles of adaptation and feedback. Within this view, teaching becomes an act of cognitive scaffolding that extends neural function into the social and cultural realm.

A neuroeducational perspective also recognizes the interdependence of cognitive and socio-emotional systems. Emotion serves as a regulatory and motivational substrate for EF, influencing how learners allocate attention, manage frustration, and persist toward goals (Cristofani et al., 2020; Tajik-Parvinchi et al., 2021). Effective pedagogical designs therefore integrate emotional literacy, mindfulness, and interpersonal connection as mediators of cognitive engagement. Such integration moves beyond the traditional separation of "academic" and "social-emotional" curricula, reflecting instead a holistic understanding of human development.

Moreover, assessment and intervention within this framework must be viewed as dynamic, context-sensitive processes rather than static evaluations of ability. As

Southon (2022) and Hamilton et al. (2025) emphasize, measuring EF should capture learners' capacity for adaptation and strategy use, not merely their performance under fixed conditions. Similarly, interventions should aim to modify the learning environment as much as the learner, promoting sustained self-regulation through iterative cycles of practice, reflection, and feedback (Barnes, 2023; Perry et al., 2022).

Crucially, this integrative paradigm positions teachers as neuroeducational mediators, professionals who translate neuroscientific knowledge into pedagogical action. Their expertise lies not in applying laboratory findings directly but in interpreting them through the lens of classroom dynamics, developmental readiness, and individual variability (Aravamuthan et al., 2024; Leifler, 2024). By cultivating metacognitive awareness in both themselves and their students, educators foster the conditions for executive growth to emerge naturally through meaningful engagement.

In practice, the neuroeducational model of EF advocates for schools that function as ecosystems of regulation, spaces where structure, empathy, and reflection converge to support cognitive and emotional balance. Such environments enable learners with and without NDDs to build the neural foundations for autonomy, resilience, and lifelong adaptability. The scientific study of executive function thus becomes inseparable from the ethical and pedagogical imperative to nurture the developing mind within responsive, equitable, and cognitively attuned learning contexts (Bombonato et al., 2024; Tao et al., 2025).

Table 1. Executive function profiles across neurodevelopmental disorders

Neurodevelopmental Disorder	Executive Functions Most Affected	Developmental Pattern	Educational and Psychosocial Implications
Specific Learning Disorders (SLD)	Working memory, planning, cognitive flexibility	Executive function difficulties often become more evident as academic demands increase	Inefficient learning strategies, difficulties with organization and task completion, need for explicit strategy instruction
Attention-Deficit/ Hyperactivity Disorder (ADHD)	Inhibitory control, sustained attention, working memory, self-monitoring	Core and persistent executive function deficits across development	Academic underachievement, behavioral regulation difficulties, increased risk for comorbid emotional problems
Autism Spectrum Disorder (ASD)	Cognitive flexibility, planning, self-regulation	Executive function profile is heterogeneous and context-dependent	Rigidity, difficulties adapting to change, challenges in independent learning and social problem-solving
Communication Disorders	Working memory, self-monitoring, cognitive flexibility	Executive difficulties often co-occur with language processing demands	Difficulties in discourse organization, narrative production, and classroom participation

continued on following page

Table 1. Continued

Neurodevelopmental Disorder	Executive Functions Most Affected	Developmental Pattern	Educational and Psychosocial Implications
Intellectual Disability (ID)	Global executive functioning, problem-solving	Executive limitations are proportional to cognitive and adaptive functioning	Need for structured environments, functional task breakdown, and adaptive learning goals
Motor and Tic Disorders	Inhibitory control, motor planning	Executive difficulties may fluctuate and increase under stress	Performance anxiety, classroom avoidance, need for reduced cognitive and motor load
Comorbid Profiles	Multiple executive domains affected simultaneously	Cumulative and interactive executive vulnerabilities	Increased functional impairment, need for integrated and interdisciplinary interventions

REFERENCES

Andersen, A. C., Thomsen, P. H., Lydersen, S., Sund, A. M., Haugan, A. L. J., & Nøvik, T. S. (2025). Behaviour ratings of executive functions in adolescents with ADHD: Correlation with core symptoms and functional impairment measured by multiple informants. *Nordic Journal of Psychiatry*, *79*(5), 347–353. DOI: 10.1080/08039488.2025.2506558 PMID: 40401509

Aravamuthan, B., Turnacioglu, S. O., & Gropman, A. L. (2024). Neurodevelopmental Disabilities: Conceptual Framework. In *Swaiman's Pediatric Neurology* (pp. 636-641). Elsevier. https://doi.org/10.1016/B978-0-443-10944-7.00068-8

Barnes, M. A. (2023). Executive functions: What are they good for? A perspective from intervention research. *Mind, Brain and Education : the Official Journal of the International Mind, Brain, and Education Society*, *17*(4), 238–245. DOI: 10.1111/mbe.12371

Bombonato, C., Del Lucchese, B., Ruffini, C., Di Lieto, M. C., Brovedani, P., Sgandurra, G., Cioni, G., & Pecini, C. (2024). Far transfer effects of trainings on executive functions in neurodevelopmental disorders: A systematic review and meta-analysis. *Neuropsychology Review*, *34*(1), 98–133. DOI: 10.1007/s11065-022-09574-z PMID: 36633797

Bowers, J. S. (2016). The practical and principled problems with educational neuroscience. *Psychological Review*, *123*(5), 600–612. https://psycnet.apa.org/buy/2016-10827-001. DOI: 10.1037/rev0000025 PMID: 26938449

Bukva, A., & Memisevic, H. (2025). The relationship between executive functions and mathematics achievements in early-grade elementary students. *Child Neuropsychology*, *31*(4), 635–648. DOI: 10.1080/09297049.2024.2421222 PMID: 39466137

Busso, D. S., & Pollack, C. (2015). No brain left behind: Consequences of neuroscience discourse for education. *Learning, Media and Technology*, *40*(2), 168–186. DOI: 10.1080/17439884.2014.908908

Capobianco, M., Puzzo, C., Di Matteo, C., Costa, A., & Adriani, W. (2025). Current virtual reality-based rehabilitation interventions in neuro-developmental disorders at developmental ages. *Frontiers in Behavioral Neuroscience*, *18*, 1441615. DOI: 10.3389/fnbeh.2024.1441615 PMID: 39882439

Carrick, A., & Hamilton, C. J. (2023). Heated behaviour in the classroom for children with FASD: The relationship between characteristics associated with ADHD, ODD and ASD, hot executive function and classroom-based reward systems. *Children (Basel, Switzerland)*, *10*(4), 685. DOI: 10.3390/children10040685 PMID: 37189934

Cortés Pascual, A., Moyano Muñoz, N., & Quílez Robres, A. (2019). The relationship between executive functions and academic performance in primary education: Review and meta-analysis. *Frontiers in Psychology, 10*, 449759. DOI: 10.3389/fpsyg.2019.01582 PMID: 31354585

Costescu, C., Ro an, A., David, C., Cozma, L., & Calota, A. (2023). The relation between cognitive and emotional processes in children and adolescents with neurodevelopmental disorders—A meta-analysis. *European Journal of Investigation in Health, Psychology and Education, 13*(12), 2811–2826. DOI: 10.3390/ejihpe13120194 PMID: 38131893

Crisci, G., Caviola, S., Cardillo, R., & Mammarella, I. C. (2021). Executive functions in neurodevelopmental disorders: Comorbidity overlaps between attention deficit and hyperactivity disorder and specific learning disorders. *Frontiers in Human Neuroscience, 15*, 594234. DOI: 10.3389/fnhum.2021.594234 PMID: 33732121

Cristofani, C., Sesso, G., Cristofani, P., Fantozzi, P., Inguaggiato, E., Muratori, P., Narzisi, A., Pfanner, C., Pisano, S., Polidori, L., Ruglioni, L., Valente, E., Masi, G., & Milone, A. (2020). The role of executive functions in the development of empathy and its association with externalizing behaviors in children with neurodevelopmental disorders and other psychiatric comorbidities. *Brain Sciences, 10*(8), 489. DOI: 10.3390/brainsci10080489 PMID: 32731515

De Felice, A., Ricceri, L., Venerosi, A., Chiarotti, F., & Calamandrei, G. (2015). Multifactorial origin of neurodevelopmental disorders: Approaches to understanding complex etiologies. *Toxics, 3*(1), 89–129. DOI: 10.3390/toxics3010089 PMID: 29056653

Filipe, M. (2021). How do executive functions issues affect writing in students with neurodevelopmental disorders. In *Executive functions and writing* (p. 160). Oxford University Press. DOI: 10.1093/oso/9780198863564.003.0007

Fogel, Y. (2022). Cognitive strategies: Moderating the relationship between executive functions and daily functioning. *International Journal of Environmental Research and Public Health, 19*(24), 16845. DOI: 10.3390/ijerph192416845 PMID: 36554722

Frazier, T. W., Crowley, E., Shih, A., Vasudevan, V., Karpur, A., Uljarevic, M., & Cai, R. Y. (2022). Associations between executive functioning, challenging behavior, and quality of life in children and adolescents with and without neurodevelopmental conditions. *Frontiers in Psychology, 13*, 1022700. DOI: 10.3389/fpsyg.2022.1022700 PMID: 36337537

Hamilton, A., Tallberg, P., Ilahi, F., Hoff, E., Ahmadi, B., & Daukantaitė, D. (2025). Behavioral manifestations of executive functioning in Swedish youth with ADHD, autism, and psychiatric comorbidity: A comparative analysis with community controls. *Child Neuropsychology*, *31*(5), 771–790. DOI: 10.1080/09297049.2024.2434736 PMID: 39655647

Harkness, K., Bray, S., Durber, C. M., Dewey, D., & Murias, K. (2025). Assessing the contribution of measures of attention and executive function to diagnosis of ADHD or autism. *Journal of Autism and Developmental Disorders*, *55*(4), 1353–1364. DOI: 10.1007/s10803-024-06275-9 PMID: 38478161

Khan, F., Amatya, B., Galea, M. P., Gonzenbach, R., & Kesselring, J. (2017). Neurorehabilitation: Applied neuroplasticity. *Journal of Neurology*, *264*(3), 603–615. DOI: 10.1007/s00415-016-8307-9 PMID: 27778158

Lane, J. M., Liu, S. H., Midya, V., Alcala, C. S., Eggers, S., Svensson, K., Martinez-Medina, S., Horton, M. K., White, R. F., Téllez-Rojo, M. M., & Wright, R. O. (2025). Childhood Pb-induced cognitive dysfunction: Structural equation modeling of hot and cold executive functions. *Journal of Exposure Science & Environmental Epidemiology*, *35*(5), 1–10. DOI: 10.1038/s41370-025-00761-7 PMID: 40033031

Lee, D., Boulton, K. A., Sun, C., Phillips, N. L., Munro, M., Kumfor, F., Demetriou, E. A., & Guastella, A. J. (2025). Attention and executive delays in early childhood: A meta-analysis of neurodevelopmental conditions. *Molecular Psychiatry*, *30*(5), 1906–1914. DOI: 10.1038/s41380-024-02802-3 PMID: 39489868

Leifler, , EBorg, , ABölte, , S. (2024). A multi-perspective study of perceived inclusive education for students with neurodevelopmental disorders. Journal of Autism and Developmental Disorders (italics), 54(4), 1611–1617. https://doi.org/10.1007/s10803-022-05643-7

Mareva, S., Astle, D., Baker, K., Gathercole, S., Holmes, J., Kievit, R., Manly, T., Akarca, D., Bathelt, J., Bettencourt, M., Bennett, M., Bignardi, G., Bishop, S., Bottacin, E., Bridge, L., Brkic, D., Bryant, A., Butterfield, S., Byrne, E., & Holmes, J.Lead InvestigatorsTeam of Researchers and PhD Students. (2024). Mapping neurodevelopmental diversity in executive function. *Cortex*, *172*, 204–221. DOI: 10.1016/j.cortex.2023.11.021 PMID: 38354470

Marini, A., Piccolo, B., Taverna, L., Berginc, M., & Ozbič, M. (2020). The complex relation between executive functions and language in preschoolers with developmental language disorders. *International Journal of Environmental Research and Public Health*, *17*(5), 1772. DOI: 10.3390/ijerph17051772 PMID: 32182903

McDermott, J. M., Troller-Renfree, S., Vanderwert, R., Nelson, C. A., Zeanah, C. H., & Fox, N. A. (2013). Psychosocial deprivation, executive functions, and the emergence of socio-emotional behavior problems. *Frontiers in Human Neuroscience*, *7*, 167. DOI: 10.3389/fnhum.2013.00167 PMID: 23675333

McFall, G. P., Sapkota, S., Thibeau, S., & Dixon, R. A. (2017). Trajectories and modifiers of executive function: Normal aging to neurodegenerative disease. In *Executive Function* (pp. 263–277). Routledge. DOI: 10.4324/9781315160719-18

Morosini, G., Cuder, A., De Vita, C., Passolunghi, M. C., & Pellizzoni, S. (2025). The relationship between executive functions and early numerical skills: A study on three- and four-year-old preschoolers. *Cognitive Development*, *74*, 101568. DOI: 10.1016/j.cogdev.2025.101568

Otterman, D. L., Koopman-Verhoeff, M. E., White, T. J., Tiemeier, H., Bolhuis, K., & Jansen, P. W. (2019). Executive functioning and neurodevelopmental disorders in early childhood: A prospective population-based study. *Child and Adolescent Psychiatry and Mental Health*, *13*(1), 38. DOI: 10.1186/s13034-019-0299-7 PMID: 31649749

Perry, V., Ellis, K., Moss, J., Beck, S. R., Singla, G., Crawford, H., Waite, J., Richards, C., & Oliver, C. (2022). Executive function, repetitive behaviour and restricted interests in neurodevelopmental disorders. *Research in Developmental Disabilities*, *122*, 104166. DOI: 10.1016/j.ridd.2021.104166 PMID: 35016127

Rosenblum, S., Sharfi, K., Elimelech, O. C., Regev, S., & Grinblat, N. (2025). Predicting emotional consequences of disorganization among adults with neurodevelopmental disorders: A structural equation model. *Disability and Rehabilitation*, *47*(8), 1954–1964. DOI: 10.1080/09638288.2024.2385733 PMID: 39119890

Ruffini, C., Osmani, F., Martini, C., Giera, W. K., & Pecini, C. (2024). The relationship between executive functions and writing in children: A systematic review. *Child Neuropsychology*, *30*(1), 105–163. DOI: 10.1080/09297049.2023.2170998 PMID: 36748722

Sadozai, A. K., Sun, C., Demetriou, E. A., Lampit, A., Munro, M., Perry, N., Boulton, K. A., & Guastella, A. J. (2024). Executive function in children with neurodevelopmental conditions: A systematic review and meta-analysis. *Nature Human Behaviour*, *8*(12), 2357–2366. DOI: 10.1038/s41562-024-02000-9 PMID: 39424962

Southon, C. (2022). The relationship between executive function, neurodevelopmental disorder traits, and academic achievement in university students. *Frontiers in Psychology*, *13*, 958013. DOI: 10.3389/fpsyg.2022.958013 PMID: 36118426

Sung, M. C., Ku, B., Leung, W., & MacDonald, M. (2022). The effect of physical activity interventions on executive function among people with neurodevelopmental disorders: A meta-analysis. *Journal of Autism and Developmental Disorders, 52*(3), 1030–1050. DOI: 10.1007/s10803-021-05009-5 PMID: 33856619

Tajik-Parvinchi, D., Farmus, L., Tablon Modica, P., Cribbie, R. A., & Weiss, J. A. (2021). The role of cognitive control and emotion regulation in predicting mental health problems in children with neurodevelopmental disorders. *Child: Care, Health and Development, 47*(5), 608–617. DOI: 10.1111/cch.12868 PMID: 33772823

Tao, R., Yang, Y., Wilson, M., Chang, J. R., Liu, C., & Sit, C. H. (2025). Comparative effectiveness of physical activity interventions on cognitive functions in children and adolescents with neurodevelopmental disorders: A systematic review and network meta-analysis of randomized controlled trials. *The International Journal of Behavioral Nutrition and Physical Activity, 22*(1), 6. DOI: 10.1186/s12966-024-01702-7 PMID: 39806448

Theodoratou, M., Gkintoni, E., & Farmakopoulou, I. (2023). Executive functions and quality of life in neurodevelopmental spectrum: An outline. *Technium Social Sciences Journal, 39*, 430–439. DOI: 10.47577/tssj.v39i1.8231

Zelazo, P. D. (2015). Executive function: Reflection, iterative reprocessing, complexity, and the developing brain. *Developmental Review, 38*, 55–68. DOI: 10.1016/j.dr.2015.07.001

Zelazo, P. D., & Carlson, S. M. (2020). The neurodevelopment of executive function skills: Implications for academic achievement gaps. *Psychology & Neuroscience, 13*(3), 273–298. DOI: 10.1037/pne0000208

Chapter 8

Side Effects of Neurodevelopmental Disorders Neurodevelopmental Disorders in the School Context:

Psychosocial Impact, Family, and Resilience

ABSTRACT

This chapter outlines how neurodevelopmental disorders intersect with school, family, and community systems to shape children's emotional, social, and academic trajectories. It describes the cumulative impact of SLDs and ADHD, where persistent learning or behavioral challenges heighten vulnerability to stress, anxiety, and reduced self-confidence. The discussion also considers how caregiver strain and societal stigma influence adjustment. Emphasizing schools as key settings for early support, the chapter highlights inclusive teaching, psychosocial programs, and family engagement as central to fostering resilience and adaptive functioning across development.

DOI: 10.4018/979-8-3693-6331-7.ch008

INTRODUCTION

The educational environment is a central context in which neurodevelopmental disorders (NDDs), such as specific learning disorders (SLDs) and attention-deficit/hyperactivity disorder (ADHD), manifest and shape developmental trajectories. Beyond academic difficulties, these disorders are associated with complex emotional, behavioral, and social consequences that influence both the child and the family system (DuPaul et al., 2013; Livingston et al., 2018; Stein et al., 2024). Recognizing the multifaceted impact of NDDs in schools is critical to fostering resilience, promoting mental well-being, and preventing long-term psychosocial challenges.

Children with SLDs and ADHD frequently experience persistent academic difficulties that, if inadequately addressed, may culminate in negative emotional outcomes such as low self-esteem, anxiety, and depression (DuPaul et al., 2013; Mugnaini et al., 2009; Nelson & Gregg, 2012). These challenges are often compounded by the presence of comorbid conditions, with approximately 40% of children with reading disorders exhibiting additional neurodevelopmental difficulties (Moll et al., 2020). The resulting psychosocial effects underscore the necessity of school-based interventions that extend beyond remediation of academic skills to encompass emotional regulation, social competence, and adaptive coping strategies.

NEURODEVELOPMENTAL DISORDERS IN THE SCHOOL CONTEXT

According to the DSM-5 (APA, 2013), dyslexia falls under the category of specific learning disorders. While prevalence estimates vary, educators report that 5–10% of children experience persistent reading difficulties consistent with dyslexia, which may correspond to approximately 2–4 children per typical classroom (Wilmot et al., 2023). Many children with dyslexia also face oral language deficits that extend beyond phonological processing (Snowling et al., 2020). ADHD affects roughly 5.29% of children worldwide and is characterized by developmentally inappropriate levels of inattention, hyperactivity, and impulsivity (Posner et al., 2020).

Comorbidity between SLDs and ADHD is common and may reflect shared genetic and neurobiological factors (Germanò et al., 2010; Lonergan et al., 2019; Nisiotou & Vlachos, 2014). Children exhibiting symptoms of both disorders often demonstrate overlapping behavioral and psychosocial challenges, including heightened academic struggles and social-emotional difficulties (Gooch et al., 2011).

In the school context, children with NDDs are at heightened risk for negative experiences that extend beyond cognitive deficits. Dyslexia, for instance, can lead to years of academic effort without commensurate success, fostering discouragement and emotional distress (Shaywitz & Shaywitz, 2020). Children with dyslexia often report higher stress levels, particularly in interactions with teachers, and may experience impaired psychological development due to repeated academic failure (Benassi et al., 2022; Deacon et al., 2022; Murray & Greenberg, 2001). Similarly, ADHD can disrupt classroom engagement, social participation, and peer relationships, reducing overall quality of life for both children and families (DuPaul et al., 2013; Leitch et al., 2019; Van Andel et al., 2022).

Internalizing problems, including depression, anxiety, and social withdrawal, are particularly prominent among children with dyslexia, while externalizing behaviors such as aggression or oppositionality are more frequent in ADHD (Brænden et al., 2023; Heiervang et al., 2001). Bullying, victimization, and societal stigmatization further exacerbate these challenges, diminishing social self-efficacy and contributing to feelings of isolation (Andreou et. al., 2015; Lisle & Wade, 2013).

Dyslexia and ADHD are closely associated with emotional dysregulation. Children with dyslexia often experience embarrassment, anger, and guilt related to academic difficulties, while ADHD frequently co-occurs with mood dysregulation, irritability, and diminished emotional control (Brænden et al., 2023; Lockiewicz et al., 2014). These internalized and externalized responses can persist into adulthood, influencing occupational performance, interpersonal relationships, and mental health outcomes (Nalavany et al., 2018; Reid & Kirk, 2001).

Research on neurodevelopmental disorders (NDDs), including ADHD, dyslexia, and autism spectrum disorders, highlights the complex interplay between cognitive difficulties, emotional well-being, and resilience. Children and adolescents with NDDs often experience academic challenges alongside heightened emotional distress, such as anxiety, depression, and reduced self-esteem, which can persist into adulthood. Studies by Bonifacci et al. (2014) and Shessel & Reiff (1999) illustrate how cognitive and behavioral profiles in dyslexia affect both learning and psychosocial outcomes, while Lee et al. (2016) and Limbers et al. (2011) show that ADHD impacts overall quality of life and family functioning. Early adverse experiences, as reported by Adaralegbe et al. (2022) and Heselton (2023), can further compromise resilience and school success, emphasizing the importance of context and environmental influences on developmental trajectories.

Family and social contexts play a pivotal role in moderating outcomes for children with NDDs. Bussing et al. (2003), Clark & Adams (2022), and Jacobs et al. (2022) highlight that caregiver stress, family support, and parent-child interactions directly shape emotional and behavioral functioning. Interventions such as parent coaching, psychoeducation, and structured family support enhance coping, foster resilience, and improve academic engagement. Moreover, stigma, labeling, and societal misconceptions, noted by Riddick (2013), Singer (2015), and Rose (1998), negatively affect self-efficacy and peer participation, reinforcing the need for inclusive educational environments. Glazzard & Dale (2013), Heselton (2023), and Palombo (2023), further demonstrate that resilience emerges from dynamic interactions between individual traits, family support, and school-based resources. Overall, integrated approaches addressing academic, emotional, and social domains are most effective in promoting adaptive functioning and long-term well-being in children with NDDs.

Building on these findings about resilience and psychosocial challenges, students with dyslexia in tertiary education frequently report heightened anxiety and depressive symptoms, partly due to limited institutional support and the challenges of re-establishing social relationships (Doikou-Avlidou, 2015; Nelson & Gregg, 2012). The compounded effects of stigma and marginalization underscore the need for school-based interventions that integrate academic support, psychosocial strategies, and community awareness (Cameron, 2016; Cortiella & Horowitz, 2014).

FAMILY SYSTEM AND PARENTAL STRESS

The psychosocial impact of NDDs extends to families, particularly parents who often experience elevated stress, anxiety, and feelings of guilt (Craig et al., 2016; Miranda et al., 2015). Parental perceptions play a critical role in children's emotional outcomes, as caregiver understanding, advocacy, and support directly influence adaptive functioning and resilience (Terras et al., 2009; Tobia et al., 2019). Parents of children with SLDs or ADHD often report strain associated with navigating educational systems, managing behavioral challenges, and coping with societal misconceptions (Earey, 2013; Serry et al., 2016).

Family-based interventions, including parent coaching, psychoeducation, and collaborative problem-solving, can mitigate these effects by equipping caregivers with strategies to support academic and emotional development (Dahl et al., 2020; Ruan et al., 2023). Adaptive family functioning and effective communication are critical mediators of resilience, buffering the child from the full impact of academic and social stressors (Montirosso et al., 2021; Rungsattatharm et al., 2025).

The dynamic interplay between child symptom severity and parental well-being underscores the systemic nature of NDDs. Parental stress can amplify child symptoms, creating feedback loops that reinforce behavioral and emotional difficulties (Breaux & Harvey, 2019; Huang et al., 2020). Gender-specific differences in child behavior and parental response further highlight the need for tailored interventions (Theule et al., 2013). Recognizing the family as an active participant in intervention planning is essential for promoting long-term resilience and holistic well-being.

EARLY IDENTIFICATION AND INCLUSIVE PRACTICES

Given that NDDs are often diagnosed during early schooling, educational settings are uniquely positioned to intervene and foster resilience. Early detection, teacher-administered screening, and individualized support plans enhance academic outcomes and reduce stress associated with repeated failure (Dahl-Leonard et al., 2023; Fletcher et al., 2021; Murphy et al., 2015). Inclusive pedagogy, flexible assessment strategies, and differentiated instruction empower students to engage with learning in ways that leverage strengths while addressing challenges (Espin et al., 2019; Levi et al., 2013).

Teacher competence and empathy are critical determinants of student well-being. Educators who understand neurodevelopmental profiles, communicate supportively, and collaborate with specialized professionals create classroom climates that reduce stigma and encourage social-emotional growth (Benassi et al., 2022; Nevill & Forsey, 2022). These positive environments are associated with improved academic engagement, self-efficacy, and overall psychological adjustment.

Complementing academic strategies, psychosocial interventions and physical activity programs have demonstrated benefits for children with ADHD and SLDs. High-intensity interval training, executive function coaching, and social-emotional learning initiatives improve attention, behavioral regulation, social competence, and quality of life (Gallego-Méndez et al., 2020; Meßler et al., 2018; Murphy et al., 2015; Zhang et al., 2024). Integrating these approaches within schools ensures that interventions address the full spectrum of developmental needs, extending support beyond the classroom to family and community contexts.

Cultural, socio-economic, and community contexts influence both the prevalence and impact of NDDs, as well as the effectiveness of interventions (Dai et al., 2024; Mercera et al., 2025). Schools that adopt culturally sensitive practices, reduce stigma, and engage families foster resilience by creating inclusive environments in which children feel valued, understood, and empowered (Nevill & Forsey, 2022; White et al., 2023).

Resilience in children with NDDs emerges from the interaction of individual, familial, and school-based factors (Slomowitz et al., 2025). Interventions that integrate academic remediation, psychosocial support, family engagement, and culturally sensitive practices provide the most effective pathway for fostering adaptive functioning (Murphy et al., 2015). Collaboration between educators, parents, school psychologists, and allied professionals ensures that interventions address the complex and interconnected challenges faced by these children (Kern et al., 2023; Weitzman et al., 2025). Practitioners are encouraged to adopt holistic strategies that consider cognitive, emotional, and social domains concurrently. Recognizing that resilience is a dynamic, systemic process underscores the importance of coordinated support that empowers children, families, and schools to work together in promoting well-being and academic success.

Despite growing research, critical gaps remain in understanding the longitudinal trajectories of children with comorbid NDDs, particularly in relation to psychosocial resilience and academic outcomes (Stein et al., 2024; White et al., 2023). Further research is needed to develop culturally adapted interventions, equity-focused policies, and models of parent-school collaboration that are responsive to diverse populations. Examining teacher attitudes, classroom climate, and peer dynamics can guide interventions that strengthen resilience within school contexts (Mercera et al., 2025; Nevill & Forsey, 2022).

Children with SLDs and ADHD navigate multifaceted challenges in the school environment that extend beyond academic difficulties. Emotional, behavioral, and social consequences intersect with family dynamics, shaping the overall trajectory of development. Schools, families, and communities collectively influence resilience, highlighting the systemic nature of intervention. Holistic, culturally informed, and family-engaged approaches provide the most effective pathway for supporting children with NDDs, fostering not only academic success but also social-emotional well-being and long-term adaptive functioning.

REFERENCES

Adaralegbe, N. J., Okobi, O. E., Omar, Z. T., Segun, E., Evbayekha, E. O., Abolurin, A., & Adedeji, A. G. (2022). Impact of adverse childhood experiences on resilience and school success in individuals with autism spectrum disorder and attention-deficit hyperactivity disorder. *Cureus*, *14*(11). Advance online publication. DOI: 10.7759/cureus.31907 PMID: 36579285

American Psychiatric Association. (2013). *Diagnostic and statistical manual of mental disorders* (5th ed.). American Psychiatric Association.

Andreou, E., Didaskalou, E., & Vlachou, A. (2015). Bully/victim problems among Greek pupils with special educational needs: Associations with loneliness and self-efficacy for peer interactions. *Journal of Research in Special Educational Needs*, *15*(4), 235–246. DOI: 10.1111/1471-3802.12028

Benassi, E., Camia, M., Giovagnoli, S., & Scorza, M. (2022). Impaired school well-being in children with specific learning disorder and its relationship to psychopathological symptoms. *European Journal of Special Needs Education*, *37*(1), 74–88. DOI: 10.1080/08856257.2020.1842975

Bonifacci, P., Montuschi, M., Lami, L., & Snowling, M. J. (2014). Parents of children with dyslexia: Cognitive, emotional and behavioural profile. *Dyslexia (Chichester, England)*, *20*(2), 175–190. DOI: 10.1002/dys.1469 PMID: 24265066

Brænden, A., Coldevin, M., Zeiner, P., Stubberud, J., & Melinder, A. (2023). Executive function in children with disruptive mood dysregulation disorder compared to attention-deficit/hyperactivity disorder and oppositional defiant disorder, and in children with different irritability levels. *European Child & Adolescent Psychiatry*, •••, 1–11. DOI: 10.1007/s00787-023-02143-6 PMID: 36680626

Breaux, R. P., & Harvey, E. A. (2019). A longitudinal study of the relation between family functioning and preschool ADHD symptoms. *Journal of Clinical Child and Adolescent Psychology*, *48*(5), 749–764. DOI: 10.1080/15374416.2018.1437737 PMID: 29578799

Bussing, R., Zima, B. T., Gary, F. A., Mason, D. M., Leon, C. E., Sinha, K., & Garvan, C. W. (2003). Social networks, caregiver strain, and utilization of mental health services among elementary school students at high risk for ADHD. *Journal of the American Academy of Child and Adolescent Psychiatry*, *42*(7), 842–850. DOI: 10.1097/01.CHI.0000046876.27264.BF PMID: 12819444

Cameron, H. E. (2016). Beyond cognitive deficit: The everyday lived experience of dyslexic students at university. *Disability & Society*, *31*(2), 223–239. DOI: 10.1080/09687599.2016.1152951

Clark, M., & Adams, D. (2022). Resilience in autism and intellectual disability: A systematic review. *Review Journal of Autism and Developmental Disorders*, *9*(1), 39–53. DOI: 10.1007/s40489-021-00239-w

Cortiella, C., & Horowitz, S. H. (2014). *The state of learning disabilities: Facts, trends and emerging issues*. National Center for Learning Disabilities.

Craig, F., Operto, F. F., De Giacomo, A., Margari, L., Frolli, A., Conson, M., Ivagnes, S., Monaco, M., & Margari, F. (2016). Parenting stress among parents of children with neurodevelopmental disorders. *Psychiatry Research*, *242*, 121–129. DOI: 10.1016/j.psychres.2016.05.016 PMID: 27280521

Dahl, V., Ramakrishnan, A., Spears, A. P., Jorge, A., Lu, J., Bigio, N. A., & Chacko, A. (2020). Psychoeducation interventions for parents and teachers of children and adolescents with ADHD: A systematic review of the literature. *Journal of Developmental and Physical Disabilities*, *32*(2), 257–292. DOI: 10.1007/s10882-019-09691-3

Dahl-Leonard, K., Hall, C., Capin, P., Solari, E. J., Demchak, A., & Therrien, W. J. (2023). Examining fidelity reporting within studies of foundational reading interventions for elementary students with or at risk for dyslexia. *Annals of Dyslexia*, *73*(2), 1–26. DOI: 10.1007/s11881-023-00279-3 PMID: 36701045

Dai, X., Cheung, Y. C. H., Zhuang, X., Ng, E. F. W., & Wong, D. F. K. (2024). Context-specific resilience through a cultural lens: Social-ecological factors among Chinese families of children with neurodevelopmental disabilities. *Journal of Autism and Developmental Disorders*, •••, 1–11. DOI: 10.1007/s10803-024-06605-x PMID: 39436517

Deacon, L., Macdonald, S. J., & Donaghue, J. (2022). "What's wrong with you, are you stupid?" Listening to the biographical narratives of adults with dyslexia in an age of 'inclusive' and 'anti-discriminatory' practice. *Disability & Society*, *37*(3), 406–426. DOI: 10.1080/09687599.2020.1815522

Doikou-Avlidou, M. (2015). The educational, social and emotional experiences of students with dyslexia: The perspective of postsecondary education students. *International Journal of Special Education*, *30*(1), 132–145.

DuPaul, G. J., Gormley, M. J., & Laracy, S. D. (2013). Comorbidity of LD and ADHD: Implications of DSM-5 for assessment and treatment. *Journal of Learning Disabilities*, *46*(1), 43–51. DOI: 10.1177/0022219412464351 PMID: 23144063

Earey, A. (2013). Parental experiences of support for pupils with dyslexia: Ignoring the effect on parents. *Support for Learning*, *28*(1), 35–40. DOI: 10.1111/1467-9604.12013

Espin, L., García, I., del Pino Sánchez, M., Román, F., & Salvador, A. (2019). Effects of psychosocial stress on the hormonal and affective response in children with dyslexia. *Trends in Neuroscience and Education*, *15*, 1–9. DOI: 10.1016/j.tine.2019.03.001 PMID: 31176466

Fletcher, J. M., Francis, D. J., Foorman, B. R., & Schatschneider, C. (2021). Early detection of dyslexia risk: Development of brief, teacher-administered screens. *Learning Disability Quarterly*, *44*(3), 145–157. DOI: 10.1177/0731948720931870 PMID: 34584341

Gallego-Méndez, J., Perez-Gomez, J., Calzada-Rodríguez, J. I., Denche-Zamorano, Á. M., Mendoza-Muñoz, M., Carlos-Vivas, J., Garcia-Gordillo, M. Á., & Adsuar, J. C. (2020). Relationship between health-related quality of life and physical activity in children with hyperactivity. *International Journal of Environmental Research and Public Health*, *17*(8), 2804. DOI: 10.3390/ijerph17082804 PMID: 32325730

Germanò, E., Gagliano, A., & Curatolo, P. (2010). Comorbidity of ADHD and dyslexia. *Developmental Neuropsychology*, *35*(5), 475–493. DOI: 10.1080/87565641.2010.494748 PMID: 20721770

Glazzard, J., & Dale, K. (2013). Trainee teachers with dyslexia: Personal narratives of resilience. *Journal of Research in Special Educational Needs*, *13*(1), 26–37. DOI: 10.1111/j.1471-3802.2012.01254.x

Gooch, D., Snowling, M., & Hulme, C. (2011). Time perception, phonological skills and executive function in children with dyslexia and/or ADHD symptoms. *Journal of Child Psychology and Psychiatry, and Allied Disciplines*, *52*(2), 195–203. DOI: 10.1111/j.1469-7610.2010.02312.x PMID: 20860755

Heiervang, A. L., Stevenson, J., & Hugdahl, K. (2001). Behavior problems in children with dyslexia. *Nordic Journal of Psychiatry*, *55*(4), 251–256. DOI: 10.1080/080394801681019101 PMID: 11839115

Heselton, G. A. (2023). Childhood adversity, resilience, and autism: A critical review of the literature. *Disability & Society*, *38*(7), 1251–1270. DOI: 10.1080/09687599.2021.1983416

Huang, Y., He, M., Li, A., Lin, Y., Zhang, X., & Wu, K. (2020). Personality, behavior characteristics, and life quality impact of children with dyslexia. *International Journal of Environmental Research and Public Health, 17*(4), 1415. DOI: 10.3390/ijerph17041415 PMID: 32098297

Jacobs, L., Parke, A., Ziegler, F., Headleand, C., & De Angeli, A. (2022). Learning at school through to university: The educational experiences of students with dyslexia at one UK higher education institution. *Disability & Society, 37*(4), 662–683. DOI: 10.1080/09687599.2020.1829553

Kern, M., Johnson, K., & Peverly, S. (2023). The need for integrated supports in schools and the role of school psychologists. In *Learning disorders across the lifespan: A mental health framework* (pp. 25–42). Springer International Publishing. DOI: 10.1007/978-3-031-21772-2_4

Lee, Y. C., Yang, H. J., Chen, V. C. H., Lee, W. T., Teng, M. J., Lin, C. H., & Gossop, M. (2016). Meta-analysis of quality of life in children and adolescents with ADHD: By both parent proxy-report and child self-report using PedsQL™. *Research in Developmental Disabilities, 51*, 160–172. DOI: 10.1016/j.ridd.2015.11.009 PMID: 26829402

Leitch, S., Sciberras, E., Post, B., Gerner, B., Rinehart, N., Nicholson, J. M., & Evans, S. (2019). Experience of stress in parents of children with ADHD: A qualitative study. *International Journal of Qualitative Studies on Health and Well-being, 14*(1), 1690091. DOI: 10.1080/17482631.2019.1690091 PMID: 31713473

Levi, U., Einav, M., Raskind, I., Ziv, O., & Margalit, M. (2013). Helping students with LD to succeed: The role of teachers' hope, sense of coherence and specific self-efficacy. *European Journal of Special Needs Education, 28*(4), 427–439. DOI: 10.1080/08856257.2013.820457

Limbers, C. A., Ripperger-Suhler, J., Boutton, K., Ransom, D., & Varni, J. W. (2011). A comparative analysis of health-related quality of life and family impact between children with ADHD treated in a general pediatric clinic and a psychiatric clinic utilizing the PedsQL. *Journal of Attention Disorders, 15*(5), 392–402. DOI: 10.1177/1087054709356191 PMID: 20065072

Lisle, K., & Wade, T. J. (2013). Does the presence of a learning disability elicit stigmatization? British *Journal of Education, Society & Behavioral Sciences, 4*(2), 211–225. https://doi.org/10.9734/BJESBS/2014/6379

Livingston, E. M., Siegel, L. S., & Ribary, U. (2018). Developmental dyslexia: Emotional impact and consequences. *Australian Journal of Learning Difficulties, 23*(2), 107–135. DOI: 10.1080/19404158.2018.1479975

Lonergan, A., Doyle, C., Cassidy, C., MacSweeney Mahon, S., Roche, R. A., Boran, L., & Bramham, J. (2019). A meta-analysis of executive functioning in dyslexia with consideration of the impact of comorbid ADHD. *Journal of Cognitive Psychology*, *31*(7), 725–749. DOI: 10.1080/20445911.2019.1669609

Mercera, G., Kalis, T., Vervoort-Schel, J., & Moonen, X. (2025). In the eye of the beholder: Visualizing strengths, burdens and desires through the lens of neurodivergent children, their parents, and professionals using the Yucel Method. *Advances in Neurodevelopmental Disorders*, *9*(2), 315–329. DOI: 10.1007/s41252-024-00411-3

Meßler, C. F., Holmberg, H. C., & Sperlich, B. (2018). Multimodal therapy involving high-intensity interval training improves the physical fitness, motor skills, social behavior, and quality of life of boys with ADHD: A randomized controlled study. *Journal of Attention Disorders*, *22*(8), 806–812. DOI: 10.1177/1087054716636936 PMID: 27013028

Miranda, A., Tárraga, R., Fernández, M. I., Colomer, C., & Pastor, G. (2015). Parenting stress in families of children with autism spectrum disorder and ADHD. *Exceptional Children*, *82*(1), 81–95. DOI: 10.1177/0014402915585479

Moll, K., Snowling, M. J., & Hulme, C. (2020). Introduction to the special issue "Comorbidities between reading disorders and other developmental disorders.". *Scientific Studies of Reading*, *24*(1), 1–6. DOI: 10.1080/10888438.2019.1702045 PMID: 35283618

Montirosso, R., Mascheroni, E., Guida, E., Piazza, C., Sali, M. E., Molteni, M., & Reni, G. (2021). Stress symptoms and resilience factors in children with neurodevelopmental disabilities and their parents during the COVID-19 pandemic. *Health Psychology*, *40*(7), 428–438. DOI: 10.1037/hea0000966 PMID: 34435794

Mugnaini, D., Lassi, S., La Malfa, G., & Albertini, G. (2009). Internalizing correlates of dyslexia. *World Journal of Pediatrics*, *5*(4), 255–264. DOI: 10.1007/s12519-009-0049-7 PMID: 19911139

Murphy, J. M., Guzmán, J., McCarthy, A. E., Squicciarini, A. M., George, M., Canenguez, K. M., Dunn, E. C., Baer, L., Simonsohn, A., Smoller, J. W., & Jellinek, M. S. (2015). Mental health predicts better academic outcomes: A longitudinal study of elementary school students in Chile. *Child Psychiatry and Human Development*, *46*(2), 245–256. DOI: 10.1007/s10578-014-0464-4 PMID: 24771270

Murray, C., & Greenberg, M. T. (2001). Relationships with teachers and bonds with school: Social-emotional adjustment correlates for children with and without disabilities. *Psychology in the Schools*, *38*(1), 25–41. DOI: 10.1002/1520-6807(200101)38:1<25::AID-PITS4>3.0.CO;2-C

Nalavany, B. A., Logan, J. M., & Carawan, L. W. (2018). The relationship between emotional experience with dyslexia and work self-efficacy among adults with dyslexia. *Dyslexia (Chichester, England)*, *24*(1), 17–32. DOI: 10.1002/dys.1575 PMID: 29230916

Nelson, J. M., & Gregg, N. (2012). Depression and anxiety among transitioning adolescents and college students with ADHD, dyslexia, or comorbid ADHD/dyslexia. *Journal of Attention Disorders*, *16*(3), 244–254. DOI: 10.1177/1087054710385783 PMID: 20978271

Nevill, T., & Forsey, M. (2022). The social impact of schooling on students with dyslexia: A systematic review of the qualitative research on the primary and secondary education of dyslexic students. *Educational Research Review*, *36*, 100507. DOI: 10.1016/j.edurev.2022.100507

Nisiotou, I., & Vlachos, F. (2014). Neurodevelopmental disorders: Is there a shared biological basis? [Translated from Greek]. *Child and Adolescent Psychiatry*, *2*, 31–41.

Palombo, J. (2023). The neuropsychodynamics of learning disorders. In *Learning disorders across the lifespan: A mental health framework* (pp. 175–194). Springer International Publishing. DOI: 10.1007/978-3-031-21772-2_11

Posner, J., Polanczyk, G. V., & Sonuga-Barke, E. (2020). Attention-deficit hyperactivity disorder. *Lancet*, *395*(10222), 450–462. DOI: 10.1016/S0140-6736(19)33004-1 PMID: 31982036

Reid, G., & Kirk, J. (2001). *Dyslexia in adults: Education and employment*. Wiley.

Riddick, B. (2013). An examination of the relationship between labeling and stigmatization with special reference to dyslexia. In *Addressing difficulties in literacy development* (pp. 317–333). Routledge.

Rose, Z. R. (1998). *A study of college graduates with dyslexia* (Vol. 9839593). UMI.

Ruan, Y., Ye, Y., & McBride, C. (2023). Effectiveness of parent coaching on the literacy skills of Hong Kong Chinese children with and without dyslexia. *Reading and Writing*, *36*, 1–22. DOI: 10.1007/s11145-022-10408-w PMID: 36687998

Rungsattatharm, L., Tasingha, P., Trairatvorakul, P., & Chonchaiya, W. (2025). Longitudinal associations between executive function and positive parenting during early childhood and resilience, self-regulation, and behavioral problems in school-age children. *Child and Adolescent Psychiatry and Mental Health*, *19*(1), 19. DOI: 10.1186/s13034-025-00875-8 PMID: 40057732

Serry, T. A., Liamputtong, P., Rose, M. L., & Bretherton, L. (2016). Parents' experiences of their children's supplementary reading intervention: A qualitative exploration. *Australian Journal of Learning Difficulties*, *21*(1), 3–20. DOI: 10.1080/19404158.2015.1128452

Shaywitz, B. A., & Shaywitz, S. E. (2020). The American experience: Towards a 21st century definition of dyslexia. *Oxford Review of Education*, *46*(4), 454–471. DOI: 10.1080/03054985.2020.1793545

Shessel, I., & Reiff, H. B. (1999). Experiences of adults with learning disabilities: Positive and negative impacts and outcomes. *Learning Disability Quarterly*, *22*(4), 305–316. DOI: 10.2307/1511264

Singer, B. (2015). *Extended stereotype threat: Parental concern that a child will confirm a negative stereotype associated with dyslexia* [Doctoral dissertation, Trinity College]. https://core.ac.uk/download/pdf/232743371.pdf

Slomowitz, R. F., Narayan, A. J., & McGrath, L. M. (2025). Resilience research in learning disabilities: Guiding principles from developmental psychopathology. *Learning and Instruction*, *97*, 102090. DOI: 10.1016/j.learninstruc.2025.102090

Snowling, M. J., Hulme, C., & Nation, K. (2020). Defining and understanding dyslexia: Past, present and future. *Oxford Review of Education*, *46*(4), 501–513. DOI: 10.1080/03054985.2020.1765756 PMID: 32939103

Stein, B., Hoeft, F., & Richter, C. G. (2024). Stress, resilience, and emotional well-being in children and adolescents with specific learning disabilities. *Current Opinion in Behavioral Sciences*, *58*, 101410. DOI: 10.1016/j.cobeha.2024.101410 PMID: 39157263

Terras, M. M., Thompson, L. C., & Minnis, H. (2009). Dyslexia and psychosocial functioning: An exploratory study of the role of self-esteem and understanding. *Dyslexia (Chichester, England)*, *15*(4), 304–327. DOI: 10.1002/dys.386 PMID: 19384920

Theule, J., Wiener, J., Tannock, R., & Jenkins, J. M. (2013). Parenting stress in families of children with ADHD: A meta-analysis. *Journal of Emotional and Behavioral Disorders*, *21*(1), 3–17. DOI: 10.1177/1063426610387433

Tobia, V., Greco, A., Steca, P., & Marzocchi, G. M. (2019). Children's well-being at school: A multi-dimensional and multi-informant approach. *Journal of Happiness Studies*, *20*(3), 841–861. DOI: 10.1007/s10902-018-9974-2

Van Andel, E., Ten Have, M., Bijlenga, D., Beekman, A. T., De Graaf, R., & Kooij, J. S. (2022). Combined impact of ADHD and insomnia symptoms on quality of life, productivity, and health care use in the general population. *Psychological Medicine*, *52*(1), 36–47. DOI: 10.1017/S0033291720001592 PMID: 32597743

Weitzman, C., Guevara, J., Curtin, M., Macias, M., Kinwa Poon, J., Joseph Smith, P. J., Christine Augustyn, M., Liu, Y. H., Anopawuia Spinks-Franklin, A. I., Zubler, J. M., Navsaria, D., Glusman, M., Uzoatu Anyigbo, C., Chen, V., Gonzales, J. L., Guevara, J. P., Nobuhide Hashikawa, A., Marie Loria, H., Miller-Fitzwater, A., & Degnon, L. (2025). Promoting optimal development: Screening for mental health, emotional, and behavioral problems: Clinical report. *Pediatrics*, *156*(3), e2025073172. DOI: 10.1542/peds.2025-073172 PMID: 40850690

White, M., Lafaille-Magnan, M. E., Roche, C., Jolicoeur-Martineau, A., Wazana, A., & Connor, K. L. (2023). Developmental origins, differential susceptibility, and resilience in neurodevelopmental disorders. In Eisenstat, D. D., Goldowitz, D., Oberlander, T. F., & Yager, J. Y. (Eds.), *Neurodevelopmental pediatrics* (pp. 1–20). Springer. DOI: 10.1007/978-3-031-20792-1_45

Wilmot, A., Pizzey, H., Leitão, S., Hasking, P., & Boyes, M. (2023). Growing up with dyslexia: Child and parent perspectives on school struggles, self-esteem, and mental health. *Dyslexia (Chichester, England)*, *29*(1), 40–54. DOI: 10.1002/dys.1729 PMID: 36349866

Zhang, H., Yu, J., Tian, Y., Ma, H., Pan, Y., Wang, C., & Zhou, C. (2024). Effects of short-term and long-term physical activity intervention on attention in children with attention deficit hyperactivity disorder: Evidence from event-related potentials. *Frontiers in Psychology*, *14*, 1301740. DOI: 10.3389/fpsyg.2023.1301740

Chapter 9
Neurodevelopmental Consequences of Pediatric Cancer

ABSTRACT

This chapter considers pediatric cancer through a neurodevelopmental lens, highlighting how treatment interacts with brain maturation to produce cognitive profiles resembling ADHD, Learning Disorders, and Executive-Function impairments. These effects shape survivors' academic performance, motivation, and emotional well-being long after medical recovery. The chapter argues that education functions as an extension of neurorehabilitation, offering structured environments where executive skills, self-regulation, and resilience can be rebuilt. By integrating insights from oncology, neuropsychology, and special education, it proposes a holistic model in which schools, families, and clinicians work together to support long-term learning and adaptation across the lifespan.

PEDIATRIC CANCER AND THE NEURODEVELOPMENTAL FRAMEWORK

Childhood cancer represents not only a medical crisis but also a profound developmental disruption. Often described metaphorically as a "mountain" that rises suddenly before the child and their family, pediatric cancer demands rapid physiological, psychological, and social adaptation (Bhakta et al., 2017). Despite major advances in treatment and survivorship, the neurodevelopmental consequences of cancer and its therapies are increasingly recognized as a critical area of inquiry for clinicians, educators, and researchers (Marusak et al., 2018; Patel et al., 2025). The

DOI: 10.4018/979-8-3693-6331-7.ch009

transition from survival to quality of life has shifted attention toward cognitive, emotional, and educational outcomes across the life span.

In the past three decades, medical progress has drastically improved survival rates in childhood cancer. However, survival often comes with a hidden cost: late effects involving cognition, executive function, and academic performance (Michel et al., 2020; Tonorezos et al., 2022). These outcomes intersect with the domain of neurodevelopmental disorders (NDDs), as both cancer-related brain injuries and traditional NDDs share overlapping neurocognitive and behavioral phenotypes (Bonti et al., 2024; Moore III, 2005). This overlap presents both diagnostic and educational challenges, calling for interdisciplinary collaboration between oncologists, neuropsychologists, and special educators.

The recognition of neurocognitive late effects has reframed survivorship as a neurodevelopmental trajectory rather than a static post-treatment phase (Erdmann et al., 2021). Children treated for leukemia, brain tumors, or solid malignancies often experience disruption in neural maturation processes, such as myelination, synaptic pruning, and executive-network development (Friedman & Robbins, 2022). These disruptions can manifest as difficulties in attention, working memory, processing speed, and emotional regulation, hallmark features that echo patterns seen in ADHD, specific learning disorders, and communication disorders (APA, 2013; Faedda et al., 2019; Krull et al., 2008).

From an educational perspective, these challenges are not merely medical side effects but developmental barriers that influence lifelong learning, motivation, and self-regulation (Fine et al., 2023; Billman et al., 2023). Cancer survivors frequently demonstrate lower academic attainment and greater special educational needs compared to their peers, including difficulties in reading fluency, language comprehension, mathematical reasoning, and organizational skills (Baughan et al., 2023; Martinez-Santos et al., 2021; Karlson et al., 2020). The educational trajectory thus becomes a lens through which clinicians and educators can assess long-term adaptation, resilience, and quality of life.

Importantly, the study of neurodevelopmental outcomes in pediatric cancer survivors intersects multiple theoretical domains. Neuropsychology provides insight into the biological mechanisms of neural disruption and recovery, while special education offers a framework for understanding learning profiles, compensatory strategies, and individualized support. Bridging these disciplines facilitates a holistic understanding of survivors as learners, not merely patients recovering from disease, but students reengaging with the cognitive and social demands of school life (Ruble et al., 2020a, 2020b; He et al., 2022).

The aim of this chapter is therefore twofold. First, it synthesizes research on the neurodevelopmental and neurocognitive sequelae of pediatric cancer, focusing on how treatments such as chemotherapy, cranial radiation, and surgery interact with brain maturation and executive function. Second, it contextualizes these findings within educational and psychosocial frameworks, emphasizing early identification, targeted intervention, and long-term support for survivors across childhood and adulthood. The goal is to align medical recovery with educational reintegration, transforming survivorship from a medical endpoint into a continuum of learning and adaptation.

Understanding cancer as a developmental event requires moving beyond traditional medical models. Survivors' trajectories often mirror those seen in neurodevelopmental conditions, where early disruptions to neural architecture have cascading effects on cognition, behavior, and academic outcomes (Lervåg, 2021; Moore III, 2005). The challenge for educators lies in differentiating between disorder-related learning disabilities and treatment-induced cognitive difficulties while applying pedagogical strategies that address both (Senter et al., 2023).

Finally, this chapter argues for an integrated clinical-educational model grounded in neuroplasticity, executive functioning, and psychosocial resilience. This approach situates educators as vital actors in rehabilitation , capable of reinforcing neural and cognitive recovery through structured learning, emotional support, and environmental adaptation (Watson et al., 2016). By situating neurodevelopmental research within the everyday realities of teaching and learning, this work contributes to a more sustainable and human-centered approach to pediatric cancer survivorship.

PEDIATRIC CANCER AND BRAIN DEVELOPMENT: MECHANISMS OF NEURODEVELOPMENTAL IMPACT

Pediatric cancer and its treatments intersect with the developing brain at a uniquely vulnerable moment in the lifespan. Neural circuits that underlie executive function, learning, memory, and emotional regulation are still undergoing refinement through processes such as synaptogenesis, myelination, and cortical pruning. When a malignancy or its treatment disrupts these processes, the resulting neurodevelopmental trajectory may diverge significantly from the expected pattern (Bhakta et al., 2017; Marusak et al., 2018). This section explores how cancer-related medical interventions, such as chemotherapy, radiation, surgery, and supportive pharmacology, affect brain structure and function, often yielding profiles that overlap with neurodevelopmental disorders (NDDs) such as ADHD, dyslexia, and specific learning disorders (Krull et al., 2018).

The neurotoxic effects of cancer treatment arise from both direct and indirect mechanisms. Chemotherapy agents, particularly those crossing the blood–brain barrier, can damage white matter integrity and disrupt oligodendrocyte development, impeding efficient neural transmission (Michel et al., 2020; Patel et al., 2025). Cranial radiation therapy compounds this damage by reducing hippocampal neurogenesis and altering prefrontal cortical maturation (Egset et al., 2024). Even non-neurotoxic treatments can produce indirect effects through inflammation, oxidative stress, and vascular dysfunction (Cheung et al., 2018).

These mechanisms converge in the prefrontal cortex, home to executive functions such as planning, inhibition, and cognitive flexibility (Friedman & Robbins, 2022). As a result, survivors frequently exhibit deficits in attention, working memory, and processing speed (Fine et al., 2023; Moore III, 2005). Such cognitive patterns resemble those observed in developmental ADHD or executive function disorder, suggesting that cancer-related brain injuries may induce "acquired neurodevelopmental syndromes" (Bonti et al., 2024). These syndromes blur the traditional clinical divide between congenital and acquired etiologies, underscoring the need for transdiagnostic frameworks in both medicine and education (Krull et al., 2008; Krull et al., 2018).

The Role of Executive Functions in Post-Cancer Adaptation

Executive functions (EFs) represent a cornerstone of adaptive learning and self-regulation. They encompass inhibition, working memory, cognitive flexibility, and goal-oriented problem solving, processes essential for academic success and independent functioning (Diamond, 2013; Cirino, 2023). Pediatric cancer survivors show significant vulnerability in these domains. Neuroimaging studies demonstrate reduced functional connectivity within prefrontal and cingulo-opercular networks, mirroring findings from children with SLDs and ADHD (Freedman et al., 2020; Faedda et al., 2019; Moore III, 2005).

Cognitive inefficiency in these domains translates to real-world learning difficulties. Survivors often struggle to sustain attention in classroom settings, organize multi-step assignments, or manage time effectively, challenges nearly identical to those reported in students with diagnosed neurodevelopmental disorders (Mulder & Cragg, 2014; Butterfuss & Kendeou, 2018; Karlson et al., 2020). These findings suggest that executive dysfunction functions as a shared mechanism linking pediatric cancer, educational underachievement, and clinical NDDs.

Moreover, the interplay between motivation and EF compounds academic challenges. Cho et al. (2023) demonstrate that intrinsic motivation improves reading comprehension and cognitive control in students with learning difficulties, whereas extrinsic pressures can impair performance. Cancer survivors, often managing anxiety, fatigue, and fear of recurrence, may rely more on external reinforcement, making motivation-based interventions crucial in re-establishing academic engagement (He et al., 2022). Teachers and school psychologists thus play a vital role in cultivating intrinsic motivation and metacognitive awareness in this population (Hebbecker et al., 2019).

Neurocognitive Profiles Across Cancer Types

Different cancer types and treatments yield distinct cognitive signatures. Survivors of brain tumors face the highest risk of neurodevelopmental impairment due to direct neural injury and localized surgical effects (Patel et al., 2025; Moore III, 2005). Common patterns include deficits in processing speed, visuospatial reasoning, and working memory, domains essential for reading and mathematical cognition. Children with leukemia treated with intrathecal or high-dose systemic chemotherapy display generalized cognitive slowing, attention problems, and fine-motor deficits (Erdmann et al., 2021; Michel et al., 2020; Krull et al., 2008).

Vogel et al. (2025) extended this perspective by linking frailty markers in congenital heart disease survivors, a population with overlapping neurodevelopmental vulnerabilities, to poorer executive and adaptive outcomes, suggesting a broader cardiovascular–neurocognitive interface in pediatric health. Similarly, Karlson et al. (2020) found that latent neurocognitive profiles in pediatric cancer survivors can predict academic and psychological outcomes, reinforcing the need for individualized interventions. Together, these findings reveal that neurodevelopmental consequences of cancer treatment are not monolithic but distributed along a spectrum of severity and domain specificity.

The convergence of medical and educational perspectives opens the path for more responsive interventions. Neuropsychological assessments can guide educators in differentiating between primary learning disabilities and secondary cognitive effects of cancer therapy. For example, a child showing slow processing speed and reduced working memory after chemotherapy may not meet diagnostic criteria for dyslexia but still require reading fluency interventions similar to those used in specific learning disorders (Berninger, 2019; He et al., 2022; Pan & Lin, 2023).

Educational supports rooted in executive function training, such as cognitive flexibility tasks, self-monitoring checklists, and goal-setting scaffolds, have shown promise in both NDDs and cancer survivorship contexts (Krull et al., 2018; Senter et al., 2023; Watson et al., 2016). Furthermore, integrating cooperative learning and peer-assisted activities enhances social cognition and sustained attention (Segundo-Marcos et al., 2022), mitigating isolation and promoting inclusion.

NEURODEVELOPMENTAL DISORDERS, EXECUTIVE FUNCTIONS, AND EDUCATIONAL PATHWAYS

Children and adolescents who survive cancer often re-enter educational systems carrying invisible neurodevelopmental burdens. Although they are medically cured, their cognitive and psychological functioning frequently remains compromised.

Research shows that survivors have significantly higher odds of requiring special educational support, including individualized instruction, accommodations for fatigue, and tailored literacy or numeracy interventions (Baughan et al., 2023; Martinez-Santos et al., 2021; Ruble et al., 2020a, b; Moore III, 2005). These educational needs align closely with the neurocognitive profiles of students diagnosed with specific learning disorders (SLDs), ADHD, and communication disorders, confirming that cancer survivorship can yield acquired neurodevelopmental-like syndromes (Bonti et al., 2024; Krull et al., 2018; Lervåg, 2021).

The similarity extends beyond behavioral symptoms. Survivors frequently exhibit reduced processing speed, working memory inefficiency, and difficulties in sustaining attention, core executive deficits also observed in developmental dyslexia and ADHD (Cirino, 2023; Karlson et al., 2020; Moll et al., 2016; Peng et al., 2018). These shared deficits raise an important theoretical question: are these difficulties manifestations of the same underlying neurobiological vulnerability, or distinct yet convergent developmental disruptions? Current evidence supports a continuum model, a neurodevelopmental spectrum on which both congenital and acquired deficits coexist and interact (Michel et al., 2020; Moore III, 2005; Patel et al., 2025).

Executive Function Deficits as a Transdiagnostic Mechanism

Executive functions (EFs) serve as the cognitive "control center" that coordinates attention, inhibition, working memory, and planning. In both SLDs and post-cancer cognitive syndromes, EF weaknesses undermine self-regulated learning (Diamond, 2013; Goldstein & Naglieri, 2014). Deficits in working memory, in particular, compromise reading fluency and mathematical reasoning (Freedman et al., 2020; He et al., 2022; Moll et al., 2016), while inhibitory-control deficits hinder task persistence and error monitoring (Butterfuss & Kendeou, 2018; Senter et al., 2023).

These patterns reveal a neuropsychological bridge between special pedagogy and medical neuropsychology: both fields describe executive dysfunction as a fundamental bottleneck in learning and adaptation. In pediatric oncology, chemotherapy-related white matter changes have been directly linked to executive-function inefficiency and attentional lapses, paralleling the neurological findings in developmental ADHD (Faedda et al., 2019; Friedman & Robbins, 2022; Krull et al., 2018).

Moreover, studies demonstrate that motivation modulates EF performance. Cho et al. (2023) found that intrinsic motivation enhances reading comprehension by facilitating executive control over attention and working memory, whereas extrinsic motivators, such as grades, praise, or external pressure, can exacerbate cognitive fatigue. Educators working with cancer survivors must therefore emphasize self-determination theory principles: autonomy, competence, and relatedness, to maintain engagement during cognitive rehabilitation (He et al., 2022; Hebbecker et al., 2019).

More specifically, reading and writing, often perceived as basic literacy skills, are in fact complex, multi-level executive processes (Berninger, 2019; Moore III, 2005). Emerging research connects orthographic and morphological awareness with domain-general executive functions, including inhibition and updating (Costa et al., 2020; Pan & Lin, 2023). Cancer survivors experiencing slowed processing and attention lapses may display writing profiles similar to dysgraphia or expressive-language disorder (Filipe, 2021).

Pedagogical Strategies and School Reintegration

Returning to school after cancer treatment represents both a milestone and a vulnerability point. Many survivors face social reintegration difficulties, feeling simultaneously "different" yet expected to perform typically (Eiser et al., 2000; Martinez-Santos et al., 2021; Ruble et al., 2020a, b). Teachers, therefore, act as mediators of belonging. Evidence suggests that structured re-entry programs, combining psychoeducation for classmates, flexible grading policies, and individualized cognitive support, significantly enhance academic and emotional outcomes (Coughtrey et al., 2018; Marchak et al., 2022).

Specific strategies include cognitive remediation targeting working memory and processing speed (Egset et al., 2024; He et al., 2022; Peng et al., 2018); executive-skills coaching, emphasizing planning, time management, and self-monitoring (Senter et al., 2023); collaborative goal setting to enhance intrinsic motivation (Cho et al., 2023); integrated multidisciplinary support, aligning educators, psychologists, and healthcare providers (Fine et al., 2023; Tonorezos et al., 2022). Such approaches situate educational intervention not as a postscript to medical recovery but as a continuation of neurorehabilitation, a critical phase in lifelong development (Krull et al., 2018; Moore III, 2005).

Cognitive Reserve and Lifespan Adaptation

Despite these challenges, many survivors display remarkable resilience, achieving academic and social success. The concept of cognitive reserve, the brain's ability to compensate for structural damage through alternative neural pathways, offers a hopeful lens for educators (Berninger, 2019; Cirino, 2023; Moore III, 2005). Cognitive reserve is strengthened through enriched environments, social engagement, and mental stimulation. Cooperative learning environments and project-based instruction can act as "educational neuroprotectants," enhancing neural plasticity and executive coordination (Segundo-Marcos et al., 2022).

Furthermore, as survivors transition into adulthood, ongoing cognitive monitoring remains essential. Longitudinal studies show that neurocognitive impairments may emerge or worsen decades after treatment due to aging, fatigue, or comorbid conditions (Cheung et al., 2018; Krull et al., 2018; Tonorezos et al., 2022). For educators and clinicians alike, this reinforces the necessity of lifespan models of support, bridging childhood, adolescence, and adulthood under a unified neurodevelopmental framework.

Ultimately, the convergence of pediatric oncology, neuropsychology, and special education highlights a paradigm shift. Cancer survivorship is no longer solely a medical triumph, it is also an educational and social responsibility. Teachers and educational leaders must perceive post-cancer cognitive and emotional adaptation as part of the broader neurodiversity continuum, deserving the same rights to accommodation, inclusion, and personalized learning as congenital NDDs (Bonti et al., 2024; Lervåg, 2021; Moore III, 2005). This integrated framework positions educators not merely as implementers of classroom strategies but as co-researchers in understanding how neurodevelopment unfolds after medical adversity. As Bhakta et al. (2017) and Marusak et al. (2018) argue, recovery is multidimensional encompassing the interplay between brain, behavior, and environment. Therefore, educational systems must adopt transdiagnostic approaches that focus on underlying cognitive mechanisms rather than categorical labels.

THE EDUCATIONAL SYSTEM AS A CONTINUUM OF NEUROREHABILITATION

For survivors of childhood cancer, education represents more than a social reintegration pathway, it is a therapeutic continuum that extends neurorehabilitation beyond the hospital setting. Advances in pediatric oncology have dramatically increased survival rates (Erdmann et al., 2021), yet the cognitive, emotional, and social sequelae of treatment persist well into adulthood (Fine et al., 2023; Krull et al., 2018; Tonorezos et al., 2022). Schools thus become not only centers of knowledge but environments of recovery, where the restoration of executive, attentional, and emotional capacities occurs in parallel with academic learning. The educational environment can serve as both a stressor and a scaffold for neurodevelopment. Survivors often return to classrooms where expectations for memory, attention, and social reciprocity are high (Martinez-Santos et al., 2021). Without systematic supports, the gap between potential and performance widens, fostering frustration and academic underachievement (Baughan et al., 2023; Moore III, 2005). However, when schools implement structured interventions targeting working memory, processing speed, and metacognition, educational participation transforms into a form of neurorehabilitative training, supporting neural reorganization and functional adaptation (Cirino, 2023; He et al., 2022; Segundo-Marcos et al., 2022).

Educational systems have historically lagged behind medical science in addressing the cognitive consequences of disease. Bridging this gap requires transforming neuropsychological findings into pedagogical models. The growing body of literature on cognitive late effects (Egset et al., 2024; Patel et al., 2025) underscores the need for instructional approaches that are adaptive, process-oriented, and neurodevelopmentally informed. For example, findings from van der Plas et al. (2024) demonstrate that adult survivors frequently experience long-term impairments in attention, emotion regulation, and memory. Translating such findings into educational language, these deficits correspond to challenges in executive self-regulation, goal maintenance, and working-memory updating, skills which are essential for independent learning (Diamond, 2013; Goldstein & Naglieri, 2014). In this context, interventions must focus not merely on content remediation but on learning-to-learn strategies that cultivate self-awareness of cognitive limitations and adaptive compensatory techniques (He et al., 2022; Krull et al., 2018; Moore III, 2005).

Cognitive strategy instruction, self-monitoring, and structured task analysis, which are common techniques drawn from special education, can be embedded within regular curricula. For instance, executive-skills training emphasizing planning, time management, and self-evaluation has been shown to enhance academic outcomes among both SLD and oncology populations (Butterfuss & Kendeou, 2018; Senter et al., 2023). Similarly, metacognitive scaffolding, such as verbalizing problem-solving steps or using visual organizers, supports working-memory constraints common among survivors (Freedman et al., 2020; Peng et al., 2018).

The psychosocial dimensions of survivorship must also be integrated into educational design. Studies document persistent anxiety, depression, and social withdrawal in cancer survivors (Bhakta et al., 2017; Eiser et al., 2000; Marusak et al., 2018; van der Plas et al., 2024). These emotional responses are not isolated symptoms but interact dynamically with executive and attentional networks, amplifying cognitive fatigue (Friedman & Robbins, 2022). Thus, emotional regulation training, teaching students to identify and manage their physiological and cognitive responses to stress, serves as both a psychological and neurocognitive intervention. Schools adopting trauma-informed approaches create predictable routines, provide emotional check-ins, and train staff to recognize the behavioral manifestations of stress (Goodman et al., 2024). Peer-support systems further promote belonging and identity reconstruction after illness, reducing social exclusion and stigma (Martinez-Santos et al., 2021). When emotional regulation and cognitive rehabilitation intersect, they yield a synergistic recovery pathway, strengthening resilience, executive control, and academic persistence simultaneously (Coughtrey et al., 2018; He et al., 2022).

Finally, fatigue and sleep disruption are among the most overlooked mediators of neurodevelopmental outcomes in pediatric cancer survivors (Goodman et al., 2024; Mogavero et al., 2020). Poor sleep impairs hippocampal functioning and prefrontal regulation, leading to deficits in memory consolidation, cognitive flexibility, and emotion modulation. These effects parallel those observed in children with ADHD or dyslexia-related attention difficulties (Bonti et al., 2024; Lervåg, 2021). Educational programs should therefore integrate fatigue-sensitive planning, e.g., distributing cognitively demanding tasks earlier in the day, providing rest breaks, and embedding movement or mindfulness activities within lessons. Sleep hygiene education can be incorporated into health curricula, promoting routines that restore circadian stability. The link between physiological and cognitive regulation underscores the importance of holistic, interdisciplinary interventions combining medical follow-up, educational design, and lifestyle management (Cheung et al., 2018).

The conceptual alignment between childhood cancer survivorship and neurodevelopmental disorders challenges the binary between medical recovery and educational inclusion. Both groups embody neurodiversity in its broadest sense: individual variability in brain function resulting from genetic, developmental, or acquired conditions (Bonti et al., 2024; Lervåg, 2021; Moore III, 2005). Framing survivorship within the neurodiversity paradigm redefines educational inclusion as the creation of environments that accommodate variability rather than remediate difference. This perspective resonates with Universal Design for Learning (UDL), which emphasizes flexible goals, materials, and assessments to support diverse cognitive pathways (Watson et al., 2016). When UDL principles are applied to post-cancer education, classrooms become adaptable ecosystems that respect fluctuating cognitive energy and attention. For example, providing multimodal representations of content (text, visuals, and oral explanations) aids students with working-memory and processing-speed limitations (Freedman et al., 2020). Similarly, choice-based learning tasks enhance intrinsic motivation and agency (Cho et al., 2023; Hebbecker et al., 2019).

Lifelong Follow-Up and Adult Functioning

Emerging evidence indicates that neurocognitive challenges can persist or even intensify decades after treatment, as the brain encounters new developmental tasks in adulthood (Cheung et al., 2018; Krull et al., 2018; Tonorezos et al., 2022). Executive dysfunctions may translate into occupational inefficiencies, fatigue intolerance, or social adaptation difficulties. These outcomes mirror adult expressions of neurodevelopmental disorders such as ADHD and dyslexia (Horowitz-Kraus, 2023; Moll et al., 2016). Thus, survivorship care must evolve into lifespan neurodevelopmental monitoring (Moore III, 2005).

Educational policy should mandate periodic cognitive screening and career guidance for adult survivors, integrating vocational rehabilitation, psychological support, and continuing education. Universities and workplaces should follow the inclusive design principles of secondary education, ensuring accessibility in learning materials, schedules, and communication channels. In this way, inclusion becomes a lifelong right, not a temporary accommodation (Krull et al., 2008; Moore III, 2005).

Integrating pediatric oncology, neuropsychology, and special education requires a systemic framework that transcends disciplinary silos. Such a framework should recognize educational environments as neurorehabilitative spaces capable of fostering executive, emotional, and social recovery in children and adolescents who have experienced cancer. It must promote interdisciplinary collaboration among educators, clinicians, and families to ensure individualized, responsive support tailored to each learner's neurocognitive profile. Within this structure, cognitive and emotional self-regulation training should be embedded into everyday instruction, allowing students to rebuild essential adaptive and academic skills in authentic contexts. Equally important is the adoption of a lifespan perspective that guarantees continuity of care and learning opportunities from childhood through adulthood. Finally, advancing comprehensive policy reform is crucial to safeguard educational rights and secure sustainable funding for all students with acquired neurodevelopmental differences, ensuring equitable access to quality education and long-term inclusion (Coughtrey et al., 2018; He et al., 2022; Moore III, 2005).

This integrative vision reframes survivorship as a journey of neurodevelopmental adaptation rather than static recovery. It situates educators not as secondary actors to medicine but as co-architects of cognitive resilience and human flourishing. The intersection between childhood cancer and neurodevelopmental disorders illuminates the porous boundaries between medical pathology and educational diversity. Both fields converge on the recognition that executive functioning, emotional regulation, and learning are intertwined processes shaped by neurobiology, experience, and environment. The future of inclusive education depends on embracing this complexity. As understanding deepens, the goal shifts from compensating for deficits to cultivating adaptive capacities. Survivors of childhood cancer, like students with SLDs or ADHD, embody the resilience and plasticity of the developing brain. Through evidence-based pedagogy, interdisciplinary collaboration, and compassionate inclusivity, educators can transform classrooms into ecosystems of recovery, where healing, learning, and growth coexist (Coughtrey et al., 2018; He et al., 2022; Moore III, 2005).

Table 1. Neurodevelopmental and educational consequences of pediatric cancer across development

Developmental Period	Neurocognitive Vulnerabilities	Psychosocial Impact	Educational and Support Needs
Early Childhood	Delays in language development, attention regulation, and emerging executive functions	Increased parental stress, early emotional vulnerability	Early developmental monitoring, language and cognitive support, family-centered interventions
Preschool and Early School Years	Difficulties in attention, memory, processing speed, and learning readiness	Adjustment difficulties, anxiety, reduced school engagement	Individualized educational planning, collaboration between medical and educational teams
Middle Childhood	Academic skill weaknesses, executive function deficits, fatigue-related cognitive fluctuations	Reduced self-esteem, peer relationship difficulties	Academic accommodations, cognitive rehabilitation strategies, psychosocial support
Adolescence	Executive dysfunction, slowed processing speed, difficulties with organization and autonomy	Identity challenges, emotional distress, increased risk for anxiety or depression	Transition-focused planning, counseling support, flexible academic pathways
Young Adulthood	Residual cognitive vulnerabilities, attentional and memory difficulties	Challenges in higher education, vocational training, and independence	Academic and workplace accommodations, mentoring, long-term follow-up services
Adulthood	Long-term neurodevelopmental imprint, adaptive coping with persistent cognitive effects	Employment instability, psychosocial vulnerability	Lifelong learning opportunities, inclusive employment practices, community-based support

REFERENCES

American Psychiatric Association. (2013). *Diagnostic and statistical manual of mental disorders* (5th ed.). Author., DOI: 10.1176/appi.books.9780890425596

Baughan, N., Pell, J. P., Mackay, D. F., Clark, D., King, A., & Fleming, M. (2023). Educational outcomes in childhood cancer survivors: A Scotland-wide record-linkage study of 766,217 schoolchildren. *PLoS One*, *18*(7), e0286840. DOI: 10.1371/journal.pone.0286840 PMID: 37494295

Berninger, V. W. (2019). *Reading and writing acquisition: A developmental neuropsychological perspective*. Routledge., DOI: 10.4324/9780429303821

Bhakta, N., Liu, Q., Ness, K. K., Baassiri, M., Eissa, H., Yeo, F., Chemaitilly, W., Ehrhardt, M. J., Bass, J., Bishop, M. W., Shelton, K., Lu, L., Huang, S., Li, Z., Caron, E., Lanctot, J., Howell, C., Folse, T., Joshi, V., & Robison, L. L. (2017). The cumulative burden of surviving childhood cancer: An initial report from the St Jude Lifetime Cohort Study (SJLIFE). *Lancet*, *390*(10112), 2569–2582. DOI: 10.1016/S0140-6736(17)31610-0 PMID: 28890157

Billman, E., Smith, S. M., & Jain, S. L. (2023). Perspectives of childhood cancer survivors as young adults: A qualitative study of illness education resources and unmet information needs. *Journal of Cancer Education*, *38*(4), 1140–1148. DOI: 10.1007/s13187-022-02240-1 PMID: 36456891

Bonti, E., Zerva, I. K., Koundourou, C., & Sofologi, M. (2024). The high rates of comorbidity among neurodevelopmental disorders: Reconsidering the clinical utility of distinct diagnostic categories. *Journal of Personalized Medicine*, *14*(3), 300. DOI: 10.3390/jpm14030300 PMID: 38541042

Butterfuss, R., & Kendeou, P. (2018). The role of executive functions in reading comprehension. *Educational Psychology Review*, *30*(3), 801–826. DOI: 10.1007/s10648-017-9422-6

Cheung, Y. T., Brinkman, T. M., Li, C., Mzayek, Y., Srivastava, D., Ness, K. K., Patel, S. K., Howell, R. M., Oeffinger, K. C., Robison, L. L., Armstrong, G. T., & Krull, K. R. (2018). Chronic health conditions and neurocognitive function in aging survivors of childhood cancer: A report from the Childhood Cancer Survivor Study. *Journal of the National Cancer Institute*, *110*(4), 411–419. DOI: 10.1093/jnci/djx224 PMID: 29088360

Cho, E., Ju, U., Kim, E. H., Lee, M., Lee, G., & Compton, D. L. (2023). Relations among motivation, executive functions, and reading comprehension: Do they differ for students with and without reading difficulties? *Scientific Studies of Reading*, *27*(4), 289–310. DOI: 10.1080/10888438.2022.2127357

Cirino, P. T. (2023). Framing executive function as a construct and its relation to academic achievement. *Mind, Brain and Education : the Official Journal of the International Mind, Brain, and Education Society*, *17*(1), 5–24. DOI: 10.1111/mbe.12360 PMID: 38093790

Costa, L. J. C., Green, J. A., & Hooper, S. R. (2020). The relations among the development of written language and executive functions for children in elementary school. In Talinn, P. (Ed.), *Writing across the lifespan* (pp. 201–221). Parlor Press., DOI: 10.37514/PER-B.2020.1053.2.10

Coughtrey, A., Millington, A., Bennett, S., Christie, D., Hough, R., Su, M. T., Constantinou, M. P., & Shafran, R. (2018). The effectiveness of psychosocial interventions for psychological outcomes in pediatric oncology: A systematic review. *Journal of Pain and Symptom Management*, *55*(3), 1004–1017. DOI: 10.1016/j.jpainsymman.2017.09.022 PMID: 28962919

Diamond, A. (2013). Executive functions. *Annual Review of Psychology*, *64*(1), 135–168. DOI: 10.1146/annurev-psych-113011-143750 PMID: 23020641

Egset, K. S., Stubberud, J., Ruud, E., Hjort, M. A., Eilertsen, M. E. B., Sund, A. M., Hjemdal, O., Weider, S., & Reinfjell, T. (2024). Neurocognitive outcome and associated factors in long-term adult survivors of childhood acute lymphoblastic leukemia, treated without cranial radiation therapy. *Journal of the International Neuropsychological Society*, *30*(2), 233–243. DOI: 10.1017/S1355617724000080 PMID: 38465668

Eiser, C., Hill, J. J., & Vance, Y. H. (2000). Examining the psychological consequences of surviving childhood cancer: Systematic review as a research method in pediatric psychology. *Journal of Pediatric Psychology*, *25*(6), 449–460. DOI: 10.1093/jpepsy/25.6.449 PMID: 10980049

Erdmann, F., Frederiksen, L. E., Bonaventure, A., Mader, L., Hasle, H., Robison, L. L., & Winther, J. F. (2021). Childhood cancer: Survival, treatment modalities, late effects, and improvements over time. *Cancer Epidemiology*, *71*, 101733. DOI: 10.1016/j.canep.2020.101733 PMID: 32461035

Faedda, N., Romani, M., Rossetti, S., Vigliante, M., Pezzuti, L., Cardona, F., & Guidetti, V. (2019). Intellectual functioning and executive functions in children and adolescents with attention deficit hyperactivity disorder (ADHD) and specific learning disorder (SLD). *Scandinavian Journal of Psychology*, *60*(5), 440–446. DOI: 10.1111/sjop.12562 PMID: 31242533

Filipe, M. (2021). How do executive function issues affect writing in students with neurodevelopmental disorders? In Limpo, T., & Olive, T. (Eds.), *Executive functions and writing* (pp. 160–180). Oxford University Press., DOI: 10.1093/oso/9780198863564.003.0007

Fine, J. R., Bafna, T., Griffith, S. C., Ransdell, J. M., Barredo, J. C., Isrow, D. M., & Isrow, D. (2023). Effects of cancer treatment on the education and cognition of South Florida pediatric cancer survivors. *Cureus*, *15*(4), e37001. DOI: 10.7759/cureus.37001 PMID: 37139042

Freedman, L., Zivan, M., Farah, R., & Horowitz-Kraus, T. (2020). Greater functional connectivity within the cingulo-opercular and ventral attention networks is related to better fluent reading: A resting-state functional connectivity study. *NeuroImage. Clinical*, *26*, 102214. DOI: 10.1016/j.nicl.2020.102214 PMID: 32092682

Friedman, N. P., & Robbins, T. W. (2022). The role of prefrontal cortex in cognitive control and executive function. *Neuropsychopharmacology : Official Publication of the American College of Neuropsychopharmacology*, *47*(1), 72–89. DOI: 10.1038/s41386-021-01132-0 PMID: 34408280

Goldstein, S., & Naglieri, J. A. (2014). *Executive functioning*. Springer., DOI: 10.1007/978-1-4614-8106-5

Goodman, P. B., Palka, J. M., Nguyen, A., Carroll, B., & Holland, A. A. (2024). Effects of screen time on emotional and behavioral functioning in pediatric cancer survivors: The potential mediating role of sleep and fatigue. *Journal of Pediatric Neuropsychology*, *10*(3), 202–212. Advance online publication. DOI: 10.1037/h0101949

He, F., Huang, H., Ye, L., Wen, X., & Cheng, A. S. (2022). Meta-analysis of neurocognitive rehabilitation for cognitive dysfunction among pediatric cancer survivors. *Journal of Cancer Research and Therapeutics*, *18*(7), 2058–2065. DOI: 10.4103/jcrt.jcrt_1429_22 PMID: 36647970

Hebbecker, K., Förster, N., & Souvignier, E. (2019). Reciprocal effects between reading achievement and intrinsic and extrinsic reading motivation. *Scientific Studies of Reading*, *23*(5), 419–436. DOI: 10.1080/10888438.2019.1598413

Horowitz-Kraus, T. (2023). The role of executive functions in fluent reading: Lessons from reading acquisition and remediation. *Mind, Brain and Education : the Official Journal of the International Mind, Brain, and Education Society*, *17*(2), 89–102. DOI: 10.1111/mbe.12357

Karlson, C. W., Sarver, D. E., Raiker, J. S., Espil, F. M., Cox, A. D., Elkin, T. D., & Annett, R. D. (2020). The contribution of neurocognitive functions to academic and psychological outcomes in pediatric cancer: A latent profile analysis. *Child Neuropsychology*, *26*(7), 881–899. DOI: 10.1080/09297049.2020.1734553 PMID: 32157961

Krull, K. R., Hardy, K. K., Kahalley, L. S., Schuitema, I., & Kesler, S. R. (2018). Neurocognitive outcomes and interventions in long-term survivors of childhood cancer. *Journal of Clinical Oncology*, *36*(21), 2181–2189. DOI: 10.1200/JCO.2017.76.4696 PMID: 29874137

Krull, K. R., Okcu, M. F., Potter, B., Jain, N., Dreyer, Z., Kamdar, K., & Brouwers, P. (2008). Screening for neurocognitive impairment in pediatric cancer long-term survivors. *Journal of Clinical Oncology*, *26*(25), 4138–4143. DOI: 10.1200/JCO.2008.16.8864 PMID: 18757327

Lervåg, A. (2021). Is there a core deficit in specific learning disabilities? *Journal of Child Psychology and Psychiatry, and Allied Disciplines*, *62*(6), 677–679. DOI: 10.1111/jcpp.13434 PMID: 34008198

Marchak, J. G., Christen, S., Mulder, R. L., Baust, K., Blom, J. M., Brinkman, T. M., Elens, I., Harju, E., Kadan-Lottick, N. S., Khor, J. W. T., Lemiere, J., Recklitis, C. J., Wakefield, C. E., Wiener, L., Constine, L. S., Hudson, M. M., Kremer, L. C. M., Skinner, R., Vetsch, J., & Michel, G. (2022). Recommendations for the surveillance of mental health problems in childhood, adolescent, and young adult cancer survivors: A report from the International Late Effects of Childhood Cancer Guideline Harmonization Group. *The Lancet. Oncology*, *23*(4), e184–e196. DOI: 10.1016/S1470-2045(21)00750-6 PMID: 35358467

Martinez-Santos, A. E., Fernandez De La Iglesia, J. D. C., Sheaf, G., & Coyne, I. (2021). A systematic review of the educational experiences and needs of children with cancer returning to school. *Journal of Advanced Nursing*, *77*(7), 2971–2994. DOI: 10.1111/jan.14784 PMID: 33598984

Marusak, H. A., Iadipaolo, A. S., Harper, F. W., Elrahal, F., Taub, J. W., Goldberg, E., & Rabinak, C. A. (2018). Neurodevelopmental consequences of pediatric cancer and its treatment: Applying an early adversity framework to understanding cognitive, behavioral, and emotional outcomes. *Neuropsychology Review*, *28*(2), 123–175. DOI: 10.1007/s11065-017-9365-1 PMID: 29270773

Michel, G., Brinkman, T. M., Wakefield, C. E., & Grootenhuis, M. (2020). Psychological outcomes, health-related quality of life, and neurocognitive functioning in survivors of childhood cancer and their parents. *Pediatric Clinics of North America*, *67*(6), 1103–1134. DOI: 10.1016/j.pcl.2020.07.005 PMID: 33131537

Mogavero, M. P., Bruni, O., DelRosso, L. M., & Ferri, R. (2020). Neurodevelopmental consequences of pediatric cancer and its treatment: The role of sleep. *Brain Sciences*, *10*(7), 411. DOI: 10.3390/brainsci10070411 PMID: 32630162

Moll, K., Göbel, S. M., Gooch, D., Landerl, K., & Snowling, M. J. (2016). Cognitive risk factors for specific learning disorder: Processing speed, temporal processing, and working memory. *Journal of Learning Disabilities*, *49*(3), 272–281. DOI: 10.1177/0022219414547221 PMID: 25124507

Moore, B. D.III. (2005). Neurocognitive outcomes in survivors of childhood cancer. *Journal of Pediatric Psychology*, *30*(1), 51–63. DOI: 10.1093/jpepsy/jsi016 PMID: 15610985

Mulder, H., & Cragg, L. (2014). Executive functions and academic achievement: Current research and future directions. *Infant and Child Development*, *23*(1), 1–3. DOI: 10.1002/icd.1836

Pan, D. J., & Lin, D. (2023). How do executive functions explain early Chinese reading and writing? *Reading and Writing*, *36*(3), 625–647. DOI: 10.1007/s11145-022-10314-1

Patel, T., Johar, P., Kanisetti, V., Talacheru, S., Avinash, V., Das, A., Sahu, S., Goyal, A., Szobody, M. W., Sayers, T., Gullapalli, S., Yallapu, M. M., Shaikh, M. H., Anand, N., Potter-Baker, K., & Gadad, B. S. (2025). Long-term neurocognitive and behavioral outcomes in survivors of pediatric brain tumors: A systematic review. *Frontiers in Neuroscience*, *19*, 1587059. DOI: 10.3389/fnins.2025.1587059 PMID: 40520501

Peng, P., Barnes, M., Wang, C., Wang, W., Li, S., Swanson, H. L., Dardick, W., & Tao, S. (2018). A meta-analysis on the relation between reading and working memory. *Psychological Bulletin*, *144*(1), 48–76. DOI: 10.1037/bul0000124 PMID: 29083201

Ruble, K., Paré-Blagoev, J., Cooper, S., & Jacobson, L. A. (2020b). Pediatric oncology provider perspectives and practices: Supporting patients and families in schooling after cancer diagnosis. *Pediatric Blood & Cancer*, *67*(4), e28166. DOI: 10.1002/pbc.28166 PMID: 31930685

Ruble, K. J., Paré-Blagoev, E. J., Cooper, S. L., & Jacobson, L. A. (2020a). Assessment of online resources for returning to school during and after treatment of childhood cancer. *Journal of Cancer Education*, *35*(5), 876–884. DOI: 10.1007/s13187-019-01537-y PMID: 31049872

Segundo-Marcos, R., Carrillo, A. M., Fernández, V. L., & González, M. T. D. (2022). Development of executive functions in late childhood and the mediating role of cooperative learning: A longitudinal study. *Cognitive Development*, *63*, 101219. DOI: 10.1016/j.cogdev.2022.101219

Senter, R., Chow, J. C., & Willis, E. C. (2023). Speech-language pathology interventions for children with executive function deficits: A systematic literature review. *Language, Speech, and Hearing Services in Schools*, *54*(1), 336–354. DOI: 10.1044/2022_LSHSS-22-00013 PMID: 36306507

Tonorezos, E. S., Cohn, R. J., Glaser, A. W., Lewin, J., Poon, E., Wakefield, C. E., & Oeffinger, K. C. (2022, April 16). Long-term care for people treated for cancer during childhood and adolescence. *Lancet*, *399*(10334), 1561–1572. DOI: 10.1016/S0140-6736(22)00460-3 PMID: 35430023

van der Plas, E., Darji, H., Srivastava, D. K., Schapiro, M., Jeffe, D., Perkins, S., Howell, R., Leisenring, W., Armstrong, G. T., Oeffinger, K., Krull, K., Edelstein, K., & Hayashi, R. J. (2024). Risk factors for neurocognitive impairment, emotional distress, and poor quality of life in survivors of pediatric rhabdomyosarcoma: A report from the Childhood Cancer Survivor Study. *Cancer*, *130*(12), 2224–2236. Advance online publication. DOI: 10.1002/cncr.35236 PMID: 38373075

Vogel, C., Marino, B. S., Cassedy, A., Natarus, M., Parikh, P., Hinkle, C., & Allen, K. Y. (2025). The novel associations between frailty and neurodevelopmental outcomes in CHD. *Cardiology in the Young*, *35*(7), 1427–1435. DOI: 10.1017/S1047951125100863 PMID: 40637129

Watson, S. M., Gable, R. A., & Morin, L. L. (2016). The role of executive functions in classroom instruction of students with learning disabilities. *International Journal of School and Cognitive Psychology*, *3*(167). Advance online publication. DOI: 10.4172/2469-9837.1000167

Chapter 10
Investigating the Interplay Between Technology Utilization, Artificial Intelligence Tools, and Neurodevelopmental Disorders

ABSTRACT

This chapter considers how the rapid spread of AI and digital tools is reshaping the experiences of learners with NDDs. These technologies can open new pathways for communication, personalized instruction, and early identification, yet they also bring challenges tied to dependency, overstimulation, and uneven access. Drawing together research from multiple disciplines and recent international guidelines, the chapter argues for a measured approach in which technological systems complement, not replace, human insight. It outlines principles for responsible, inclusive use of AI that strengthens learning while safeguarding developmental and ethical considerations.

INTRODUCTION

Recent advances in Artificial Intelligence (AI) and digital technologies have reshaped approaches to learning, communication, and therapy for individuals with neurodevelopmental disorders (NDDs). Tools based on machine learning, natural-

DOI: 10.4018/979-8-3693-6331-7.ch010

language processing, and robotics offer unprecedented opportunities for personalized diagnosis and intervention. Yet, these same technologies can also reinforce dependence, distraction, or inequity if applied without careful ethical and pedagogical consideration. This chapter explores the dynamic relationship between technology utilization, AI systems, and NDDs, synthesizing research from computer science, psychology, and education. It examines supportive uses of AI in inclusive education, potential developmental risks of excessive screen exposure, and strategies for integrating AI tools responsibly within school environments. Drawing on recent international guidelines (Li et al., 2023; OECD, 2024; Polydoros et al., 2025; UNESCO, 2021) and empirical studies, it proposes a balanced model of "intelligent inclusion," where technology augments rather than replaces human support.

Artificial Intelligence has moved from an abstract theoretical concept to a pervasive presence in daily life. Its integration into education and child development has been especially rapid, influencing how teachers assess, how students learn, and how special-education professionals support children with diverse needs. Neurodevelopmental disorders, including autism spectrum disorder (ASD), attention-deficit/hyperactivity disorder (ADHD), specific learning disorders (SLDs), developmental coordination disorder (DCD), and communication impairments represent conditions where AI's potential is particularly visible (APA, 2015; Thapar et al., 2017).

While digital tools can scaffold learning and communication, growing evidence highlights new challenges: increased screen dependency, reduced face-to-face interaction, and the risk of over-stimulation (Torous et al., 2021). The question is no longer whether technology affects neurodevelopment, but how to direct this impact constructively. This chapter investigates the interplay between technology use and NDDs through an interdisciplinary lens. It situates AI historically, reviews its cognitive and clinical applications, evaluates associated risks, and discusses educational integration guided by international ethical frameworks (European Commission, 2023b; UNESCO, 2021). The goal is to synthesize current evidence into a coherent understanding of how intelligent systems can both support and challenge neurodiverse learners.

Artificial Intelligence as a scientific field was formally born with Turing's seminal question "Can machines think?" in 1950 (Turing, 1950). Since then, developments such as expert systems (Buchanan & Shortliffe, 1984), symbolic reasoning (Brachman & Levesque, 2004), and neural networks (LeCun et al., 2015) have transformed computational modeling of human cognition. The resurgence of deep learning (Goodfellow et al., 2016) and reinforcement learning (Sutton & Barto, 2018) in the 2010s led to powerful systems like AlphaGo (Silver et al., 2016) and GPT-4 (Bubeck et al., 2023). AI's success in pattern recognition and natural-language understanding has direct implications for neurodevelopmental research, where detecting subtle behavioral or linguistic cues can improve diagnosis and intervention (Jordan & Mitchell, 2015; Jurafsky & Martin, 2023).

Early rule-based systems such as MYCIN (Buchanan & Shortliffe, 1984) relied on explicit knowledge representation. Today's transformer architectures (Vaswani et al., 2017; Wolf et al., 2020) enable contextual language modeling and multimodal reasoning. Generative AI models, capable of producing coherent text, images, or speech, open new pathways for assistive communication technologies, especially relevant for individuals with speech and language disorders. However, these systems also raise ethical and developmental questions: do they enhance children's expressive capacities or risk replacing authentic communication? Scholars argue that AI should remain a complement to human interaction rather than a substitute (Floridi & Chiriatti, 2020; UNESCO, 2021).

Governments and organizations have recognized both the promise and peril of educational AI. The European Commission (2023a) emphasizes inclusion and data ethics; OECD (2024) stresses system-level strategies for equitable adoption; UNESCO (2021) advocates human-centered AI that respects learners' rights. Together these frameworks encourage adaptive technologies that personalize learning without reinforcing bias or dependency.

NEURODEVELOPMENTAL DISORDERS IN THE DIGITAL ERA

Neurodevelopmental disorders are conditions of early onset that affect cognition, communication, motor control, or behavior (APA, 2015; Thapar et al., 2017). They share overlapping genetic and neural pathways (Astle et al., 2022) and require long-term multidisciplinary management. With the proliferation of digital tools, children with NDDs increasingly interact with screens, apps, and online learning platforms. This creates both opportunities for scaffolding deficits and risks of technological overexposure.

Digital devices can serve as cognitive prostheses, external supports that extend memory, attention, and communication (Topol, 2019). For autistic learners, visual schedules and AI-driven social-skills coaches can facilitate routine and social understanding (Li et al., 2023). In ADHD, gamified feedback systems improve motivation and self-regulation (Polydoros & Antoniou, 2025b). For dyslexia or SLDs, text-to-speech and adaptive reading software help overcome decoding barriers (Protopapas & Parrila, 2018). Yet, heavy reliance on digital feedback loops can narrow attention spans or reduce intrinsic motivation (Holmes et al., 2022). Balanced exposure and guided mediation by educators are essential to ensure that technology strengthens rather than supplants executive-function training.

AI-based analytics are increasingly used to identify atypical developmental trajectories. Digital phenotyping, the passive collection of behavioral data from smartphones or wearables, offers real-time insights into mental-health patterns (Melcher et al., 2020; Torous et al., 2021). Machine-learning algorithms can detect early markers of autism through gaze tracking or vocal prosody (Jurafsky & Martin, 2023). In ADHD, AI systems analyze motor activity and response times to refine diagnostic precision (Obermeyer & Emanuel, 2016). However, ethical debates persist about data privacy, consent, and potential stigmatization (Floridi, 2019; UNESCO, 2021).

The benefits of AI-driven tools are unevenly distributed. Socioeconomic disparities affect access to reliable internet, up-to-date devices, and trained educators (Crawford, 2021; OECD, 2024; Eubanks, 2018). Without inclusive policy frameworks, technological innovation may exacerbate rather than alleviate educational inequality.

Supportive Roles Of AI And Technology In NDDs

AI systems can individualize content delivery based on real-time learner data. Adaptive platforms adjust difficulty, pacing, and modality to match cognitive profiles (Holmes et al., 2022; Luckin et al., 2016). For students with SLD in mathematics, computer-assisted mathematics environments optimize fraction learning and enhance motivation (Polydoros & Antoniou, 2025a). In geometry, AI-integrated applications improve spatial reasoning while reducing anxiety (Polydoros et al., 2025).

Similarly, natural-language processing tools such as speech recognizers and chatbots support communication practice for children with language impairments (Jurafsky & Martin, 2023). Virtual-reality simulations assist motor coordination and sensory integration in developmental coordination disorder (Wilson et al., 2020). These systems embody "augmented intelligence", the synergy between human and artificial cognition that amplifies learning potential (Holmes et al., 2022).

Machine-learning models can analyze multimodal data, such as eye-tracking, speech, and physiological signals, to customize therapeutic feedback (Chen et al., 2020). For instance, emotion-recognition algorithms help autistic children interpret facial cues, while reinforcement-learning frameworks adapt difficulty levels to sustain engagement (Sutton & Barto, 2018). In special education, AI tutors capable of conversational interaction can simulate social scenarios, enabling safe rehearsal of communication skills. Importantly, these systems should be embedded within teacher-guided pedagogies to preserve empathy and contextual judgment (Luckin, 2018).

Teachers generally perceive AI tools as beneficial for differentiating instruction but express concerns about data ethics and workload (Zawacki-Richter et al., 2019). Parental attitudes vary: some value technology's motivational appeal, others fear dependency or privacy risks (Goodall, 2016). Training in digital literacy for both educators and caregivers is therefore vital to sustain informed, balanced adoption (European Commission, 2023a).

Technology Dependence in Children with NDDs

Although digital tools can scaffold communication and cognition, excessive screen exposure can provoke attentional dysregulation and compulsive behaviors, vulnerabilities already present in many neurodevelopmental profiles (Torous et al., 2021). Prolonged interactive stimulation activates dopaminergic circuits similarly to gaming or social-media addiction, producing cycles of overstimulation and withdrawal (Topol, 2019). For children with autism, repetitive media consumption can narrow interests and diminish spontaneous social exploration, while those with ADHD may experience intensified distractibility and sleep disruption. Neuroscientific findings suggest that early, unmediated exposure to rapidly changing digital stimuli may hinder the maturation of executive functions, e.g. planning, inhibition, and cognitive flexibility, essential for adaptive learning (Holmes et al., 2019; OECD, 2024).

At the emotional level, continuous connectivity can heighten anxiety and social comparison (Crawford, 2021). For children with NDDs, whose self-regulation systems are still developing, such feedback loops can be particularly harmful. Bostrom (2014) reminds that every technological gain entail opportunity costs. Time spent interacting with devices displaces unstructured play, physical activity, and face-to-face communication. UNESCO (2021) emphasizes guided mediation: adults should co-view, co-play, and interpret digital experiences with children.

Educational institutions increasingly adopt "digital hygiene" curricula that teach self-regulation and mindful technology use (European Commission, 2023b). Cognitive-behavioral interventions integrated with AI monitoring apps can track screen time and prompt breaks (Melcher et al., 2020). Balanced exposure, structured sessions of purposeful use combined with offline activity, helps preserve the benefits of technology without reinforcing dependency (Eubanks, 2018; Floridi, 2019).

Children with NDDs exhibit heterogeneity in sensory thresholds and cognitive tempo (Astle et al., 2022). Adaptive systems must therefore calibrate stimulation intensity, visual complexity, and feedback frequency. Excessive animation or rapid pacing may overwhelm those with hypersensitivity, whereas under-stimulation may disengage others. Machine-learning frameworks can learn these profiles, but only when developers collaborate closely with clinicians and educators (Luckin et al., 2016).

The ideal model is not substitution but symbiosis. Teachers remain interpreters of context, empathy, and ethical nuance; AI provides analytic precision and scalability (Holmes et al., 2022). Polydoros et al. (2025) demonstrated that AI-driven math environments yielded optimal outcomes only when teachers mediated interpretation of feedback. Hence, effective practice requires hybrid intelligence, human insight amplified, not replaced, by computation (Haenlein & Kaplan, 2019).

Inclusive education benefits from Universal Design for Learning (UDL), emphasizing flexible pathways that accommodate diverse learners. AI enhances UDL by dynamically adjusting presentation, engagement, and assessment (OECD, 2024; UNESCO, 2021). Adaptive systems analyze learner analytics to identify struggling students early and deliver targeted scaffolds. In mathematics, computer-assisted modules employing reinforcement learning adjust task difficulty based on performance trends (Polydoros & Antoniou, 2025a; Sutton & Barto, 2018).

For children with communication disorders or ASD, natural-language processing (NLP) interfaces, such as speech recognizers, text-to-speech tools, conversational agents, enable practice in pragmatic language within controlled contexts (Jurafsky & Martin, 2023). Virtual-reality platforms replicate social environments, allowing rehearsal of eye contact, turn-taking, and emotional inference (Wilson et al., 2020). These technologies exemplify the merger of cognitive neuroscience and computational modeling first envisioned by Turing (1950) and extended by modern deep-learning architectures (LeCun et al., 2015).

Teacher readiness determines whether AI becomes an equalizer or a source of confusion. Professional-development programs in digital pedagogy enhance confidence and ethical awareness (Mintz et al., 2023; Zawacki-Richter et al., 2019). The European Commission (2023a) guidelines stress continuous training on data protection, bias recognition, and accessibility. Without such grounding, educators may either over-trust algorithmic recommendations or reject them entirely. Balanced competence ensures critical yet constructive adoption.

Home use of assistive applications reinforces classroom learning when caregivers understand the principles behind AI adaptation (Goodall, 2016). Parental workshops on safe use and digital boundaries mitigate overreliance while maintaining engagement.

ETHICAL, SOCIETAL, AND POLICY CONSIDERATIONS

AI's opacity, the so-called black-box problem, poses challenges for accountability in education (Russell & Norvig, 2021). Explainable-AI methods aim to make algorithmic decisions interpretable (Bussmann et al., 2021). Educational authorities must mandate transparency, particularly when algorithms influence grading or placement of students with disabilities (Bench-Capon & Ashley, 2017). Children's data are uniquely sensitive. The UNESCO (2021) Recommendation on the Ethics of Artificial Intelligence and the OECD & GPAI (2024) Recommendation on AI both underscore informed consent, minimal data collection, and contextual integrity. Systems used in special education should operate on local servers or encrypted clouds, minimizing exposure to third-party analytics (European Commission, 2023b).

Machine-learning models reflect the biases of their training data (Crawford, 2021). Under-representation of neurodiverse users can produce misclassification or exclusion. Developers must include diverse datasets and participatory design processes with neurodivergent individuals and educators (Li et al., 2023). Ethical AI in education thus demands inclusivity not only in application but also in creation. The neurodiversity framework interprets NDDs not as deficits but as natural variations of human cognition (Miranda-Ojeda et al., 2025). AI tools aligned with this paradigm should aim to amplify strengths (e.g. pattern recognition, visual thinking, persistence) rather than normalize difference. Educational policy must ensure that automation enhances agency and dignity rather than enforcing conformity (UNESCO, 2021).

Artificial Intelligence offers transformative possibilities for supporting individuals with neurodevelopmental disorders. When guided by ethical design and pedagogical insight, AI can function as an equalizing force, diagnosing earlier, personalizing learning, and expanding communicative opportunities. Yet, the same systems can deepen inequalities or foster dependency if applied without developmental sensitivity. Future research should pursue three converging paths: first, interdisciplinary co-design, involving collaboration among engineers, educators, psychologists, and neurodiverse communities to ensure relevance and fairness; second, longitudinal evaluation through large-scale studies assessing cognitive and emotional outcomes of sustained AI use in NDD populations; and third, policy alignment, which entails continuous revision of standards to reflect evolving technologies, with global coordination through UNESCO and OECD frameworks.

As Floridi (2019) argues, the success of AI depends not on surpassing human intelligence but on designing systems that co-flourish with it. For children with neurodevelopmental disorders, this means building technologies that are not only intelligent but also compassionate tools that teach machines to understand humans, and humans to understand each other better through machines.

Table 1. Opportunities and ethical considerations of AI-supported approaches in neurodevelopmental disorders

Application Domain	Potential Benefits	Associated Risks	Ethical and Educational Considerations
Assessment and Screening	Early identification, fine-grained profiling, data-driven decision-making	Overreliance on algorithms, false positives/negatives, bias in datasets	Transparency, human oversight, culturally responsive assessment
Personalized Learning	Adaptive instruction, individualized pacing, increased engagement	Data privacy concerns, reduced pedagogical flexibility	Data protection, informed consent, teacher-mediated implementation
Intervention and Skill Training	Targeted cognitive and academic support, increased accessibility	Limited generalization, unequal access to technology	Equity of access, integration with evidence-based practices
Assistive Technologies	Enhanced communication, support for literacy and organization	Dependence on technology, stigmatization	Universal design principles, normalization of assistive tools
Monitoring and Progress Tracking	Continuous feedback, early detection of emerging difficulties	Surveillance concerns, misuse of sensitive data	Ethical data governance, purpose-limited data use

REFERENCES

American Psychiatric Association. (2015). *Neurodevelopmental disorders: DSM-5® selections*. American Psychiatric Publishing.

Astle, D. E., Holmes, J., Kievit, R., & Gathercole, S. E. (2022). Annual research review: The transdiagnostic revolution in neurodevelopmental disorders. *Journal of Child Psychology and Psychiatry, and Allied Disciplines*, *63*(4), 397–417. DOI: 10.1111/jcpp.13481 PMID: 34296774

Bench-Capon, T., & Ashley, K. D. (2017). Argumentation in artificial intelligence and law. *The Knowledge Engineering Review*, *32*, e10.

Bostrom, N. (2014). *Superintelligence: Paths, dangers, strategies*. Oxford University Press.

Brachman, R. J., & Levesque, H. J. (2004). *Knowledge representation and reasoning*. Morgan Kaufmann.

Bubeck, S., Chandrasekaran, V., Eldan, R., . . . Zhang, Y. (2023). Sparks of artificial general intelligence: Early experiments with GPT-4. *arXiv preprint arXiv:2303.12712.*

Buchanan, B. G., & Shortliffe, E. H. (1984). *Rule-based expert systems: The MYCIN experiments of the Stanford heuristic programming project*. Addison-Wesley. https://doi.org//arXiv.2303.12712DOI: 10.48550

Bussmann, N., Giudici, P., Marinelli, D., & Papenbrock, J. (2021). Explainable machine learning in credit risk management: A fintech perspective. *Frontiers in Artificial Intelligence*, *3*, 26. Advance online publication. DOI: 10.3389/frai.2020.00026 PMID: 33733145

Chen, L., Chen, P., & Lin, Z. (2020). Artificial intelligence in education: A review. *IEEE Access : Practical Innovations, Open Solutions*, *8*, 75264–75278. DOI: 10.1109/ACCESS.2020.2988510

Crawford, K. (2021). *Atlas of AI: Power, politics, and the planetary costs of artificial intelligence*. Yale University Press., DOI: 10.1109/ACCESS.2020.2988510

Eubanks, V. (2018). *Automating inequality: How high-tech tools profile, police, and punish the poor*. St. Martin's Press.

European Commission. (2023a). *Artificial intelligence in education: Guidelines for policy and practice*. Publications Office of the EU.

European Commission. (2023b). *Ethical guidelines on the use of artificial intelligence in education.* Publications Office of the EU.

Floridi, L. (2019). *The logic of information: A theory of philosophy as conceptual design.* Oxford University Press. DOI: 10.1093/oso/9780198833635.001.0001

Floridi, L., & Chiriatti, M. (2020). GPT-3: Its nature, scope, limits, and consequences. *Minds and Machines, 30*(4), 681–694. DOI: 10.1007/s11023-020-09548-1

Goodall, J. S. (2016). Technology and school-home communication. *International Journal of Pedagogies and Learning, 11*(2), 118–131. DOI: 10.1080/22040552.2016.1227252

Goodfellow, I., Bengio, Y., & Courville, A. (2016). *Deep learning.* MIT Press.

Haenlein, M., & Kaplan, A. (2019). A Brief History of Artificial Intelligence: On the Past, Present, and Future of Artificial Intelligence. *California Management Review, 61*(4), 5–14. DOI: 10.1177/0008125619864925

Holmes, W., Bialik, M., & Fadel, C. (2022). *Artificial intelligence in education: Promises and implications for teaching and learning* (2nd ed.). Center for Curriculum Redesign.

Jordan, M. I., & Mitchell, T. M. (2015). Machine learning: Trends, perspectives, and prospects. *Science, 349*(6245), 255–260. DOI: 10.1126/science.aaa8415 PMID: 26185243

Jurafsky, D., & Martin, J. H. (2023). *Speech and language processing* (3rd ed.). Stanford University.

LeCun, Y., Bengio, Y., & Hinton, G. (2015). Deep learning. *Nature, 521*(7553), 436–444. DOI: 10.1038/nature14539 PMID: 26017442

Li, X., Chen, Y., & Wang, J. (2023). *Artificial intelligence and inclusive education: Emerging trends and ethical challenges.* ISBN: 978-981-13-8160-7DOI: 10.1007/978-981-13-8161-4

Luckin, R. (2018). *Machine learning and human intelligence: The future of education for the 21st century.* UCL IOE Press.

Luckin, R., Holmes, W., Griffiths, M., & Forcier, L. B. (2016). *Intelligence unleashed: An argument for AI in education.* Pearson Education.

Melcher, J., Hays, R., & Torous, J. (2020). Digital phenotyping for mental health of college students. *Evidence-Based Mental Health, 23*(4), 161–166. DOI: 10.1136/ebmental-2020-300180 PMID: 32998937

Mintz, J., Holmes, W., Liu, L., & Perez-Ortiz, M. (2023). Artificial Intelligence and K-12 education: Possibilities, pedagogies and risks. *Computers in the Schools, 40*(4), 325–333. DOI: 10.1080/07380569.2023.2279870

Miranda-Ojeda, R., Wickramasinghe, A., Ntolkeras, G., Castanho, I., & Yassin, W. (2025). The neurodiversity framework in medicine: On the spectrum. *Developmental Neurobiology, 85*(1), e22960. DOI: 10.1002/dneu.22960 PMID: 39874176

Obermeyer, Z., & Emanuel, E. J. (2016). Predicting the future—Big data, machine learning, and clinical medicine. *The New England Journal of Medicine, 375*(13), 1216–1219. DOI: 10.1056/NEJMp1606181 PMID: 27682033

OECD. (2024). *AI and the futures of education: Systems, skills and strategies.* OECD Publishing.

Polydoros, G., & Antoniou, A. (2025). Optimization of fraction learning for students with learning difficulties in mathematics: Computer-assisted educational environments. *International Journal of Educational Methodology, 11*(2), 127–141. DOI: 10.12973/ijem.11.2.127

Polydoros, G., & Antoniou, A. S. (2025b). Empowering students with learning disabilities: Examining serious digital games' potential. *Behavioral Sciences (Basel, Switzerland), 15*(3), 282–296. DOI: 10.3390/bs15030282 PMID: 40150177

Polydoros, G., Galitskaya, V., Antoniou, A. S., & Drigas, A. (2025). AI technology integration in elementary geometry. *Scientific Electronic Archives, 18*(2), 1–11. DOI: 10.36560/18220252050

Protopapas, A., & Parrila, R. (2018). Is dyslexia a brain disorder? *Brain Sciences, 8*(4), 61. DOI: 10.3390/brainsci8040061 PMID: 29621138

Russell, S. J., & Norvig, P. (2021). *Artificial intelligence: A modern approach* (4th ed.). Pearson., https://library.giadinh.edu.vn/handle/GDU/2783

Silver, D., Huang, A., Maddison, C. J., Guez, A., Sifre, L., van den Driessche, G., Schrittwieser, J., Antonoglou, I., Panneershelvam, V., Lanctot, M., Dieleman, S., Grewe, D., Nham, J., Kalchbrenner, N., Sutskever, I., Lillicrap, T., Leach, M., Kavukcuoglu, K., Graepel, T., & Hassabis, D. (2016). Mastering the game of Go with deep neural networks and tree search. *Nature, 529*(7587), 484–489. DOI: 10.1038/nature16961 PMID: 26819042

Sutton, R. S., & Barto, A. G. (2018). Temporal-difference learning. *Reinforcement learning: an introduction*, 131-132.

Thapar, A., Cooper, M., & Rutter, M. (2017). Neurodevelopmental disorders. *The Lancet. Psychiatry*, *4*(4), 339–346. DOI: 10.1016/S2215-0366(16)30376-5 PMID: 27979720

Topol, E. (2019). *Deep medicine: How artificial intelligence can make healthcare human again*. Basic Books.

Torous, J., Bucci, S., Bell, I. H., Kessing, L. V., Faurholt-Jepsen, M., Whelan, P., Carvalho, A. F., Keshavan, M., Linardon, J., & Firth, J. (2021). The growing field of digital psychiatry. *World Psychiatry; Official Journal of the World Psychiatric Association (WPA)*, *3*(3), 318–335. DOI: 10.1002/wps.20883 PMID: 34505369

UNESCO. (2021). *AI and education: Guidance for policymakers*. UNESCO Publishing.

Wilson, P., Ruddock, S., Rahimi-Golkhandan, S., Piek, J., Sugden, D., Green, D., & Steenbergen, B. (2020). Cognitive and motor function in developmental coordination disorder. *Developmental Medicine and Child Neurology*, *62*(11), 1317–1323. DOI: 10.1111/dmcn.14646 PMID: 32770756

Wolf, T., Debut, L., Sanh, V., Chaumond, J., Delangue, C., Moi, A., & Rush, A. M. (2020, October). Transformers: State-of-the-art natural language processing. In *Proceedings of the 2020 conference on empirical methods in natural language processing: system demonstrations* (pp. 38-45). DOI:DOI: 10.18653/v1/2020.emnlp-demos.6

Zawacki-Richter, O., Marín, V. I., Bond, M., & Gouverneur, F. (2019). Systematic review of research on AI applications in higher education. *International Journal of Educational Technology in Higher Education*, *16*(1), 1–27. DOI: 10.1186/s41239-019-0171-0

Chapter 11
Navigating Neurodiversity in the Transition to Higher Education

ABSTRACT

This chapter considers how students with SLDs and ADHD experience the shift from school to higher education, a period marked by lingering literacy difficulties, shifting support structures, and growing academic expectations. Many arrive with histories of anxiety, uneven accommodations, or late diagnosis, making the university environment both an opportunity and a source of strain. By drawing on research in inclusive practice, teacher training, and educational policy, the chapter highlights the need for flexible assessment, informed staff, and early, empathetic intervention. It argues that genuine inclusion depends on coordinated institutional commitment, allowing neurodiverse learners to develop confidence, belonging, and sustainable academic pathways.

INTRODUCTION

Many contemporary studies focus on Specific Learning Difficulties (SLD) and/or Attention Deficit Hyperactivity Disorder (ADHD) within the school context, especially in the early grades of primary school, reflecting the need for both early detection and diagnosis of neurodevelopmental disorders (NDDs) as well as timely intervention aimed at promoting the well-being and academic development of children. This journey often continues into higher education. For example, having completed secondary education, individuals with specific learning difficulties, such as dyslexia, face new challenges (Gibson, 2012; Stampoltzis & Polychronopoulou, 2009). Research demonstrates that literacy problems often persist into

DOI: 10.4018/979-8-3693-6331-7.ch011

adulthood, confirming the clinical continuum of NDDs and their association with lower educational attainment, higher rates of unskilled employment, and periods of unemployment (Bishop et al., 2020; Le Cunff et al., 2025).

Specific challenges, such as financial limitations and inconsistent support between secondary and tertiary education, along with low self-confidence, can negatively affect access to higher education (Quigley & Gallagher, 2025; Gibson, 2012; O'Byrne et al., 2019). In university settings, dyslexic students often struggle with reading and writing, which may lead to declines in academic performance and negative psychological consequences, including anxiety, poor academic self-concept, and low self-esteem (Jacobs et al., 2022; Kattari & Erickson, 2025; Mortimore & Crozier, 2006; Novita, 2016).

Some European universities, for example, are beginning to recognize that their student populations include dyslexic individuals and have introduced individualized measures such as oral examinations (Stampoltzis & Polychronopoulou, 2009; Stampoltzis et al., 2015). However, despite favorable legal frameworks for supporting dyslexic students, institutionalized uniform policies are lacking, and students must actively request accommodations for them to be implemented.

Special education is a complex area that educational leaders must manage (Pazey & Cole, 2012). Providing equal opportunities for all students, including those with SLDs, should be a core ethical concern of educational leadership (McBeath, 2003, p. 2). Ethical leadership is linked to school improvement, shaping student behavior and responses to adaptive changes proposed by leadership with the aim of including individuals with unique learning profiles (Ainscow, 1999; Ainscow et al., 2006; Ainscow & Sandill, 2010; Corbett, 2001). Achieving this requires interventions targeting the value systems, mindsets, beliefs, and perceptions of all members within educational organizations, where moral leadership plays a key role.

Inclusive education integrates neurotypical and neurodiverse individuals in the same environment, valuing the diversity and contributions of each student. A truly inclusive environment promotes safety, belonging, and shared decision-making in learning. Staff at all levels should receive sufficient training, support, and resources to cater to the needs of all students, regardless of neurodiversity (Le Cunff et al., 2025).

Neurodevelopmental disorders persist throughout life, with symptoms and affected domains varying with age, highlighting the ongoing educational challenge posed by specific learning disorders.

THE ROLE OF EDUCATORS

Kern et al. (2017) suggest that improving the school environment, daily routines, and core activities can enhance the well-being of neurodiverse individuals, particularly those with SLDs. However, Jacobs et al. (2022) note that many educators, counsellors, and support staff lack the necessary knowledge to support students with neurodevelopmental disorders, negatively affecting the overall educational experience. Barriers to learning are often embedded in institutional practices. For example, literacy is considered a central aspect of academic competence, and written assessments such as essays or exams can inadvertently disadvantage students with dyslexia or other SLDs. Alternative assessment methods may therefore be more inclusive and accessible.

Despite challenges in primary and secondary education, many students remain motivated to pursue higher education. Educators should avoid framing NDDs solely as cognitive deficits and should recognize that inflexible learning environments can hinder academic potential (Guzzardo et al., 2021; Jacobs et al., 2022). Additionally, students with NDDs may experience lower positive affect due to learned anxiety from prior academic failures, approaching new situations with fear and anticipation of failure (Espin et al., 2019).

The literature consistently highlights the negative impact of SLDs on both children and adults (Doikou-Avlidou, 2015; Kargiotidis & Manolitsis, 2023; Lievore et al., 2024; Nevill & Forsey, 2022; Seiradakis, 2023; Siegel, 2020; Snowling & Hulme, 2024). Many conceal their difficulties due to fear of negative attitudes, causing stress that can persist into adulthood. Nevertheless, longitudinal studies suggest that individuals can demonstrate resilience and empathy, and with appropriate educator support, learning difficulties need not lead to negative emotional outcomes (Elliott & Grigorenko, 2024; Glazzard & Dale, 2013; Le Cunff et al., 2025; Nevill & Forsey, 2022; Riddick, 2013; Shessel & Reiff, 1999; Wolf et al., 2024).

Special education teachers report higher levels of expectation and self-efficacy in supporting students with learning difficulties (Levi et al., 2013; Michael, 2024; Reyes et al., 2024). Early identification and diagnosis are critical in preventing misconceptions and traumatic experiences (Livingston et al., 2018). Research consistently shows that early diagnosis and intervention improve both academic and emotional outcomes for students with dyslexia (Dahl-Leonard et al., 2023; de Bree et al., 2022; Fletcher et al., 2021; Liu et al., 2024). Despite rehabilitation programs, some individuals show limited academic improvement (Siegel, 2020).

The role of teachers is therefore pivotal in achieving inclusive education. Their training and specialization directly influence interactions with neurodiverse students, justifying the need for professional development in special education. Properly trained educators can support the academic success and overall development of neurodiverse students, helping them navigate both school and higher education contexts (Kattari & Erickson, 2025; Le Cunff et al., 2025; Quigley & Gallagher, 2025).

EDUCATIONAL POLICIES FOR SPECIAL EDUCATION

The European Agency for Development in Special Needs and Inclusive Education (EASNIE) serves as a cooperation platform for its 31 member countries in the field of special needs and inclusive education. Its aim is to improve educational policy and practice for students with disabilities and special educational needs. Recent findings (European Agency of Special Needs Education and Inclusive Education, 2018) indicate that attendance in inclusive education contexts, when accompanied by adequate support, increases the likelihood of pursuing higher education. The link between inclusive education and higher education attendance depends on effective transition planning, which should begin in secondary education and involve the broader community.

Barriers to higher education enrollment remain, including lack of financial resources, insufficient assistance with applications, limited identification of accommodations, restricted curriculum variation, and low-quality transition programs. Education in inclusive contexts can enhance employment prospects for individuals with disabilities, though social factors such as local labor markets, employer attitudes, politics, and accessibility also play critical roles. The nature of the curriculum is influential: access to general curricula leads to qualifications that improve employment opportunities, whereas specialized curricula may restrict them. High-quality transition programs implemented during secondary education have been shown to increase employment prospects, while short-term programs conducted by specialist teachers without general education collaboration may limit access to the open labor market. This underscores the importance of training all teachers, including general education staff, in inclusive and special education strategies (European Agency of Special Needs Education and Inclusive Education, 2018; Le Cunff et al., 2025).

Inclusive education provides students with academic and professional qualifications that increase opportunities for supported employment, open employment, and self-employment. Conversely, education in segregated or special education settings is linked to limited qualifications, employment in sheltered workshops, financial dependency, reduced independent living opportunities, and smaller social networks. Attendance in inclusive schools improves the likelihood of higher education participation and can increase chances of securing paid employment in the open labor market. However, other barriers such as lack of guidance, absence of transition programs, politics, and inaccessible employment frameworks may still impede success. EASNIE emphasizes the importance of quality teacher training in inclusion. Both initial and in-service programs should cover special education and co-educational topics, enabling teachers, psychologists, and non-teaching staff to develop the knowledge, skills, and attitudes necessary for inclusive pedagogy. Professional development opportunities should focus on collaborative curriculum design, teaching, and assessment, supported by adequate resources (European Agency of Special Needs Education and Inclusive Education, 2018; Le Cunff et al., 2025).

Teachers require ongoing education on inclusion and co-education to benefit neurodiverse students across their academic and professional trajectories. Literature reviews indicate that many teachers initially hold neutral or negative attitudes toward inclusive education due to insufficient knowledge, competence, and confidence (Finkelstein et al., 2021; Saloviita, 2020; Soldevila-Pérez et al., 2022). Teachers must reconceptualize their roles, acquire new skills, and develop fresh pedagogical knowledge to adapt to inclusive models (Emmers et al., 2020; Saloviita, 2020).

The UNESCO Open Dossier on Inclusive Education (2018, 2021) outlines critical requirements for teachers adopting inclusive curricula, including engagement in local curriculum development, curriculum adaptation, managing complex classroom activities, supporting diverse learners, and working flexibly within cultural contexts (Bremner et al., 2022). Some researchers note that mainstream teachers are increasingly expected to perform functions similar to those of special educators, despite not receiving comparable intensive training (Emmers et al., 2020; Saloviita, 2020).

Most teachers recognize that co-education promotes both academic and social development, and inclusive models are gaining acceptance as viable options for many students (Kamran et al., 2023; Schwab et al., 2024). School communities and educational programs must equip teachers with skills, knowledge, and practical experience for co-teaching. Ultimately, responsibility for implementing inclusive education, impacting academic development, cognitive progress, higher education enrollment, and labor market integration for neurodiverse students—rests with state policies and education authorities (Kattari & Erickson, 2025; Le Cunff et al., 2025; Quigley & Gallagher, 2025).

NEURODIVERSITY IN SCHOOL YEARS

The concept of "school" is closely associated with learning and represents the foundational level where individuals acquire knowledge. In this paper, it is considered equivalent to any structure that promotes lifelong learning in contemporary society. Globally, there is an increasing demand for schools that are democratic, humane, social, inclusive, digitally adept, and supportive of diverse learners. Inclusive education for children and adolescents in mainstream schools has been a topic of scholarly discussion since the early 1990s. Significant progress has been made with the support of UNESCO and other international bodies, but substantial challenges remain. Gewirtz (2006) defined social justice in education as the provision of equal opportunities, interventions to prevent disadvantages, and promotion of equitable living conditions for all members of society. In alignment with lifelong learning legislation, inclusive education aims to broaden access to education and training, ensuring that all students can acquire higher qualifications and, consequently, enhance employment opportunities.

From a European perspective, the Maastricht Treaty (Title VIII, Articles 117–127) outlines the EU's general educational objectives, including the preparation of young people for seamless integration into the single European labor market. The Lisbon Strategy (2000) emphasized education's role in contributing to global economic competitiveness (Flink & Kaldewey, 2018). Subsequently, the European Agenda 2030 identified inclusive access to education as a key objective (SDG4) (ECDPM, 2016; OECD, 2023). The concept of Global Citizenship Education (GCE) emerged to train students to engage in global society and economy, integrating values such as knowledge, inclusion, and sustainability into curricula (Goren & Yemini, 2017; Yoshida & Van der Walt, 2018).

Equal access to education is a fundamental right, as stated in the United Nations General Assembly Convention on the Rights of the Child (1989; UNO, 2024), with initiatives such as the "No Child Left Behind" framework emphasizing inclusive education. Specific Learning Difficulties (SLDs) and ADHD (APA, 2013) are often displayed in mainstream classrooms, necessitating empathy and support from educators. While school staff may not possess specialized knowledge of neurodevelopmental disorders (NDDs), they can implement strategies informed by social and humanitarian principles, seeking support from specialists when necessary (Le Cunff et al., 2025; Quigley & Gallagher, 2025).

Recent studies indicate that 5–10% of the global population is affected by dyslexia, yet teachers often have only a basic understanding of behavioral manifestations, lacking awareness of neurobiological and cognitive dimensions (Knight, 2018). Consequently, support for non-neurotypical students in primary and secondary education can be superficial, leaving gaps in the preparation for higher education, where academic and social demands evolve. Psychoeducation and awareness of familiar difficulties by both children and parents have positive impacts on behavioral management and overall well-being for children with ADHD (Dahl et al., 2020). Additionally, physical activity has been shown to alleviate ADHD symptoms, improve executive functions, and enhance social and emotional skills, benefiting both students and their families (Gallego-Méndez et al., 2020; Meßler et al., 2018; Zhang et al., 2020).

Both SLDs and ADHD are commonly observed in mainstream classrooms, and the educational community must foster greater empathy toward affected children and families. While staff may lack specialized expertise, they can implement strategies based on moral and social considerations and engage relevant specialists when needed. By promoting inclusive education from the earliest stages, schools can improve academic outcomes, social participation, and well-being for neurodiverse students, laying the foundation for successful progression into higher education (Kattari & Erickson, 2025; Le Cunff et al., 2025).

NEURODIVERSITY IN HIGHER EDUCATION

Neurodiversity is often perceived as a deficit in one's studies, a "personal tragedy" that hinders productivity and timely graduation from higher education (Gibson, 2012; Nieminen & Pesonen, 2022). The growing number of students with neurodiverse learning differences entering higher education presents an increasing challenge for educators and university administrators worldwide. Students with SLDs and ADHD require tailored services and technologies to meet their specific learning needs. Despite the availability of support services, many higher education institutions remain inflexible in addressing neurodiversity, resulting in a continued need for substantial support throughout the students' academic careers. Low levels of staff awareness, ambivalence, and rigid teaching and assessment approaches have been documented (Clouder et al., 2020; Gibson, 2012; Le Cunff et al., 2025; Quigley & Gallagher, 2025).

Research indicates that individuals with dyslexia often face negative academic and psychological outcomes, including anxiety and depression, which may not be obvious to others despite difficulties in reading and writing (Bazen et al., 2023). Educational experiences strongly influence self-concept; students with dyslexia may perceive themselves as less capable than their peers, resulting in low self-esteem and self-efficacy (Ben-Naim et al., 2019; Brunswick & Bargary, 2022; Kattari & Erickson, 2025). Due to the clinical continuum of NDDs, these difficulties may intensify in higher education, where previously established supports, such as individualized education programs and parental involvement, are less frequent. Students must also navigate the rebuilding of social relationships in new peer and academic environments (Nelson & Gregg, 2012).

Longitudinal studies have highlighted the contrast between academic demands in higher education and previous educational experiences. Students with dyslexia often encounter more complex tasks without specialist support, resulting in heightened emotional strain. Cameron (2016) reported that students frequently experienced stigmatization from lecturers and peers, corroborating earlier findings (Denhart, 2008). Consequently, students may experience persistent low self-esteem and negative emotions, which can extend into adulthood (Hansen & Dawson, 2020; Livingstone et al., 2018).

A further challenge is the high rate of undiagnosed dyslexia. In the UK, studies indicate that many students are assessed only in adulthood, with over 80% not diagnosed at school (All-Party Parliamentary Group for Dyslexia and Other Specific

Learning Difficulties, 2019; Brunswick & Bargary, 2022). In Denmark, the lack of standardized diagnostic protocols similarly results in delayed identification (Ottosen et al., 2022). Early and accurate identification is essential not only for academic support but also for fostering self-concept, resilience, and emotional well-being (Bonti et al., 2024; Carter & Sellman, 2013; Pino & Mortari, 2014).

Effective support extends beyond assessment. Students with dyslexia show deficits in error detection and higher-order writing skills even when revising texts, emphasizing the need for differentiated and multimodal support in higher education (Mazur & Chenu, 2023; Sumner & Connelly, 2020; Zerva, 2023). Providing flexible testing, multimodal notes, and allowances for pacing in reading and writing tasks can significantly improve outcomes (Brèthes et al., 2022; Cavalli et al., 2019; Moojen et al., 2020; Petersen, 2015; Reis et al., 2020; Tops et al., 2012).

Studies also emphasize the importance of psychosocial and emotional support. Psychological interventions, self-awareness of the disorder, and self-acceptance can reduce stress associated with writing and other academic activities (Austin & Pisano, 2017; Bonti et al., 2024; Brunswick & Bargary, 2022; Mazur-Palandre et al., 2021; Witko, 2019). University-level support programs that integrate study skills, compensatory strategies, and digital tools have been shown to improve academic performance, self-image, and emotional resilience (Faísca et al., 2023; Johnson et al., 2019; Kelly & Erwin, 2022). Additionally, fostering a sense of community and belonging through student clubs and peer networks positively impacts self-esteem and academic engagement (Ilaria et al., 2022; Podsiadlik, 2023; Vaccaro et al., 2018).

Despite these recommendations, many institutions fail to fully implement inclusive practices, and universal design strategies remain fragmented (Clouder et al., 2020; Le Cunff et al., 2025). Promoting trust, inclusivity, and tolerance without over-reliance on labels or special measures can enable students to thrive academically and emotionally. Individualized interventions, early diagnosis, digital supports, and collaborative networks between student affairs and academic departments are essential to ensuring equitable educational experiences for neurodiverse students (Murray, 2025; Quigley & Gallagher, 2025). Ultimately, universities must prioritize inclusive policies, proactive support, and training for educators to enhance academic outcomes, social integration, and lifelong employability for neurodiverse learners.

REFERENCES

All-Party Parliamentary Group for Dyslexia and Other Specific Learning Difficulties. (2019). *Educational cost of dyslexia, 2019.* All Party Parliamentary Group on Specific Learning Difficulties. British Dyslexia Association.

American Psychiatric Association (APA). (2013). *Diagnostic and statistical manual of mental disorders* (5th ed.). American Psychiatric Association.

Austin, R. D., & Pisano, G. P. (2017). Neurodiversity as a competitive advantage. *Harvard Business Review*, *95*(3), 96–103. Retrieved April 1, 2024, from https://www.hbs.edu/faculty/Pages/item.aspx?num=52624&=undefined

Bazen, L., de Bree, E. H., van den Boer, M., & de Jong, P. F. (2023). Perceived negative consequences of dyslexia: The influence of person and environmental factors. *Annals of Dyslexia*, *73*(2), 214–234. DOI: 10.1007/s11881-022-00274-0 PMID: 36449221

Ben-Naim, S., Laslo-Roth, R., Einav, M., Biran, H., & Margalit, M. (2019). Academic self-efficacy, sense of coherence, hope and tiredness among college students with learning disabilities. In *Postsecondary educational opportunities for students with special education needs* (pp. 18–34). Routledge. DOI: 10.4324/9781351107570-3

Bonti, E., Zerva, I. K., Koundourou, C., & Sofologi, M. (2024). The high rates of comorbidity among neurodevelopmental disorders: Reconsidering the clinical utility of distinct diagnostic categories. *Journal of Personalized Medicine*, *14*(3), 300. DOI: 10.3390/jpm14030300 PMID: 38541042

Brèthes, H., Cavalli, E., Denis-Noël, A., Melmi, J. B., El Ahmadi, A., Bianco, M., & Colé, P. (2022). Text reading fluency and text reading comprehension do not rely on the same abilities in university students with and without dyslexia. *Frontiers in Psychology*, *13*, 866543. DOI: 10.3389/fpsyg.2022.866543 PMID: 35615197

Brunswick, N., & Bargary, S. (2022). Self-concept, creativity, and developmental dyslexia in university students: Effects of age of assessment. *Dyslexia (Chichester, England)*, *28*(3), 293–308. DOI: 10.1002/dys.1722 PMID: 35818173

Cameron, H. E. (2016). Beyond cognitive deficit: The everyday lived experience of dyslexic students at university. *Disability & Society*, *31*(2), 223–239. DOI: 10.1080/09687599.2016.1152951

Carter, C., & Sellman, E. (2013). A view of dyslexia in context: Implications for understanding differences in essay writing experience amongst higher education students identified as dyslexic. *Dyslexia (Chichester, England), 19*(3), 149–164. DOI: 10.1002/dys.1457 PMID: 23780811

Cavalli, E., Colé, P., Brèthes, H., Lefevre, E., Lascombe, S., & Velay, J. L. (2019). E-book reading hinders aspects of long-text comprehension for adults with dyslexia. *Annals of Dyslexia, 69*(2), 243–259. DOI: 10.1007/s11881-019-00182-w PMID: 31313046

Clouder, L., Karakus, M., Cinotti, A., Ferreyra, M. V., Fierros, G. A., & Rojo, P. (2020). Neurodiversity in higher education: A narrative synthesis. *Higher Education, 80*(4), 757–778. DOI: 10.1007/s10734-020-00513-6

Dahl, V., Ramakrishnan, A., Spears, A. P., Jorge, A., Lu, J., Bigio, N. A., & Chacko, A. (2020). Psychoeducation interventions for parents and teachers of children and adolescents with ADHD: A systematic review of the literature. *Journal of Developmental and Physical Disabilities, 32*(2), 257–292. DOI: 10.1007/s10882-019-09691-3

Dahl-Leonard, K., Hall, C., Capin, P., Solari, E. J., Demchak, A., & Therrien, W. J. (2023). Examining fidelity reporting within studies of foundational reading interventions for elementary students with or at risk for dyslexia. *Annals of Dyslexia, 73*(2), 1–26. DOI: 10.1007/s11881-023-00279-3 PMID: 36701045

de Bree, E. H., van den Boer, M., Toering, B. M., & de Jong, P. F. (2022). A stitch in time…: Comparing late-identified, late-emerging and early identified dyslexia. *Dyslexia (Chichester, England), 28*(3), 276–292. DOI: 10.1002/dys.1712 PMID: 35586881

Denhart, H. (2008). Deconstructing barriers: Perceptions of students labeled with learning disabilities in higher education. *Journal of Learning Disabilities, 41*(6), 483–497. DOI: 10.1177/0022219408321151 PMID: 18931016

Doikou-Avlidou, M. (2015). The educational, social and emotional experiences of students with dyslexia: The perspective of postsecondary education students. *International Journal of Special Education, 30*(1), 132–145. https://eric.ed.gov/?id=EJ1094794

ECDPM. (2016). *Implementation of the 2030 Agenda in the European Union.* Retrieved March 6, 2024, from https://ecdpm.org/download_file/429/1387

Elliott, J. G., & Grigorenko, E. L. (2024). *The dyslexia debate revisited.* Cambridge University Press. DOI: 10.1017/9781009083140

Emmers, E., Baeyens, D., & Petry, K. (2020). Attitudes and self-efficacy of teachers towards inclusion in higher education. *European Journal of Special Needs Education, 35*(2), 139–153. DOI: 10.1080/08856257.2019.1628337

Espin, L., García, I., del Pino Sánchez, M., Román, F., & Salvador, A. (2019). Effects of psychosocial stress on the hormonal and affective response in children with dyslexia. *Trends in Neuroscience and Education, 15*, 1–9. DOI: 10.1016/j.tine.2019.03.001 PMID: 31176466

European Agency of Special Needs Education and Inclusive Education. (2018). *EASIE. European Agency Statistics on Inclusive Education. Data Set Cross-country Data Report*. Electronic.

Faísca, L., Reis, A., & Araújo, S. (2023). Cognitive subtyping of university students with dyslexia in a semi-transparent orthography: What can weaknesses and strengths tell us about compensation? *Journal of Cultural Cognitive Science, 7*(2), 1–16. DOI: 10.1007/s41809-023-00124-4

Finkelstein, S., Sharma, U., & Furlonger, B. (2021). The inclusive practices of classroom teachers: A scoping review and thematic analysis. *International Journal of Inclusive Education, 25*(6), 735–762. DOI: 10.1080/13603116.2019.1572232

Fletcher, J. M., Francis, D. J., Foorman, B. R., & Schatschneider, C. (2021). Early detection of dyslexia risk: Development of brief, teacher-administered screens. *Learning Disability Quarterly, 44*(3), 145–157. DOI: 10.1177/0731948720931870 PMID: 34584341

Flink, T., & Kaldewey, D. (2018). The language of science policy in the twenty-first century. *Basic and Applied Research: The Language of Science Policy in the Twentieth Century, 4*, 251–284. DOI: 10.2307/j.ctv8bt0z7.15

Gallego-Méndez, J., Perez-Gomez, J., Calzada-Rodríguez, J. I., Denche-Zamorano, Á. M., Mendoza-Muñoz, M., Carlos-Vivas, J., Garcia-Gordillo, M. Á., & Adsuar, J. C. (2020). Relationship between health-related quality of life and physical activity in children with hyperactivity. *International Journal of Environmental Research and Public Health, 17*(8), 2804. DOI: 10.3390/ijerph17082804 PMID: 32325730

Gewirtz, S. (2006). Towards a contextualized analysis of social justice in education. *Educational Philosophy and Theory, 38*(1), 69–81. DOI: 10.1111/j.1469-5812.2006.00175.x

Gibson, S. (2012). Narrative accounts of university education: Socio-cultural perspectives of students with disabilities. *Disability & Society, 27*(3), 353–369. DOI: 10.1080/09687599.2012.654987

Glazzard, J., & Dale, K. (2013). Trainee teachers with dyslexia: Personal narratives of resilience. *Journal of Research in Special Educational Needs, 13*(1), 26–37. DOI: 10.1111/j.1471-3802.2012.01254.x

Goren, H., & Yemini, M. (2017). Global citizenship education redefined – A systematic review of empirical studies on global citizenship education. *International Journal of Educational Research, 82*, 170–183. DOI: 10.1016/j.ijer.2017.02.004

Guzzardo, M. T., Khosla, N., Adams, A. L., Bussmann, J. D., Engelman, A., Ingraham, N., Gamba, R., Jones-Bey, A., Moore, M. D., Toosi, N. R., & Taylor, S. (2021). "The ones that care make all the difference": Perspectives on student-faculty relationships. *Innovative Higher Education, 46*(1), 41–58. DOI: 10.1007/s10755-020-09522-w PMID: 33012971

Hansen, K. D., & Dawson, D. L. (2020). "We can do better": Community college faculty preparedness for teaching students with learning disabilities. *Journal of Diversity in Higher Education, 13*(4), 309–319. DOI: 10.1037/dhe0000142

Ilaria, B., Marcella, B., Valentina, P., Juri, T., Tony, U., Andrea, Z., & Calabrò, G. (2022). Clustering analysis of factors affecting academic career of university students with dyslexia in Italy. *Scientific Reports, 12*(1), 9010. DOI: 10.1038/s41598-022-12985-w PMID: 35637234

Jacobs, L., Parke, A., Ziegler, F., Headleand, C., & De Angeli, A. (2022). Learning at school through to university: The educational experiences of students with dyslexia at one UK higher education institution. *Disability & Society, 37*(4), 662–683. DOI: 10.1080/09687599.2020.1829553

Johnson, C., Rossiter, H., Cartmell, B., Domingos, M., & Svanaes, S. (2019). *Evaluation of disabled students' allowances.* London: DfE.

Kamran, M., Siddiqui, S., & Adil, M. S. (2023). Breaking barriers: The influence of teachers' attitudes on inclusive education for students with mild learning disabilities (MLDs). *Education Sciences, 13*(6), 606. DOI: 10.3390/educsci13060606

Kargiotidis, A., & Manolitsis, G. (2023). Are children with early literacy difficulties at risk for anxiety disorders in late childhood? *Annals of Dyslexia,* •••, 1–15. DOI: 10.1007/s11881-023-00291-7 PMID: 37935991

Kattari, S. K., & Erickson, D. (2025). "If you're asking for my diversity, you're asking for all of me": Challenging experiences of neurodivergent higher education faculty in the United States. *Journal of Diversity in Higher Education.* Advance online publication. DOI: 10.1037/dhe0000639

Kelly, D. R., & Erwin, V. M. (2022). Specific learning difficulty tutors: Direct supports for navigating disabilities and the university environment. *Disability & Society*, •••, 1–24. DOI: 10.1080/09687599.2022.2114882

Kern, L., Mathur, S. R., Albrecht, S. F., Poland, S., Rozalski, M., & Skiba, R. (2017). Recommendations for school-based mental health services. *Behavioral Disorders*, *43*, 223–226. DOI: 10.1177/0198742917728355

Knight, C. (2018). What is dyslexia? An exploration of the relationship between teachers' understandings of dyslexia and their training experiences. *Dyslexia (Chichester, England)*, *24*(3), 207–219. DOI: 10.1002/dys.1593 PMID: 30019501

Le Cunff, A. L., Ross, F., Westwood, S. J., Koya, S., Caldwell, D. M., Russell, A. E., & Dommett, E. J. (2025). Key research questions to support neurodiversity in higher education: A participatory priority setting exercise. *Education Sciences*, *15*(7), 839. DOI: 10.3390/educsci15070839

Levi, U., Einav, M., Raskind, I., Ziv, O., & Margalit, M. (2013). Helping students with LD to succeed: The role of teachers' hope, sense of coherence and specific self-efficacy. *European Journal of Special Needs Education*, *28*(4), 427–439. DOI: 10.1080/08856257.2013.820457

Lievore, R., Caviola, S., & Mammarella, I. C. (2024). How trait and state mathematics anxiety could affect performance: Evidence from children with and without Specific Learning Disorders. *Learning and Individual Differences*, *112*, 102459. DOI: 10.1016/j.lindif.2024.102459

Livingston, E. M., Siegel, L. S., & Ribary, U. (2018). Developmental dyslexia: Emotional impact and consequences. *Australian Journal of Learning Difficulties*, *23*(2), 107–135. DOI: 10.1080/19404158.2018.1479975

Mazur-Palandre, A., Quignard, M., & Witko, A. (2021). Confronting lexical choice and error distribution in written French: New insights into the linguistic insecurity of students with dyslexia. *Brain Sciences*, *11*(7), 922. DOI: 10.3390/brainsci11070922 PMID: 34356156

Michael, K. (2024). *An Examination of Methods for Increasing Teachers' Treatment Fidelity for Special Education Students* (Order No. 31295138). Available from ProQuest Dissertations & Theses Global. (3099989987). https://www.proquest.com/dissertations-theses/examination-methods-increasing-teachers-treatment/docview/3099989987/se-2

Murray, F. (2025). Neurodiversity in higher education: Reflections from lived experience. Neurodivergent Education and Lifelong Learning, 49. ISBN 9798369380468

Nelson, J. M., & Gregg, N. (2012). Depression and anxiety among transitioning adolescents and college students with ADHD, dyslexia, or comorbid ADHD/dyslexia. *Journal of Attention Disorders, 16*(3), 244–254. DOI: 10.1177/1087054710385783 PMID: 20978271

Nevill, T. P., & Forsey, M. (2022). We are all thrown into one basket": Dyslexia, schools and the (non) enactment of policies of inclusion. *Disability Studies Quarterly, 42*(1). Advance online publication. DOI: 10.18061/dsq.v42i1.7649

Novita, S. (2016). Secondary symptoms of dyslexia: A comparison of self-esteem and anxiety profiles of children with and without dyslexia. *European Journal of Special Needs Education, 31*(2), 279–288. DOI: 10.1080/08856257.2015.1125694

Ottosen, H. F., Bønnerup, K. H., Weed, E., & Parrila, R. (2022). Identifying dyslexia at the university: Assessing phonological coding is not enough. *Annals of Dyslexia, 72*(3), 566–567. DOI: 10.1007/s11881-022-00259-z PMID: 35672600

Petersen, A. (2015). Perspectives of Special Education Teachers on General Education Curriculum Access: Preliminary Results: Preliminary Results. *Research and Practice for Persons with Severe Disabilities : the Journal of TASH, 41*(1), 19–35. DOI: 10.1177/1540796915604835

Pino, M., & Mortari, L. (2014). The inclusion of students with dyslexia in higher education: A systematic review using narrative synthesis. *Dyslexia (Chichester, England), 20*(4), 346–369. DOI: 10.1002/dys.1484 PMID: 25293652

Quigley, E., & Gallagher, T. (2025). Neurodiversity and higher education: Double masking by neurodivergent students. *European Journal of Special Needs Education*, •••, 1–17. DOI: 10.1080/08856257.2025.2511369

Reis, A., Araújo, S., Morais, I. S., & Faísca, L. (2020). Reading and reading-related skills in adults with dyslexia from different orthographic systems: A review and meta-analysis. *Annals of Dyslexia, 70*(3), 339–368. DOI: 10.1007/s11881-020-00205-x PMID: 32918699

Reyes, J. M., Gentry, J. E., Atchley, S., & Phillips, J. (2024). Special education directors and special education teachers' self-efficacy for serving students with English learner and special education needs. *International Journal of Disability Development and Education, 71*(4), 573–589. DOI: 10.1080/1034912X.2022.2146068

Riddick, B. (2013). Researching the social and emotional consequences of dyslexia. In *Addressing difficulties in literacy development* (pp. 282–302). Routledge.

Saloviita, T. (2020). Teacher attitudes towards the inclusion of students with support needs. *Journal of Research in Special Educational Needs*, *20*(1), 64–73. DOI: 10.1111/1471-3802.12466

Schwab, S., Resch, K., & Alnahdi, G. (2024). Inclusion does not solely apply to students with disabilities: Pre-service teachers' attitudes towards inclusive schooling of all students. *International Journal of Inclusive Education*, *28*(2), 214–230. DOI: 10.1080/13603116.2021.1938712

Seiradakis, E. V. (2023). Giving students a voice: Dyslexia and language learning experiences from childhood to adulthood in Greece. *European Journal of Education and Pedagogy*, *4*(1), 43–48. DOI: 10.24018/ejedu.2023.4.1.561

Shessel, I., & Reiff, H. B. (1999). Experiences of adults with learning disabilities: Positive and negative impacts and outcomes. *Learning Disability Quarterly*, *22*(4), 305–316. DOI: 10.2307/1511264

Siegel, L. S. (2020). Early identification and intervention to prevent reading failure: A response to intervention (RTI) initiative. *The Educational and Developmental Psychologist*, *37*(2), 140–146. DOI: 10.1017/edp.2020.21

Snowling, M., & Hulme, C. (2024). Do we really need a new definition of dyslexia? A commentary. *Annals of Dyslexia*, *74*(3), 1–8. DOI: 10.1007/s11881-024-00305-y PMID: 38526759

Soldevila-Pérez, J., Naranjo, M., & Collet, J. (2022). Global inclusive education: Challenges for the future. In *Global inclusive education: Lessons from Spain* (pp. 157–167). Springer International Publishing. DOI: 10.1007/978-3-031-11476-2_11

Sumner, E., & Connelly, V. (2020). Writing and revision strategies of students with and without dyslexia. *Journal of Learning Disabilities*, *53*(3), 189–198. DOI: 10.1177/0022219419899090 PMID: 31941395

Tops, W., Callens, M., Lammertyn, J., Van Hees, V., & Brysbaert, M. (2012). Identifying students with dyslexia in higher education. *Annals of Dyslexia*, *62*(3), 186–203. DOI: 10.1007/s11881-012-0072-6 PMID: 22815104

UNESCO. (2018). *Handbook on measuring equity in education*. UNESCO Institute for Statistics.

UNESCO/The World Bank. (2021). *Cities, culture, creativity: Leveraging culture and creativity for sustainable urban development and inclusive growth.* Retrieved March 6, 2024, from https://documents1.worldbank.org/curated/en/104121621556036559/pdf/Cities-Culture-Creativity-Leveraging-Culture-and-Creativity-for-Sustainable-Urban-Development-and-Inclusive-Growth.pdf

United Nations Organization. (n.d.). *Promotion and protection of the rights of children; Follow-up to the outcome of the special session on children.* Retrieved March 22, 2024, from https://www.un.org/en/ga/62/plenary/children/bkg.shtml

Vaccaro, A., Kimball, E. W., Moore, A., Newman, B. M., & Troiano, P. F. (2018). Narrating the self: A grounded theory model of emerging purpose for college students with disabilities. *Journal of College Student Development*, *59*(1), 37–54. DOI: 10.1353/csd.2018.0003

Wilmot, A., Pizzey, H., Leitao, S., Hasking, P., & Boyes, M. (2023). Growing up with dyslexia: Child and parent perspectives on school struggles, self-esteem, and mental health. *Dyslexia (Chichester, England)*, *29*(1), 40–54. DOI: 10.1002/dys.1729 PMID: 36349866

Witko, A. (2019). La dyslexie à l'âge adulte: Comment révéler son potentiel pour dépasser son trouble. *L'Orthophoniste, 391,* 18–20. https://hal.archives-ouvertes.fr/hal-02264404

Wolf, M., Gotlieb, R. J., Kim, S. A., Pedroza, V., Rhinehart, L. V., Tempini, M. L. G., & Sears, S. (2024). Towards a dynamic, comprehensive conceptualization of dyslexia. *Annals of Dyslexia*, *74*(3), 1–22. DOI: 10.1007/s11881-023-00297-1 PMID: 38217783

Yoshida, K., & Van der Walt, J. L. (2018). The policy-implementation-results linkage for education development and aid effectiveness in the Education 2030 era. *Compare: A Journal of Comparative Education*, *48*(1), 39–55. DOI: 10.1080/03057925.2017.1283583

Zerva, I. K. (2023). Metalinguistic abilities: The contribution to writing. *Social Sciences & Humanities Open*, *8*(1), 100609. DOI: 10.1016/j.ssaho.2023.100609

Zhang, M., Liu, Z., Ma, H., & Smith, D. M. (2020). Chronic physical activity for attention deficit hyperactivity disorder and/or autism spectrum disorder in children: A meta-analysis of randomized controlled trials. *Frontiers in Behavioral Neuroscience*, *14*, 564886. DOI: 10.3389/fnbeh.2020.564886 PMID: 33192364

Chapter 12
The Impact of Neurodevelopmental Disorders on Labor Market Participation and Employment Opportunities

ABSTRACT

This chapter considers how neurodevelopmental disorders continue to influence adults long after schooling, particularly in their access to stable and fulfilling employment. Many carry early experiences of stigma and uneven support into the workplace, shaping confidence and opportunities. Drawing on international policy efforts and emerging corporate models, the chapter illustrates how targeted accommodations, informed educators, and inclusive hiring practices can transform outcomes. It argues that meaningful labor market participation arises when education and employment systems work together to recognize neurodiverse strengths, reduce structural barriers, and create environments where varied ways of thinking can thrive.

NEURODIVERSITY AND EMPLOYMENT

Neurodiversity, encompassing conditions such as autism, ADHD, dyslexia, and other neurodevelopmental disorders, reflects the natural variation in human neurological structure and function (APA, 2013). Despite growing recognition of the unique strengths that neurodiverse individuals can bring to the workplace, many

DOI: 10.4018/979-8-3693-6331-7.ch012

adults with neurodevelopmental disorders continue to face significant challenges in labor market participation. Research indicates that negative or uncomfortable experiences associated with specific learning disorders (SLDs) are linked to reduced work self-efficacy and overall work ability, as well as heightened work-related stress (Nalavany et al., 2018).

Stigma and exclusion are pervasive, often beginning in childhood and continuing into adulthood. Adults who were diagnosed with an NDD in their formative years frequently report discrimination, marginalization, and internalized perceptions of deficit (Deacon et al., 2022; Nalavany et al., 2018). Importantly, this deficit-focused narrative is not only imposed by neurotypical groups but may also be internalized by neurodiverse individuals themselves, potentially leading to self-limiting behaviors and what Deacon et al. (2022) describe as a "self-fulfilling prophecy."

European policy frameworks have increasingly emphasized the importance of inclusive education as a means to address such long-term inequalities. The European Agency for Special Needs and Inclusive Education prioritize interventions for students at risk of marginalization, including those with disabilities or special educational needs (Pappas et al., 2018). The Agency provides evidence-based recommendations across key domains, such as early intervention, inclusive assessment practices, teacher training, professional development, accessible technology, and data collection. By sharing information on the status of inclusive education across member states, the Agency seeks to promote systemic practices that facilitate equitable opportunities for neurodiverse students, laying the groundwork for future labor market inclusion.

In recent years, neurodiversity has increasingly been framed as an asset in the workplace rather than a deficit (Austin & Pisano, 2017). Several European companies have implemented neurodiversity hiring programs and tailored support structures to enable neurodiverse employees to thrive. For example, SAP's Autism at Work program provides mentorship, training, and workplace accommodations to support employees on the autism spectrum. Microsoft has developed a similar Autism Hiring Program, focusing on recruiting neurodiverse individuals for roles in software development and testing, coupled with individualized support. Ernst & Young (EY) operates Neuro-Diverse Centers of Excellence, supporting employees with autism, ADHD, and dyslexia through specialized training and mentorship. Similarly, JP-Morgan Chase has established a Neurodiversity Program targeting technology and other sectors, offering accommodations and guidance to optimize employee success. Collectively, these initiatives demonstrate that when employers proactively address the specific needs of neurodiverse workers, meaningful participation in the labor market becomes achievable.

EDUCATIONAL FOUNDATIONS AND LABOR MARKET OUTCOMES

Educational experiences significantly influence labor market participation for neurodiverse adults. The OECD (2023) reports that although special education in Greece absorbs substantial state and European resources, it has not yet translated into effective labor market integration (Lambrechts et al., 2024; OECD, 2023). Evaluative and self-assessment tools are crucial for improving educational outcomes rather than merely controlling performance, supporting continuous enhancement of special education systems (Lemoine et al., 2024). European policies such as the "School for All" initiative utilize financial, technical, and human resources to ensure students with special educational needs have equal educational opportunities, with the ultimate aim of improving employability and social participation (OECD, 2023).

Despite these efforts, gaps remain. Neurodiverse students often face inadequate academic support during both primary and higher education, which can negatively impact confidence, academic achievement, and subsequent employment prospects. Teacher competence and awareness are pivotal in this context. Professional development programs that enhance educators' understanding of neurodiversity and provide strategies for inclusive pedagogy can improve educational engagement and equip neurodiverse students with skills and confidence to pursue higher education and employment opportunities (Pappas et al., 2018).

OPPORTUNITIES FOR WORKFORCE INCLUSION

Improving labor market participation for neurodiverse individuals requires strategic collaboration among employers, educational institutions, and governmental bodies. Partnerships can foster initiatives that support recruitment, retention, and career development. By providing specialized accommodations, mentorship, and targeted professional development, organizations can ensure neurodiverse employees thrive in diverse roles. These programs presume that equitable access to education and skill acquisition mitigates psychological stress and enhances workforce readiness. Furthermore, initiatives to raise awareness of neurodiversity among employers and colleagues promote inclusive work cultures that value diverse cognitive approaches.

Evidence suggests that a neurodiverse-inclusive approach to employment benefits not only individuals but also organizations. Tailored accommodations and structured support systems facilitate higher productivity, innovation, and problem-solving capacities within the workforce (Austin & Pisano, 2017). Inclusive education and workplace practices are therefore mutually reinforcing well-prepared, confident, and supported neurodiverse graduates enter employment ready to contribute meaningfully, while organizations equipped to accommodate diversity can leverage this untapped potential.

Neurodiversity presents both challenges and opportunities within the European labor market. While stigma, inadequate educational support, and systemic barriers continue to impede full workforce participation, emerging policies and corporate initiatives demonstrate promising strategies for inclusion. Critical interventions include high-quality, inclusive education, teacher training, workplace accommodations, mentorship programs, and cross-sector partnerships. By addressing structural and attitudinal barriers, the European Union can foster labor markets that value neurodiverse talent, enabling adults with NDDs to translate their unique skills and perspectives into meaningful employment. Ultimately, the competence and preparedness of educators, alongside the commitment of employers and policymakers, constitute the foundational mechanisms by which neurodiverse individuals can achieve equitable and sustainable labor market participation.

REFERENCES

American Psychiatric Association. (2013). *Diagnostic and statistical manual of mental disorders* (5th ed.). American Psychiatric Association.

Austin, R. D., & Pisano, G. P. (2017). Neurodiversity as a competitive advantage. *Harvard Business Review*, *95*(3), 96–103. Retrieved April 1, 2024, from https://www.hbs.edu/faculty/Pages/item.aspx?num=52624

Deacon, L., Macdonald, S. J., & Donaghue, J. (2022). "What's wrong with you, are you stupid?" Listening to the biographical narratives of adults with dyslexia in an age of 'inclusive' and 'anti-discriminatory' practice. *Disability & Society*, *37*(3), 406–426. DOI: 10.1080/09687599.2020.1815522

Lambrechts, A. A., Cavallaro, M., & Lepori, B. (2024). The European Universities initiative: Between status hierarchies and inclusion. *Higher Education*, *88*(4), 1–21. DOI: 10.1007/s10734-023-01167-w PMID: 39391862

Lemoine, L., Bernier, T., Peter, L., Noël, Y., & Besançon, M. (2024). Teachers' attitudes toward inclusive education for children with disabilities. *European Journal of Psychology of Education*, *39*(3), 1–34. DOI: 10.1007/s10212-024-00812-x

Nalavany, B. A., Logan, J. M., & Carawan, L. W. (2018). The relationship between emotional experience with dyslexia and work self-efficacy among adults with dyslexia. *Dyslexia (Chichester, England)*, *24*(1), 17–32. DOI: 10.1002/dys.1575 PMID: 29230916

OECD. (2023). *Education at a glance*. OECD. Retrieved April 13, 2024, from https://www.oecd.org/education/education-at-a-glance/

Pappas, M. A., Papoutsi, C., & Drigas, A. S. (2018). Policies, practices, and attitudes toward inclusive education: The case of Greece. *Social Sciences (Basel, Switzerland)*, *7*(6), 90. DOI: 10.3390/socsci7060090

Chapter 13
The Imprint of Neurodevelopmental Disorders in Adulthood

ABSTRACT

This extended chapter, presented in two complementary subchapters, examines the lifelong presence of neurodevelopmental differences and their implications for adult life. The first section traces how early educational experiences leave enduring psychological and social patterns that shape identity, confidence, and modes of coping in adulthood. The second shifts to the urban scale, showing how contemporary cities, often designed around neurotypical expectations, can either support or hinder autonomy for neurodivergent adults. Bringing together perspectives from developmental psychology, inclusive pedagogy, and urban studies, the chapter introduces the concept of the "neurodiver-city," arguing that sensory-aware planning, accessible public information, and participatory governance are essential for building civic environments where diverse cognitive styles can thrive.

THE PSYCHOSOCIAL CONTINUUM OF NEURODEVELOPMENTAL DISORDERS IN ADULTHOOD

This subchapter considers how the experiences of neurodivergent pupils' echo across adulthood, shaping how individuals see themselves and navigate relationships, work, and community life. Early school environments, whether inclusive or exclusionary, leave emotional traces that often outlast the academic challenges themselves. Patterns such as masking, self-doubt, or resilience emerge early and continue to guide adult coping and identity. By bringing together research from education, neuroscience, and mental health, the chapter shows that adult outcomes are

DOI: 10.4018/979-8-3693-6331-7.ch013

not fixed by diagnosis but by the quality of support and recognition offered across the lifespan. It argues for systems that honor neurodiversity and sustain growth long after schooling ends.

Neurodevelopmental disorders (NDDs) are no longer conceptualized as discrete, childhood-bound clinical phenomena, but as lifelong conditions that shape the individual's trajectory from early education to adult identity and psychosocial wellbeing. The school environment, as the primary developmental and social setting of childhood, plays a critical role in mediating this trajectory, forming the foundation for cognitive, emotional, and social patterns that persist well into adulthood (Antolini & Colizzi, 2023; Bosia et al., 2022). This continuity reflects the clinical and psychosocial imprint of early experience, where educational contexts, peer interactions, and institutional expectations intertwine with neurobiological predispositions to construct what might be called the "lifelong signature" of NDDs.

From early childhood, the classroom operates as both a social microcosm and a regulatory space in which neurodivergent pupils encounter their first structured expectations of conformity, communication, and performance. In this context, difficulties in executive functioning, social cognition, or sensory regulation often become visible, marking the child as atypical in systems designed around neurotypical norms (Landgren, 2022). Yet the impact of these early challenges extends far beyond academic performance. They influence self-concept, social belonging, and emotional security, core components of later psychosocial health (Zoccante et al., 2022). When schools fail to scaffold these areas effectively, the experience of exclusion or misunderstanding can leave enduring emotional traces that shape adulthood, manifesting as anxiety, low self-worth, or social withdrawal (Stewart et al., 2025; Zapata & Worrell, 2023).

Educational systems that prioritize standardization often magnify these challenges. The psychosocial cost of fitting into rigid pedagogical structures is particularly high for neurodivergent pupils, who may spend years learning to "mask" their differences to avoid punishment or ostracism. Such early masking behaviors, learned adaptations to maintain social acceptance, can later solidify into internalized patterns of self-concealment and perfectionism in adulthood, frequently linked to depression, burnout, or identity confusion (Quigley & Gallagher, 2025). In this sense, the school is not only a site of cognitive learning but also a crucible in which emotional strategies are forged, some protective, others maladaptive. These learned responses underpin the psychosocial imprint of NDDs, creating continuities between childhood experiences of visibility or invisibility and adult modes of coping with social and occupational demands.

Research increasingly highlights how the psychosocial implications of NDDs begin not at diagnosis, but within the educational environment's interactional dynamics (Augustine et al., 2022). Teachers and peers shape neurodivergent pupils' perceptions of competence and belonging; positive support and adaptive learning environments can foster resilience and motivation, whereas chronic misunderstanding or stigma compounds developmental vulnerability. For instance, Augustine and colleagues (2022) demonstrate that adolescents with self-reported NDDs show divergent trajectories of participation and mental health depending on the inclusivity of their school context. When inclusion is meaningful, embedding flexibility, predictability, and recognition of individual strengths, the risk of later psychosocial distress diminishes. Conversely, environments that emphasize deficits over capabilities reinforce internalized stigma, setting a template for later difficulties in relationships and employment.

This link between school experiences and adult outcomes is also grounded in neurobiological and developmental continuity. Allegra et al. (2021) argue that early environmental stressors may interact with neurodevelopmental vulnerabilities through epigenetic and neuro-immunological pathways, affecting emotional regulation systems that persist into adulthood. These findings lend biological weight to psychosocial observations: the school environment is not merely a social structure but a setting that can modulate neural development. Repeated exposure to stress, rejection, or over-stimulation during schooling can alter the maturation of stress-response systems, influencing how adults with NDDs manage anxiety, sensory input, and social cues later in life (Malatesta et al., 2021). Hence, the psychosocial imprint is both psychological and physiological, a bidirectional process where environment and neurobiology shape each other across developmental time.

The implications of this continuity are particularly visible during the transition from primary to secondary education, a phase often marked by increased academic pressure and social complexity. For many neurodivergent young people, this period represents a critical inflection point where coping resources are tested and identity formation begins to crystallize (da Silva & Eisenstein, 2026). Adolescents with NDDs frequently encounter intensifying expectations of autonomy and executive function, which, without sufficient support, may lead to frustration, disengagement, and emotional exhaustion (Kuwahara et al., 2025). These difficulties, if unaddressed, may evolve into chronic psychosocial stressors that colour adult relationships, occupational choices, and mental health. The sense of competence or failure established during these educational transitions often echoes into adulthood, influencing how individuals approach further education, employment, and social participation.

A central question, therefore, concerns how schools can operate as preventive rather than reactive environments. While medical and therapeutic interventions remain essential, the psychosocial imprint of NDDs is often determined less by the presence of the condition itself than by how educational systems respond to it (Zoccante et al., 2022). Inclusive pedagogical practices, such as multimodal instruction, flexible pacing, and co-designed learning experiences, are now recognized as key determinants of long-term psychosocial outcomes (Le Cunff et al., 2025). When neurodivergent pupils are invited to participate in shaping their learning processes, they develop a sense of agency that translates into stronger self-efficacy and adaptive functioning in adulthood. This participatory model not only improves educational engagement but also fosters identity integration, countering the fragmentation and alienation that many adults with NDDs describe retrospectively. However, despite growing awareness, the gap between policy and practice remains substantial. Many school systems continue to rely on medicalized models of difference, emphasizing remediation over acceptance. This approach inadvertently pathologizes neurodivergent ways of thinking and learning, reinforcing early experiences of inadequacy that carry into adult life (Bonti et al., 2024). The persistence of comorbidity between NDDs and mood or anxiety disorders in adulthood can thus be traced, in part, to these cumulative psychosocial experiences (Bosia et al., 2022; Gur, 2025). The emotional toll of early educational marginalization may heighten vulnerability to later psychopathology, particularly when combined with ongoing social exclusion or employment difficulties.

At the same time, research offers hope by illustrating how neurodivergent strengths, when nurtured in supportive school environments, contribute positively to adult adaptation. Creativity, hyperfocus, and pattern recognition, often regarded as liabilities in rigid educational systems, can become sources of professional excellence and social contribution in adulthood when properly channeled (Pagán & Loveland, 2025). The school years thus represent both risk and opportunity: a developmental crucible where psychosocial resilience can either erode or take root. When inclusion is genuine and affirming, the lifelong imprint of NDDs becomes not one of deficit but of adaptive difference, a foundation for diverse forms of intelligence, empathy, and innovation.

Transition To Adulthood And Early Adult Psychosocial Outcomes

As young people with neurodevelopmental disorders progress through adolescence into early adulthood, the psychosocial imprint of their educational experience becomes increasingly visible. The shift from school to higher education, work, or independent living brings with it new demands for autonomy, executive functioning, and social adaptability areas that frequently intersect with the enduring cognitive and emotional profiles shaped during school years (Thanalingam et al., 2022). For many, this transition represents a second critical developmental juncture where the protective structures of school dissolve and the individual must negotiate identity and belonging within broader social and occupational systems. The patterns of resilience or vulnerability established in earlier educational environments often re-emerge here, influencing mental health, self-efficacy, and social integration (Pagán & Loveland, 2025; Zorzi & Berteotti, 2025).

Universities and vocational settings frequently reproduce, in subtler ways, the same systemic barriers first encountered in school. Students with neurodevelopmental differences often describe feeling required to conceal their needs, navigating institutional cultures that valorize productivity and conformity over diversity of cognition and learning (Quigley & Gallagher, 2025). The phenomenon of "double masking", in which neurodivergent individuals manage both the expectations of the environment and the internal strain of suppressing authentic expression, is particularly pronounced in higher education and early career contexts. This behavioral adaptation, while socially protective in the short term, carries psychological costs over time, contributing to fatigue, anxiety, and identity diffusion (Kattari & Erickson, 2025). The educational system's latent emphasis on normative achievement thus extends its influence into adult life, shaping not only professional trajectories but also self-perception and emotional wellbeing.

At the same time, the transition to adulthood offers opportunities for transformation when appropriate supports are in place. Pagán and Loveland (2025) describe structured group therapy and psychoeducational programs designed to assist young adults with NDDs in navigating this complex stage, focusing on self-advocacy, social communication, and life-skills development. These interventions, when grounded in participatory approaches, enable individuals to reinterpret their neurodivergence as a form of diversity rather than deficiency. This reframing process is vital, as it challenges the internalized stigma that often originates in early educational contexts and provides a bridge toward more coherent adult identity formation. The emphasis shifts from remediation to empowerment, enabling adults with NDDs to reclaim authorship of their developmental narrative.

However, the persistence of systemic gaps in service provision often undermines these efforts. Thanalingam et al. (2022) have highlighted the absence of cohesive transition frameworks between pediatric and adult services, noting that many individuals with NDDs "fall through the cracks" during late adolescence. The discontinuity of care reflects a broader societal difficulty in recognizing neurodevelopmental diversity as a lifelong condition rather than a pediatric diagnosis. Without consistent support, the psychosocial gains achieved during school years may dissipate, leaving adults to reconstruct coping mechanisms independently. This fragmentation can lead to heightened stress and disengagement, particularly when combined with the social isolation frequently reported among adults with neurodevelopmental conditions (Schiltz et al., 2024).

Loneliness, a recurrent theme across developmental stages, often intensifies during the early adult years. Schiltz and colleagues (2024) show that individuals with autism and related conditions report persistent feelings of social disconnection despite increased cognitive maturity. These experiences are not simply products of interpersonal deficits but are shaped by cumulative environmental and cultural factors, systems that fail to accommodate alternative modes of social engagement or communication. The psychosocial imprint of exclusion internalized over years of subtle and overt marginalization, manifests here as emotional fatigue and reduced sense of belonging. Yet, it also reflects a structural issue: adulthood, much like schooling, remains configured for neurotypical norms of participation. The continuity of these barriers illustrates how early educational dynamics resonate across the lifespan, linking classroom experience to workplace integration and community inclusion.

From a clinical perspective, these adult outcomes cannot be disentangled from the developmental processes that precede them. Longitudinal studies increasingly demonstrate that NDDs are not static disorders, but dynamic trajectories influenced by the interaction of genetic, neurobiological, and psychosocial factors over time (Bosia et al., 2022; Indelicato et al., 2022). When early interventions focus narrowly on symptom management without addressing social context, they may fail to prevent later psychosocial distress. For example, Mishra et al. (2024) illustrate how disruptions in early life rhythms, such as circadian desynchrony, can lead to enduring behavioral and emotional alterations into adulthood. Such findings emphasize the necessity of holistic approaches that integrate biological, psychological, and environmental dimensions, aligning with emerging transdiagnostic perspectives in the field (Bonti et al., 2024). The adult presentation of an NDD thus reflects not only intrinsic neurodevelopmental variation but also the cumulative psychosocial pressures exerted across schooling and societal interaction.

The significance of this developmental continuity becomes especially apparent when considering mental health outcomes. Gur (2025) identifies early-life factors associated with heightened vulnerability to psychosis spectrum disorders, suggesting that certain neurodevelopmental trajectories predispose individuals to psychiatric comorbidities under conditions of chronic stress or social adversity. In this sense, the psychosocial imprint is a mediator between neurobiological predisposition and clinical manifestation. Environments that fail to recognize or support neurodiversity effectively amplify risk through sustained emotional dysregulation, social rejection, and self-stigmatization. These processes can gradually erode resilience, leading to depression, anxiety, or maladaptive coping patterns in adulthood (Stewart et al., 2025). Conversely, environments that validate neurodivergent identity and provide structured autonomy can mitigate these risks, reinforcing positive self-concept and social engagement.

The educational-to-adulthood continuum therefore represents a crucial frontier for intervention. Recent research emphasizes the value of participatory and co-designed educational reforms that empower neurodivergent learners to become active agents in their developmental trajectory (Le Cunff et al., 2025). When neurodivergent voices are included in the design of learning environments, support policies, and transition programs, the resulting systems are more adaptive and equitable. These collaborative processes foster a sense of shared ownership, which extends into adulthood as civic participation and self-advocacy. They also bridge the gap between academic and psychosocial development, recognizing that learning is not solely cognitive but profoundly emotional and relational. In effect, inclusive education serves as a rehearsal for inclusive citizenship.

Technology also emerges as a mediating factor in this transition. As digital tools increasingly permeate education and work, they offer new means of accommodation and expression for neurodivergent individuals. Tan et al. (2022) demonstrate how virtual and augmented reality applications can enhance psychosocial rehabilitation, supporting the acquisition of social and emotional skills through experiential learning. When integrated ethically and creatively, such technologies can extend the supportive qualities of inclusive schooling into adult life, promoting autonomy, confidence, and belonging. However, these interventions must be grounded in respect for cognitive diversity and ethical design principles that avoid pathologizing difference. As Al-Raeei (2025) and Perry et al. (2024) argue in broader discussions on technological ethics, equity in access and representation is essential to ensure that digital innovation supports, rather than constrains, neurodiverse lives.

Ultimately, the psychosocial imprint of NDDs in adulthood is not a fixed outcome but an evolving dialogue between the individual and their environment. Each phase of education and life adds new layers to this dialogue, reconfiguring earlier experiences within emerging contexts of meaning. Adults who once struggled in school may, with support and understanding, reinterpret their neurodivergence as a source of strength, developing unique professional and creative identities. Others may continue to grapple with the residues of exclusion or misunderstanding, their adult psychosocial landscape echoing the emotional climates of their schooling years. Recognizing this continuity is essential for clinicians, educators, and policymakers alike, as it calls for integrated frameworks that bridge childhood education, adolescent transition, and adult inclusion.

Neurodevelopmental Disorders (NDDs) From School In Mature Life

By adulthood, the enduring psychosocial imprint of neurodevelopmental disorders becomes evident in the multiple domains that define participation and quality of life. The school environment, once the primary site of social learning and self-comparison, leaves a trace that shapes the adult's sense of capability, worth, and relational belonging. Experiences of early academic marginalization or affirmation continue to inform how individuals interpret success and failure, how they engage in relationships, and how they respond to structural barriers in work and community life (Landgren, 2022). These imprints operate through subtle internalized narratives, beliefs about competence, social acceptance, and the legitimacy of one's difference, which either enable or constrain adult flourishing.

For many adults with NDDs, residual difficulties in executive functioning, communication, or social reciprocity are compounded by the emotional sequelae of earlier educational exclusion. Feelings of inferiority or inadequacy that originated from comparison with neurotypical peers often persist, influencing occupational engagement and interpersonal confidence (Nalavany et al., 2018). In this sense, adult psychosocial outcomes cannot be viewed solely through a medical lens but must be understood as products of cumulative educational and societal experience. The concept of "lifelong imprinting" proposed by Allegra et al. (2021) in relation to neurobiological development finds its psychosocial analogue here: early environments program emotional and cognitive expectations that endure across decades. Whether that programming fosters resilience or vulnerability depends largely on the quality of school relationships and the inclusivity of learning cultures encountered in childhood and adolescence.

Adults who have experienced inclusive educational contexts, where difference was recognized and accommodated, often demonstrate higher self-efficacy and adaptive coping in later life. These individuals tend to frame their neurodivergence as a distinctive mode of cognition rather than a defect (Austin & Pisano, 2017; Zoccante et al., 2022). This reframing aligns with the neurodiversity paradigm, which conceptualizes neurological variation as an element of human biodiversity, contributing to creative and societal enrichment. Yet such positive identification is not automatic. It requires sustained opportunities for self-understanding, peer support, and narrative reconstruction, opportunities that are too often absent in adult services and workplaces. As Le Cunff et al. (2025) note, higher education can serve as a crucial space for redefining self-perception, but only when institutional structures are participatory and sensitive to neurodivergent voices. Without that, adults may continue to internalize the same ableist logics encountered during their school years, perceiving themselves as perpetually compensating for deficits rather than contributing unique perspectives.

The persistence of these internalized patterns partly explains why psychosocial outcomes among adults with NDDs vary widely despite similar clinical profiles. Emotional adjustment depends less on the severity of symptoms than on the social meaning attributed to them over time. Malatesta et al. (2021) highlight the importance of maternal and environmental influences in shaping early neural asymmetries, but the social environment remains equally formative in adulthood. When workplaces and communities remain rigidly standardized, they replicate the exclusivity of traditional schooling, translating educational barriers into occupational ones. Adults with autism, ADHD, or dyslexia frequently report underemployment and workplace stigma, not due to lack of competence but to inadequate systemic accommodation (Deacon et al., 2022; Zapata & Worrell, 2023). In these contexts, the psychosocial imprint of educational exclusion evolves into a professional one: learned caution, emotional suppression, and chronic stress.

Conversely, when adult environments mirror the principles of inclusive education, emphasizing flexibility, communication, and mutual respect neurodivergent individuals often thrive. Programs such as SAP's *Autism at Work* and similar initiatives in the European Union demonstrate that neurodiversity-oriented employment practices can transform both individual outcomes and organizational cultures (Austin & Pisano, 2017). These initiatives provide mentorship, structured feedback, and sensory-friendly workspaces, extending the logic of differentiated instruction from the classroom to the professional sphere. Adults participating in such programs report not only higher job satisfaction but also a renewed sense of belonging, a psychosocial corrective to earlier experiences of isolation. Zorzi and Berteotti (2025) argue that this kind of structural inclusion supports independent living by enhancing both competence and confidence, reinforcing the principle that psychosocial development is relational rather than individual.

Still, inclusion in adulthood remains unevenly distributed, reflecting disparities in policy implementation, social attitudes, and access to continuing education. In many European contexts, including Greece, the legacy of special education systems that prioritized segregation continues to hinder integration into the labor market (OECD, 2023; Lambrechts et al., 2024). Educational evaluation frameworks, while increasingly data-driven, have yet to align fully with the lived realities of neurodiverse learners. Lemoine et al. (2024) observe that teacher attitudes toward inclusion remain a decisive factor in shaping both immediate learning experiences and long-term self-concept. If educators perceive neurodiversity primarily through a deficit model, their students carry those perceptions into adulthood as internalized stigma. Conversely, teachers who practice reflective, adaptive pedagogy foster not just academic achievement but emotional security, a resource that later protects against psychosocial decline.

The continuity of support from education to adulthood thus becomes a cornerstone of resilience. Gregory & Murray (2026), reflecting on lived experience in higher education, underscores that neurodivergent adults often depend on the early development of self-advocacy and self-awareness skills, capacities nurtured through inclusive schooling and mentorship. These skills form the psychological infrastructure for later independence, allowing individuals to navigate the complexities of employment, relationships, and health systems. Without them, adults may find themselves re-enacting patterns of dependency or avoidance first formed in childhood educational contexts. Effective transition planning, as advocated by

Thanalingam et al. (2022), bridges this gap by connecting pedagogical practices with adult-oriented services and community supports. In this way, the psychosocial imprint of NDDs can be transformed from a residue of exclusion into a foundation for autonomy.

At the population level, the diversity of adult outcomes underscores the need for multidimensional research frameworks that integrate clinical, psychological, and sociological perspectives. Bosia et al. (2022) and Indelicato et al. (2022) both argue for a re-evaluation of diagnostic categories, suggesting that the boundaries between developmental and psychiatric disorders are porous and shaped by environmental feedback. This perspective resonates with the notion that adulthood represents not the end of neurodevelopment but its social continuation. Lifelong neuroplasticity ensures that environments (e.g. educational, occupational, or relational) continue to sculpt cognitive and emotional capacities well beyond childhood. Consequently, interventions targeting adult wellbeing must engage with the same principles of early inclusion: consistency, participation, and empathy.

In recent years, emerging research has explored how technological and psychosocial innovations can enhance adult adaptation. Tan et al. (2022) describe how virtual and augmented reality tools facilitate experiential learning for social and emotional skill-building among adults with NDDs. Such technologies replicate the immersive learning qualities of inclusive classrooms, allowing individuals to rehearse social interactions in controlled environments that reduce anxiety. Similarly, community-based rehabilitation initiatives draw from the pedagogical ethos of differentiated instruction, offering peer mentoring and structured autonomy as pathways toward social participation. These models illustrate that inclusion, when grounded in ethical and creative use of technology, can sustain the developmental gains of childhood across the lifespan.

Adulthood introduces challenges that no technology or policy alone can resolve. The psychosocial imprint of early marginalization interacts with biological ageing, social expectations, and economic pressures, often manifesting as burnout or social withdrawal. Stewart et al. (2025) document the persistence of worry and depressive symptoms among adolescents with NDDs, which, when unaddressed, tend to consolidate into chronic affective patterns in adulthood. The continuity of these mental health concerns reaffirms the necessity of early psychosocial support and lifelong monitoring. Similarly, Kuwahara et al. (2025) demonstrate that behavioral difficulties in adolescence often persist into young adulthood, shaped by environmental responses and self-concept. Without adaptive frameworks for self-regulation and social belonging, the adult landscape can become one of quiet exhaustion, where the cognitive effort required to “pass” as neurotypical depletes emotional resources over time.

Against this backdrop, a more compassionate understanding of adulthood as a continuation, not a resolution, of neurodevelopmental experience becomes essential. The psychosocial imprint of NDDs is not merely the shadow of childhood difficulty but the evolving record of how individuals and systems have related across time. Education, in its broadest sense, extends beyond the classroom into every domain where learning, adaptation, and identity construction occur. When society views neurodiversity as integral to its collective intelligence rather than as deviation from it, adulthood becomes a stage of fulfilment rather than deficit management. In this vision, the imprint left by schooling is not erased but reinterpreted, an enduring memory that, when held within inclusive systems, generates wisdom, empathy, and innovation.

Across the lifespan, neurodevelopmental disorders reveal themselves not as static diagnostic entities but as evolving interactions between biology, environment, and meaning. The psychosocial imprint that emerges from this interaction is neither wholly determined by neural circuitry nor entirely constructed by society; rather, it is co-authored by both. Adulthood represents the cumulative expression of these formative dialogues, the negotiation between early schooling experiences, societal structures, and the individual's ongoing effort to make sense of their own neurodivergence. For many, this lifelong process involves transforming the residues of educational exclusion into sources of insight, resilience, and self-definition.

The school environment, as the first structured social world most individuals encounter, establishes the templates through which later relationships with work, learning, and authority are filtered. Where schools emphasize competition over collaboration or treat difference as deficiency, they unwittingly reproduce a hierarchy of minds that continues to govern adult inclusion. The emotional scars of this hierarchy are subtle yet enduring: the avoidance of risk for fear of failure, the hesitancy to disclose difference in professional settings, or the chronic effort to camouflage one's authentic way of thinking and feeling (Quigley & Gallagher, 2025). These behaviors, adaptive in constrained environments, can become maladaptive when internalized as self-concepts. Adults may thus find themselves both desiring and fearing visibility, a paradox born of formative years spent negotiating belonging on unequal terms.

Conversely, inclusive schooling plants the seeds of lifelong psychosocial stability. When educators recognize that learning differences do not signal lesser potential but alternate pathways to understanding, they cultivate resilience and curiosity that transcend academic achievement. Lemoine et al. (2024) highlight that teacher attitudes are decisive not only for students' immediate engagement but for their long-term sense of self-worth. The transition to adulthood, then, is less a rupture than a recontextualization: the same adaptive strategies once used to survive the classroom are repurposed to navigate employment, relationships, and civic life.

Yet, the conditions under which this transformation occurs remain deeply uneven. Structural inequities in educational provision, as documented by Lambrechts et al. (2024) and the OECD (2023), ensure that some young people enter adulthood already burdened by fragmented support systems and constrained expectations.

The persistence of these inequities underscores the importance of reconceptualizing education as a continuum that extends beyond childhood. Learning, in the case of neurodevelopmental difference, is inseparable from self-understanding and social negotiation. Gregory & Murray (2025) also observes that neurodivergent adults often experience education as a cyclical process, a lifelong project of reconciling internal identity with external demands. This insight aligns with participatory research by Le Cunff et al. (2025), who argue that higher education can become a site of empowerment when neurodivergent individuals help shape its research priorities and pedagogical practices. Such co-creation not only enhances institutional inclusivity but also symbolically reverses the historical erasure of neurodivergent voices. Through collaboration, academia itself becomes a model for the inclusive societies it seeks to inform.

The translation of inclusive ideals into adult reality demands more than goodwill; it requires systemic coherence. Thanalingam et al. (2022) emphasize that the transition between pediatric and adult services remains one of the weakest links in the care continuum. Many young adults, having outgrown school-based support, encounter fragmented medical and occupational systems that treat adulthood as a period of closure rather than development. This institutional discontinuity contrasts sharply with neuroscientific evidence of ongoing neuroplasticity and psychosocial growth throughout life (Bosia et al., 2022; Indelicato et al., 2022). Bridging this gap entails designing intersectoral frameworks that follow individuals across educational, health, and community domains, ensuring that identity development and functional support remain synchronized. Technology offers both opportunities and ethical challenges in this respect. Virtual and augmented reality platforms, as described by Tan et al. (2022), provide immersive tools for social learning and psychosocial rehabilitation. Yet their efficacy depends on ethical design that respects cognitive diversity and avoids reductionist assumptions. When implemented thoughtfully, digital environments can extend the ethos of inclusive schooling into adulthood, offering adaptive pathways for emotional regulation, communication, and participation. However, as Al-Raeei (2025) and Perry et al. (2024) remind us, technological inclusion without ethical reflexivity risks replicating the same exclusions it seeks to redress. The challenge, therefore, lies not only in innovation but in cultivating empathy through design, a reminder that technology, like education, reflects the values of its creators.

The psychosocial imprint of neurodevelopmental disorders also has profound implications for mental health practice. Gur (2025) and Stewart et al. (2025) both emphasize that the relationship between neurodevelopmental variation and psychiatric vulnerability is mediated by environmental adversity. Chronic social rejection, lack of autonomy, and internalized stigma serve as stress multipliers, transforming manageable differences into sources of psychological suffering. Effective clinical care, therefore, must move beyond symptom reduction to include narrative reconstruction helping individuals reinterpret past experiences through frameworks of acceptance and growth. Pagán and Loveland (2025) exemplify this approach in their work with young adults transitioning to independence, demonstrating how group-based interventions can transform deficit-oriented narratives into strengths-based identities. Such practices echo educational principles of co-design and self-advocacy, bridging the artificial divide between pedagogy and therapy.

Adult life also reveals the complexity of comorbidity and intersectionality in neurodevelopmental conditions. Bonti et al. (2024) challenge the clinical utility of discrete diagnostic categories, arguing that overlapping traits across ADHD, autism, and specific learning disorders reflect shared developmental mechanisms. This insight reinforces the need for flexible, person-centered interventions that account for both cognitive and psychosocial diversity. Allegra et al. (2021) further demonstrate that early neurobiological imprinting processes have lifelong consequences, suggesting that psychosocial and biological trajectories are intertwined from the outset. The implication for adult care is clear: effective support must integrate the clinical and the contextual, acknowledging that the architecture of the brain and the architecture of experience co-evolve.

From a societal perspective, these insights demand a shift in how success and independence are defined. Zorzi and Berteotti (2025) propose that independent living for adults with NDDs should not be equated with isolation or total self-sufficiency but understood as interdependence the ability to participate meaningfully in community life with appropriate supports. This reconceptualization parallels the ethos of inclusive education: success is relational, not solitary. Similarly, Zapata and Worrell (2023) highlight that adults who find personal meaning in their neurodivergence experience higher psychosocial functioning, particularly when they perceive their work as aligned with their values. Employment, therefore, is not merely economic participation but a continuation of the educational journey, a site where the lessons of inclusion or exclusion are enacted daily.

These perspectives converge on a central principle: the psychosocial imprint of neurodevelopmental disorders is modifiable. It is not destiny but dialogue between individual agency and societal structure, between the legacies of the past and the possibilities of the present. Adulthood offers the potential to rewrite these narratives, but only when social systems provide the scaffolding for reinterpretation. Inclusive education, ethical technology, participatory research, and compassionate clinical practice are all expressions of the same philosophy: that difference enriches, rather than diminishes, the collective human story.

Finally, the challenge for educators, clinicians, and policymakers is to design environments that remember. Systems must retain the memory of their youngest participants, the children once told they were too slow, too distracted, too intense, and ensure that these memories are transmuted into understanding rather than exclusion. The adult who thrives despite a difficult start does so not because they have outgrown their neurodivergence but because they have found spaces that honor it. Such spaces, informed by empathy and evidence alike, represent the true measure of societal maturity. For in recognizing that neurodevelopment is lifelong, we affirm that humanity itself is a continuous work of becoming a collective learning process in which every mind, however it develops, belongs.

NEURODIVER-CITY IN ADULTHOOD: DESIGNING INCLUSIVE AND SUSTAINABLE URBAN FUTURES

This subchapter situates neurodevelopmental diversity within the broader pursuit of sustainable and socially cohesive urban life. It argues that many contemporary cities, despite technological progress, continue to assume neurotypical patterns of attention, perception, and communication, creating barriers for autistic, dyslexic, and otherwise neurodivergent adults. Drawing on work in urban studies, inclusive pedagogy, and emerging AI-based systems, the chapter outlines how sensory-aware design, accessible public information, and participatory planning can transform urban environments into supportive ecosystems. It introduces the idea of the "neurodiver-city," a model that treats cognitive diversity as essential to resilience and equitable civic participation.

Discussions of sustainable urban development must increasingly account for social inclusion, particularly for individuals with neurodevelopmental disorders (NDDs) such as autism and dyslexia. Although significant progress has been made in special education, urban spaces often fail to provide the structures necessary for functional independence and full civic participation in adulthood.

A truly inclusive city guarantees equal opportunities, autonomy, and dignity for all residents, regardless of neurotypicality. When principles from special education intersect with universal design, they have the potential to guide cities toward greater accessibility and equity.

Sustainability in the twenty-first century cannot be defined solely in environmental or economic terms. Social and cognitive inclusion across the lifespan is equally critical. Adults with NDDs, including autism spectrum disorder (ASD), ADHD, dyslexia, developmental coordination disorder (DCD), and intellectual disability, experience urban life in ways that either enhance independence or reinforce exclusion.

While inclusive pedagogy has grown within schools, translating these frameworks to adult urban life remains limited (André et al., 2012; Kenna, 2023).

Urban environments profoundly shape daily independence, social participation, and wellbeing. Concepts like the "smart city" or "sustainable city," often celebrated for technological sophistication, have seldom been examined through a neurodiversity lens (Vanolo, 2023). Notably, Kenna (2022) and Tokarek et al. (2024) suggest that the future requires neurodiver-cities: urban ecosystems that embrace cognitive and sensory diversity as integral to civic resilience.

This chapter synthesizes literature on creating accessible cities for neurodiverse individuals, who face challenges ranging from sensory overload and social navigation to difficulties interpreting public information, signage, and official forms. Drawing on international practices, it proposes a policy framework emphasizing accessible information, trained frontline staff, neurodiversity-friendly urban design (e.g., quiet zones, dyslexia-friendly fonts, symbolic signage), and cross-sector collaboration. Social sustainability, it argues, must integrate ecological, economic, and cognitive considerations.

Urban life is an ongoing negotiation between structure and lived experience. For neurodivergent adults, the built environment is often designed with implicit assumptions about attention, communication, and perception that align with neurotypical norms (Kenna, 2022; Vanolo, 2023). Research indicates that sensory overload, unpredictable social interactions, and complex infrastructures can provoke stress and disorientation for autistic and dyslexic individuals (Alfonso et al., 2023; Chan, 2018).

City sensory landscapes such as crowds, lights, noise create barriers that inhibit participation (Holt-Damant et al., 2013). Alfonso et al. (2023) highlight that overstimulation affects mental and physiological health, shaping behaviors such as avoidance, social withdrawal, and spatial exclusion. Shiue (2016) similarly reports that adults with dyslexia experience lower neighborhood satisfaction, reflecting the combined impact of inaccessible information and environmental stressors.

At the same time, cities concentrate resources, cultural institutions, transport systems, healthcare networks, that, if inclusively designed, promote autonomy (Mirzoev et al., 2022). Sustainable urbanism therefore requires integrating environmental, social, and cognitive dimensions, what Mazalán and Mazalán (2021) describe as preparing cities for neurodiversity.

From Education to Citizenship: Lifelong Inclusion

Special education has historically prioritized childhood and school-based interventions. Yet adulthood, the longest life stage, often exposes the absence of inclusive infrastructures (André et al., 2012). Education must therefore expand beyond classrooms toward the civic scale of the "learning city," where urban spaces themselves function as arenas for continuous skill development. André et al. (2012) argue that social cohesion in education is reciprocal: citizens learn *for* the city, developing civic competence, and *from* it, acquiring skills through lived experience. Applied to neurodivergent populations, this perspective calls for embedding inclusive pedagogy into public communications, transport, and digital systems. For instance, dyslexia-friendly forms, visual signage, and accessible libraries serve as extensions of literacy education (Kennecke et al., 2022; Lamont et al., 2013).

Transitional services bridging childhood education and adult civic life are critical. Lifelong learning centers, community hubs, and public institutions should employ neuro-inclusive design, incorporating clear spatial sequencing, multimodal information, and sensory zoning to reduce anxiety and bolster confidence (Durrett, 2025; Verona & Carpenter, 2023).

Each NDD interacts uniquely with urban affordances. Autistic adults may perceive cities as unpredictable or overwhelming (Vanolo, 2023). DCD can constrain mobility in visually complex, crowded spaces, while dyslexia complicates reading-based navigation like signage and schedules (Lamont et al., 2013). ADHD introduces challenges in interactions requiring sustained focus (Kennecke et al., 2022). Kenna's (2023) research across Europe underscores how neurodivergent adults oscillate between belonging and exclusion depending on spatial context. Predictable transport lines or quiet parks foster autonomy, whereas chaotic streets and poorly marked services increase stress. Accessibility, therefore, must encompass sensory and cognitive dimensions alongside physical mobility (Mazalán & Mazalán, 2021; Verona & Carpenter, 2023).

Smart Technologies and Neuro-Inclusive Cities

AI, IoT, and real-time data offer promising tools for creating adaptive urban environments. Dahmane et al. (2025) and Zhou et al. (2024) identify AI-driven services as underused mechanisms for neurodiversity inclusion. Perry et al. (2024) review AI systems that scaffold adaptive functioning in daily life, such as wearable sensors, voice assistants, and predictive navigation tools. Basak et al. (2025) note that AI systems originally designed for autistic children can translate to adult contexts, enabling real-time crowd mapping, route optimization, and sensory intensity prediction, transforming cities into responsive ecosystems.

However, AI also poses risks. Al-Raeei (2025) cautions that algorithmic models may reinforce exclusion if built on neurotypical assumptions. Özkaynak et al. (2024) stress that neuro-inclusive smart cities must integrate neuroscience, engineering, and policy to anticipate both cognitive and infrastructural needs. Environmental sustainability intersects with neurological wellbeing. King (2025) links air pollution to NDDs, reinforcing the importance of green spaces and clean air in inclusive urban design.

Social systems remain central. Irkey and Wiater (2023) report uneven municipal service access for neurodiverse residents, and Wiater and Firszt (2025) show that even innovative smart-city initiatives can inadvertently exclude neurodivergent citizens if accessibility is not embedded from the start. Kenna (2023) describes "neurodiversity deserts" urban zones lacking inclusive services or sensory-safe spaces. Conversely, Sólheimar Ecovillage demonstrates that predictable, participatory, and ecologically sensitive communities can foster neurodivergent wellbeing (Durrett, 2025; Verona & Carpenter, 2023).

Toward the Neurodiver-City

Inclusive urbanism requires more than architecture, it demands recognition of cognitive diversity as a valid dimension of urban planning (Kenna, 2022; Vanolo, 2023). Frameworks such as Design for Inclusivity (Mostafa et al., 2023) emphasize biophilic and neuro-aware design: lighting modulation, tactile navigation surfaces, and predictable spatial arrangements. Alfonso et al. (2023) propose "soft, deployable environments," adaptable safe spaces responsive to sensory needs. Redundant design, as advocated by Verona and Carpenter (2023), provides multiple perceptual channels, reducing dependence on any single sensory modality. Mazalán and Mazalán (2021) highlight predictability and optionality, structured layouts reduce anxiety, while optional quiet zones allow self-regulation.

Information accessibility is equally critical. Dyslexia-friendly designs, such as sans-serif fonts, clear hierarchy, pictorial supports, and generous spacing, enhance legibility and reduce exclusion (Kennecke et al., 2022; Lamont et al., 2013). AI-driven text-to-speech, translation tools, and multimodal travel information further support participation (Lei & Yueyi, 2022; Zhou et al., 2024).

Ethical AI integration is crucial. Predictive algorithms can empower or marginalize (Al-Raeei, 2025; Dahmane et al., 2025). Gaines and Rodrigues (2024) advocate human-centered AI with transparency, accountability, and participatory design, echoing Floridi's (2019) principle of infospheric stewardship. Neurodivergent co-design committees can guide implementation, ensuring systems reflect lived cognitive diversity (Tokarek et al., 2024).

Social structures, employment, and housing also mediate wellbeing. Inclusive workplaces with sensory modulation enhance productivity and mental health (Hutson & Hutson, 2023). Participatory governance, like neurodiversity-focused budgeting, empowers marginalized residents (Wiater & Firszt, 2025). Interdepartmental collaboration ensures neurodiversity is a core urban criterion (Mirzoev et al., 2022). Environmental stressors exacerbate cognitive barriers. Clean air, noise reduction, and green corridors promote both physical and cognitive resilience (King, 2025;

Telaumbanua et al., 2024). IoT-enabled waste management further reduces sensory pollution, supporting neurodivergent citizens (Palagan et al., 2025).

Educators and therapists serve as crucial mediators between clinical insight and civic application. Schools prepare neurodivergent individuals for urban life while informing municipal planning. Collaborative urban literacy programs teach navigation, digital skills, and self-advocacy, transforming education into a tool for spatial justice (André et al., 2012; Lei & Yueyi, 2022).

The principles of sustainable neuro-inclusion emphasize the creation of urban environments that are both cognitively and socially accessible. Sensory legibility requires that cities communicate through multimodal channels, accommodating diverse perceptual styles and reducing stress for neurodivergent adults (Alfonso et al., 2023; Chan, 2018). Cognitive accessibility ensures that public information, forms, and signage are simplified, symbol-rich, and time-flexible, thereby enabling independent navigation and participation (Kennecke et al., 2022; Lamont et al., 2013). Central to these efforts is participatory co-design, in which neurodivergent individuals actively contribute to policy development and urban planning, ensuring that infrastructures reflect lived cognitive diversity (Kenna, 2022; Wiater & Firszt, 2025). Equally critical are technological ethics, which mandate that AI and data systems deployed in cities respect cognitive differences, maintain equity, and safeguard privacy (Al-Raeei, 2025; Perry et al., 2024). Finally, educational continuity underscores the need for inclusive education to extend beyond schools into civic spaces, bridging learning with real-world urban experiences and fostering lifelong competence (André et al., 2012). When implemented collectively, these principles, aligned with UNESCO (2021) frameworks and the UN Sustainable Development Goals, position neurodiversity as a central component of sustainable cities.

The Neurodiver-City paradigm thus reframes neurodiversity as a civic asset rather than a clinical exception, emphasizing empathy, accessibility, and belonging. In this vision, architecture provides safety, technology ensures adaptability, and pedagogy offers meaning, creating urban systems where all cognitive styles can flourish. Adult neurodevelopmental disorders, therefore, serve not as limitations but as guiding frameworks for building more resilient, humane, and intelligent cities.

Table 1. Transition, inclusion, and participation from higher education to adulthood

Life Domain	Key Challenges	Protective Factors	Support and Policy Implications
Higher Education	Academic demands, executive function difficulties, reduced access to individualized support	Self-advocacy skills, inclusive university services, flexible learning environments	Disability services, reasonable accommodations, mentoring and transition planning
Vocational Training	Mismatch between skills and training demands, limited guidance	Strength-based orientation, experiential learning opportunities	Career counseling, individualized vocational pathways, collaboration with training institutions
Employment	Workplace inflexibility, stigma, difficulties with organization and social communication	Supportive employers, job matching, workplace accommodations	Inclusive employment policies, supported employment models, employer education
Independent Living	Executive functioning demands, financial and organizational challenges	Gradual autonomy, community support, adaptive strategies	Life-skills training, community-based services, long-term follow-up
Social Participation	Social isolation, limited peer networks	Social competence, community inclusion initiatives	Community programs, peer-support networks, inclusive social policies
Mental Health and Well-being	Increased vulnerability to anxiety, depression, and burnout	Resilience, access to mental health services, stable support networks	Integrated mental health care, continuity of support across adulthood

REFERENCES

Al-Raeei, M. (2025). The smart future for sustainable development: Artificial intelligence solutions for sustainable urbanization. *Sustainable Development, 33*(1), 508–517. DOI: 10.1002/sd.3131

Alfonso, S., Tsafoulia, L., & Obeid, I. (2023). Frameworks for Dynamic Environments and Neurodiversity. Soft, a Deployable, Stress-Relief and, Adaptive Safe Space. In Mostafa, M., Baumeister, R., Thomsen, M. R., & Tamke, M. (Eds.), *Design for Inclusivity. UIA 2023. Sustainable Development Goals Series*. Springer. DOI: 10.1007/978-3-031-36302-3_31

Allegra, A., Giarratana, R. M., Scola, L., & Balistreri, C. R. (2021). The close link between the fetal programming imprinting and neurodegeneration in adulthood: The key role of "hemogenic endothelium" programming. *Mechanisms of Ageing and Development, 195*, 111461. DOI: 10.1016/j.mad.2021.111461 PMID: 33600833

André, I., Carmo, A., Abreu, A., Estevens, A., & Malheiros, J. (2012). Learning for and from the city: The role of education in urban social cohesion. *Belgeo. Revue Belge de Geographie*, (4). Advance online publication. DOI: 10.4000/belgeo.8587

Antolini, G., & Colizzi, M. (2023, April). Where do neurodevelopmental disorders go? Casting the eye away from childhood towards adulthood. *Health Care, 11*(7), 1015. DOI: 10.3390/healthcare11071015 PMID: 37046942

Augustine, L., Lygnegård, F., & Granlund, M. (2022). Trajectories of participation, mental health, and mental health problems in adolescents with self-reported neurodevelopmental disorders. *Disability and Rehabilitation, 44*(9), 1595–1608. DOI: 10.1080/09638288.2021.1955304 PMID: 34353177

Austin, R. D., & Pisano, G. P. (2017). Neurodiversity as a competitive advantage. *Harvard Business Review, 95*(3), 96–103. Retrieved April 1, 2024, from https://www.hbs.edu/faculty/Pages/item.aspx?num=52624&=undefined

Basak, D., Kaul, V., Ananthanarayan, S., Kaul, A., & Chacko, N. K. (2025). Building 6 AI-Enabled for Smart Cities Children with Autism Spectrum Disorder. *Handbook of Artificial Intelligence for Smart City Development: Management Systems and Technology Challenges*, 106. https://doi.org/10.1201/9781003225317

Bonti, E., Zerva, I. K., Koundourou, C., & Sofologi, M. (2024). The high rates of comorbidity among neurodevelopmental disorders: Reconsidering the clinical utility of distinct diagnostic categories. *Journal of Personalized Medicine, 14*(3), 300. DOI: 10.3390/jpm14030300 PMID: 38541042

Bosia, M., ... (2022). Adult consequences of neurodevelopmental disorders. In R. Cavallaro & C. Colombo (Eds.), *Fundamentals of psychiatry for health care professionals* (pp. 113–130). Springer. DOI: 10.1007/978-3-031-07715-9_7

Chan, E. R. L. (2018). Neurodivergent themed neighbourhoods as A strategy to enhance the liveability of cities: The blueprint of an autism village, its benefits to neurotypical environments. *Urban Science (Basel, Switzerland)*, *2*(2), 42. DOI: 10.3390/urbansci2020042

da Silva, E. J. C., & Eisenstein, E. (2026). Transition to adolescence in neurodevelopmental disorders: Challenges and clinical perspectives. *Jornal de Pediatria*, *102*, 101463. DOI: 10.1016/j.jped.2025.101463 PMID: 41120046

Dahmane, W. M., Ouchani, S., & Bouarfa, H. (2025). Smart cities services and solutions: A systematic review. *Data and Information Management*, *9*(2), 100087. DOI: 10.1016/j.dim.2024.100087

Durrett, C. (2025). *Neuro-Inclusive Community Design: Lessons from Sólheimar Ecovillage*. Taylor & Francis. DOI: 10.4324/9781003642336

Floridi, L. (2019). *The logic of information: A theory of philosophy as conceptual design*. Oxford University Press. DOI: 10.1093/oso/9780198833635.001.0001

Gaines, K., & Rodrigues, R. (2024). Smart Design and AI for Developmental Disabilities: Characteristics of People with Autism Spectrum Disorders and Other Developmental Disabilities. In *The Routledge Companion to Smart Design Thinking in Architecture & Urbanism for a Sustainable, Living Planet* (pp. 333-344). Routledge. DOI: 10.4324/9781003384113

Gregory, D., & Murray, F. (2026). Neurodiversity in Higher Education: Reflections From Lived Experience. In *Neurodivergent Education and Lifelong Learning* (pp. 49-72). IGI Global Scientific Publishing.

Gur, R. E. (2025). Pathways to psychosis: Factors associated with risk for psychosis spectrum disorders. *The American Journal of Psychiatry*, *182*(11), 984–990. DOI: 10.1176/appi.ajp.20250950 PMID: 41174893

Holt-Damant, K., Guaralda, M., Taylor Gomez, M., & Nicollet, C. (2013). Urban jungles: Making cities healthy places for australians with neurodiversity. In *Proceedings of the 6th Making Cities Liveable Conference in conjunction with Sustainable Transformation Conference* (pp. 116–132). AST Management Pty Ltd.

Hutson, J., & Hutson, P. (2023). *Neuroinclusive workplaces and biophilic design: Strategies for promoting occupational health and sustainability in smart cities*. Glob Health Econ Sustain. DOI: 10.36922/ghes.0549

Indelicato, E., Zech, M., Amprosi, M., & Boesch, S. (2022). Untangling neurodevelopmental disorders in adulthood: A movement disorder is the clue. *Orphanet Journal of Rare Diseases*, *17*(1), 55. DOI: 10.1186/s13023-022-02218-8 PMID: 35172867

Irkey, T., & Wiater, M. (2023). Availability of Municipal Services for Neurodiverse People. *Journal of Psychosocial Rehabilitation and Mental Health*, *10*(4), 473–493. DOI: 10.1007/s40737-023-00336-5

Kenna, T. (2022). Cities of neurodiversity: New directions for an urban geography of neurodiversity. *Area*, *54*(4), 646–654. DOI: 10.1111/area.12803

Kenna, T. (2023). Neurodiversity in the city: Exploring the complex geographies of belonging and exclusion in urban space. *The Geographical Journal*, *189*(2), 370–382. DOI: 10.1111/geoj.12512

Kennecke, A. K., Wessel, D., & Heine, M. (2022). Dyslexia and accessibility guidelines–how to avoid barriers to access in public services. *i-com, 21*(1), 139-155. DOI: 10.1515/icom-2021-0040

King, C. (2025). Commentary: Air pollution and neurodevelopmental disorders: a cause for concern in an urbanising world. *Global Public Health: An International Journal for Research, Policy and Practice*, *20*(1), 2492234. DOI: 10.1080/17441692.2025.2492234 PMID: 40249163

Kuwahara, Y., Yonezawa, M., Miya, H., Moroto, M., & Iehara, T. (2025). Behavioral problems in adolescents and young adults with neurodevelopmental disorders. *Pediatrics International*, *67*(1), e70115. DOI: 10.1111/ped.70115 PMID: 40536041

Lambrechts, A. A., Cavallaro, M., & Lepori, B. (2024). The European Universities initiative: Between status hierarchies and inclusion. *Higher Education*, *88*(4), 1227–1247. Advance online publication. DOI: 10.1007/s10734-023-01167-w PMID: 39391862

Lamont, D., Kenyon, S., & Lyons, G. (2013). Dyslexia and mobility-related social exclusion: The role of travel information provision. *Journal of Transport Geography*, *26*, 147–157. DOI: 10.1016/j.jtrangeo.2012.08.013

Landgren, V. (2022). *Neurodevelopmental disorders (ESSENCE): Early detection and outcome in adulthood.* [Doctoral dissertation, Lund University]. ISBN 978-91-8009-861-8 (Print); 978-91-8009-862-5 (PDF).

Le Cunff, A.-L., Ross, F., Westwood, S. J., Koya, S., Caldwell, D. M., Russell, A. E., & Dommett, E. J. (2025). Key research questions to support neurodiversity in higher education: A participatory priority setting exercise. *Education Sciences*, *15*(7), 839. DOI: 10.3390/educsci15070839

Lei, W., & Yueyi, L. (2022). Shenzhen, China: The innovation and technology for sustainable development in dyslexia services. In *The Routledge international handbook of dyslexia in education* (pp. 47-54). Routledge. DOI: 10.4324/9781003162520

Lemoine, L., Bernier, T., Peter, L., Noël, Y., & Besançon, M. (2024). Teachers' attitudes toward inclusive education for children with disabilities. *European Journal of Psychology of Education, 39*(3), 2867–2900. Advance online publication. DOI: 10.1007/s10212-024-00812-x

Malatesta, G., Marzoli, D., Prete, G., & Tommasi, L. (2021). Human lateralization, maternal effects, and neurodevelopmental disorders. *Frontiers in Behavioral Neuroscience, 15*, 668520. DOI: 10.3389/fnbeh.2021.668520 PMID: 33828467

Mazalán, P., & Mazalán, P. (2021). The city as a place prepared for neurodiversity. *Architektúra & Urbanizmus, 55*(1-2), 18-31. ISSN: 0044-8680

Mirzoev, T., Tull, K. I., Winn, N., Mir, G., King, N. V., Wright, J. M., & Gong, Y. Y. (2022). Systematic review of the role of social inclusion within sustainable urban developments. *International Journal of Sustainable Development and World Ecology, 29*(1), 3–17. DOI: 10.1080/13504509.2021.1918793

Mishra, A., Lin, H., Singla, R., Le, N., Oraebosi, M., Liu, D., & Cao, R. (2024). Circadian desynchrony in early life leads to enduring autistic-like behavioral changes in adulthood. *Communications Biology, 7*(1), 1485. DOI: 10.1038/s42003-024-07131-3 PMID: 39528720

Mostafa, M., Baumeister, R., Thomsen, M. R., & Tamke, M. (Eds.). (2023). *Design for inclusivity*. Springer. DOI: 10.1007/978-3-031-36302-3

Nalavany, B. A., Logan, J. M., & Carawan, L. W. (2018). The relationship between emotional experience with dyslexia and work self-efficacy among adults with dyslexia. *Dyslexia (Chichester, England), 24*(1), 17–32. DOI: 10.1002/dys.1575 PMID: 29230916

Özkaynak, B., Aras, N., Daloğlu Çetinkaya, İ., Ersoy, C., Durmaz İncel, Ö., Koca, M., Nalça, İ., Onay, T. T., Öncü, S., Ülger Vatansever, B., Yücesoy, E., & Yücesoy, C. A. (2024). Neurochallenges in smart cities: State-of-the-art, perspectives, and research directions. *Frontiers in Neuroscience, 18*, 1279668. DOI: 10.3389/fnins.2024.1279668 PMID: 39744330

Pagán, A. F., & Loveland, K. A. (2025). Launching! to adulthood: An updated group therapy program for young adults with neurodevelopmental differences transitioning to adulthood. *Advances in Neurodevelopmental Disorders, 9*(1), 115–129. DOI: 10.1007/s41252-024-00402-4 PMID: 41445725

Palagan, C. A., Joe, S. S. A., Mary, S. J., & Jijo, E. E. (2025). Predictive analysis-based sustainable waste management in smart cities using IoT edge computing and blockchain technology. *Computers in Industry*, *166*, 104234. DOI: 10.1016/j.compind.2024.104234

Perry, N., Sun, C., Munro, M., Boulton, K. A., & Guastella, A. J. (2024). AI technology to support adaptive functioning in neurodevelopmental conditions in everyday environments: A systematic review. *NPJ Digital Medicine*, *7*(1), 370. DOI: 10.1038/s41746-024-01355-7 PMID: 39702672

Perry, N., Sun, C., Munro, M., Boulton, K. A., & Guastella, A. J. (2024). AI technology to support adaptive functioning in neurodevelopmental conditions in everyday environments: A systematic review. *NPJ Digital Medicine*, *7*(1), 370. DOI: 10.1038/s41746-024-01355-7 PMID: 39702672

Quigley, E., & Gallagher, T. (2025). Neurodiversity and higher education: Double masking by neurodivergent students. *European Journal of Special Needs Education*, 1–17. Advance online publication. DOI: 10.1080/08856257.2025.2511369

Schiltz, H., Gohari, D., Park, J., & Lord, C. (2024). A longitudinal study of loneliness in autism and other neurodevelopmental disabilities: Coping with loneliness from childhood through adulthood. *Autism*, *28*(6), 1471–1486. DOI: 10.1177/13623613231217337 PMID: 38153097

Shiue, I. (2016). People with dyslexia and heart, chest, skin, digestive, musculoskeletal, vision, learning, speech and mental disorders were more dissatisfied with neighbourhoods: Scottish Household Survey, 2007–2008. *Environmental Science and Pollution Research International*, *23*(23), 23840–23853. DOI: 10.1007/s11356-016-7585-1 PMID: 27628699

Stewart, S., Houghton, S. J., Glasgow, K., & Macqueen, L. (2025). Worry and depressive symptoms in adolescents with neurodevelopmental disorders. *International Journal of Environmental Research and Public Health*, *22*(2), 185. DOI: 10.3390/ijerph22020185 PMID: 40003411

Tan, B. L., Shi, J., Yang, S., Loh, H., Ng, D., Choo, C., & Medalia, A. (2022). The use of virtual reality and augmented reality in psychosocial rehabilitation for adults with neurodevelopmental disorders: A systematic review. *Frontiers in Psychiatry*, *13*, 1055204. DOI: 10.3389/fpsyt.2022.1055204 PMID: 36590624

Telaumbanua, E., Harsono, I., & Soegiarto, I. (2024). Urbanisation in Indonesia: The relationship between income inequality, urban infrastructure, access to education, and population growth with social cohesion, environmental resilience, and housing quality to look at urbanisation in Indonesia. *International Journal of Business, Law, and Education, 5*(1), 603–614. https://ijble.com/index.php/journal/index. DOI: 10.56442/ijble.v5i1.443

Thanalingam, Y., Langridge, F., Gordon, I., Russell, J., Muir, C., & Hamm, L. M. (2022). Strategies and tools to aid the transition between paediatric and adult health services for young adults with neurodevelopmental disorders: A scoping review protocol. *BMJ Open, 12*(11), e065138. DOI: 10.1136/bmjopen-2022-065138 PMID: 36446454

Tokarek, A., Ufel, W., & Grabkowska, M. NeurodiverCITY: Leveraging the potential for neurodiverse deliberation in urban life (June 11, 2024). Available at http://dx.doi.org/DOI: 10.2139/ssrn.4891374

UNESCO. (2021). *AI and education: Guidance for policymakers.* UNESCO Publishing.

Vanolo, A. (2023). Autistic cities: Critical urbanism and the politics of neurodiversity. *City (London, England), 27*(1,2), 190–208. DOI: 10.1080/13604813.2023.2178273

Verona, I., & Carpenter, J. (2023). Neurodiversity, Sustainability, and Inclusion: A Case for Redundancy in Architecture. In Mostafa, M., Baumeister, R., & Thomsen, M.R. *Design for inclusivity.* Springer.

Wiater, M., & Firszt, D. (2025). Innovations in the public sector in response to the challenges of neurodiversity–findings from Krakow. *Scientific Papers of Silesian University of Technology. Organization & Management/Zeszyty Naukowe Politechniki Slaskiej. Seria Organizacji i Zarzadzanie*, (218). DOI: 10.29119/1641-3466.2025.218.38

Zapata, M. A., & Worrell, F. C. (2023). Disability self-worth and personal meaning relate to psychosocial functioning among employed U.S. adults with LD and ADHD. *Rehabilitation Psychology, 68*(2), 184–195. DOI: 10.1037/rep0000484 PMID: 36951738

Zhou, S., Loiacono, E. T., & Kordzadeh, N. (2024). Smart cities for people with disabilities: A systematic literature review and future research directions. *European Journal of Information Systems, 33*(6), 845–862. DOI: 10.1080/0960085X.2023.2297974

Zoccante, L., Antolini, G., Passarella, L., Puttini, E., Rizzo, V., & Colizzi, M. (2022). Neurodevelopmental disorders and psychosocial issues later in life. In *Prevention in mental health: From risk management to early intervention* (pp. 125–141). Springer International Publishing. DOI: 10.1007/978-3-030-97906-5_7

Zorzi, S., & Berteotti, L. (2025). *Independent living for persons with neurodevelopmental disorders*. IntechOpen. DOI: 10.5772/intechopen.1008380

Conclusions

As this volume draws to a close, the contours of a unifying idea become clear: human development unfolds as a continuous, relational process, and education is both the mirror and the medium of that continuity. The chapters gathered here have traced how neurodevelopmental differences emerge, transform, and interact with experience across the lifespan. From early childhood interventions to the challenges of adulthood, each section has illustrated that neurodevelopmental disorders cannot be confined to clinical categories alone, they are lived experiences shaped by context, learning, and social participation.

The early chapters established the theoretical and clinical foundations of this journey. *Neurodevelopmental Disorders in Educational Contexts* and *Understanding and Exploring the Spectrum of Neurodevelopmental Disorders and Comorbidity* presented these conditions not as fixed impairments but as dynamic variations within human neurobiology. They encouraged educators and clinicians alike to regard differences in cognition, language, or behavior as invitations to adapt environments and methods. *The Clinical Continuum Hypothesis* then wove these insights together, framing development as a seamless process in which early vulnerabilities, educational opportunities, and social experiences continually interact. This perspective supports the work of researchers such as Vargo (2015), Nadeau et al. (2020), and Leifler (2022), who view educational inclusion as a lifelong dialogue between neuroscience and pedagogy.

The subsequent chapters followed this dialogue through the phases of life. *Neurodevelopmental Disorders During Early Childhood* stressed the crucial role of early detection and intervention, aligning with Gutiérrez-Ruiz and Santoya Montes (2025), who emphasize that preschool inclusion sets the foundation for later participation. *The School Experience of Students with Neurodevelopmental Disorders: Symptomatology and Support Mechanisms* highlighted classrooms as both diagnostic and transformative spaces. Teachers, as shown by Van der Merwe et al. (2020) and Berglund Melendez et al. (2020), function as frontline observers and agents of inclusion, using observation, empathy, and adaptive strategies to guide learning.

The chapters addressing bilingual learners and "twice-exceptional" students revealed how neurodevelopmental and cultural diversity intertwine. Misunderstanding these intersections can lead to marginalization, underscoring the need for relational competence (Ewe & Aspelin, 2025) and educator self-efficacy (Bettcher Erickson, 2025). Inclusion, therefore, is not merely a matter of resources but of professional awareness and sensitivity. As the focus shifted to adolescence, *The Relationship Among Neurodevelopmental Disorders and Executive Functions* demonstrated how self-regulation, working memory, and emotion control bridge the gap between neurological mechanisms and everyday functioning. Findings by Ferreira et al. (2025) and Tajik-Parvinchi et al. (2021) illustrate that enhancing these capacities can improve both academic and emotional outcomes.

The following analyses of psychosocial impact, family resilience, and environmental influences broadened the scope beyond individual pathology. Families and schools create the relational ecosystems that enable or hinder development (Beckman et al., 2016; Salminen et al., 2025). *Neurodevelopmental Consequences of Pediatric Cancer* reminded readers that neurocognitive vulnerabilities may arise in medical contexts too, demanding equally nuanced educational and psychosocial support. The discussion on technology and artificial intelligence brought the continuum into the present: digital tools can amplify inclusion when used ethically and collaboratively (Frei-Landau et al., 2023). Yet, they must never replace the human relationships that sustain growth.

In later chapters, the focus turned toward higher education and employment. *Neurodiversity in Higher Education* and *The Impact of Neurodevelopmental Disorders on Labor Market Participation and Employment Opportunities* showed that adulthood is not an endpoint but another developmental phase requiring adaptation and advocacy. Leifler et al. (2024) demonstrated that perceived inclusion in universities correlates with institutional culture, while Edin (2025) and Justus et al. (2025) illustrated how workplace structures still pose barriers to equity. Recognizing neurodiversity at these levels redefines competence and productivity in more humane terms. The final analytical chapter, *The Imprint of Neurodevelopmental Disorders in Adulthood*, closed the empirical circle, showing that long-term outcomes depend less on early severity than on sustained support and opportunities for self-determination (Hadders-Algra, 2022; Van Herwegen et al., 2019). Education, in its broadest sense, thus remains the thread that binds these experiences together, a lifelong act of self-construction and social negotiation.

Taken together, these insights converge on several central principles. First, neurodevelopmental disorders cannot be separated from their educational dimensions: learning is both affected by and a response to neurological difference. Second, inclusion is not a fixed goal but a developmental process reflecting the same continuity that characterizes human growth. Third, resilience arises through relationships between families, teachers, and communities, that affirm belonging. And finally, the recognition of neurodiversity transforms practice from remediation toward empowerment. To study neurodevelopmental disorders, therefore, is to engage in an ethical and creative exploration of what it means to educate and to understand humanity itself.

Looking ahead, the path forward lies in deepening this integration of science, education, and lived experience. Teacher education must evolve to include neuroscience and inclusive pedagogy, cultivating the empathy and reflective competence that inclusive classrooms require (Bettcher Erickson, 2025; Van der Merwe et al., 2020). Interdisciplinary collaboration should become standard practice, ensuring that health and education systems support individuals consistently across life stages (Salminen et al., 2025). Emerging technologies and artificial intelligence can enhance accessibility but must always remain guided by ethics and human connection (Ferreira et al., 2025; Güneş Özler & Yıldız, 2025). Employment policies, too, must extend the spirit of inclusive education into adult life, transforming workplaces into environments of learning and belonging (Edin, 2025; Justus et al., 2025).

More broadly, the future demands a collective embrace of the neurodiversity paradigm, a recognition, as Bölte et al. (2021) and Leifler et al. (2024) propose, that variation in thinking and relating is a natural condition of humanity. Inclusion, in this sense, becomes a civic project through which societies learn to live intelligently with difference. This redefinition of inclusion calls for ethical leadership (Angelides et al., 2012) and intercultural responsibility (Sorkos et al., 2025), shifting education from mere accommodation toward shared transformation.

Ultimately, the measure of an educational system lies not in how efficiently it standardizes, but in how generously it responds to uniqueness. The clinical continuum described throughout this book reminds us that development is ongoing; it changes form but never ceases. In the same way, inclusion is not a static policy but a living dialogue, between individuals and communities, between knowledge and compassion, between what is and what might yet be. It is in this dialogue that education fulfills its highest purpose: to cultivate the conditions under which every mind can learn, contribute, and belong. Human nature is endless…

All works cited in this conclusion are included in the reference lists of the preceding chapters.

Compilation of References

Abutbul-Oz, H., & Armon-Lotem, S. (2022). Parental questionnaires and assessment of bilingual children with developmental language disorder. *Journal of Speech, Language, and Hearing Research: JSLHR*, *65*(4), 1023–1038. DOI: 10.1044/2022_JSLHR-21-00567

Açar, D., & Gül, M. D. (2025). Academic and socio-emotional experiences of a twice-exceptional student. *Behavioral Sciences (Basel, Switzerland)*, *15*(10), 1349. DOI: 10.3390/bs15101349 PMID: 41153139

Adaralegbe, N. J., Okobi, O. E., Omar, Z. T., Segun, E., Evbayekha, E. O., Abolurin, A., & Adedeji, A. G. (2022). Impact of adverse childhood experiences on resilience and school success in individuals with autism spectrum disorder and attention-deficit hyperactivity disorder. *Cureus*, *14*(11). Advance online publication. DOI: 10.7759/cureus.31907 PMID: 36579285

Agnafors, S., Norman Kjellström, A., Torgerson, J., & Rusner, M. (2019). Somatic comorbidity in children and adolescents with psychiatric disorders. *European Child & Adolescent Psychiatry*, *28*(11), 1517–1525. DOI: 10.1007/s00787-019-01313-9 PMID: 30895480

Albayrak, Z. S., Serdengeçti, N., Ateş, B. Ö., Seçinti, D. D., Kadak, M. T., & Doğangün, B. (2025). *Autistic traits and their relationship with internalization-externalization symptoms in ADHD and SLD*. Turkish Journal of Child and Adolescent Mental Health. DOI: 10.4274/tjcamh.galenos.2024.60252

Aldharman, S. S., Al-Jabr, K. H., Alharbi, Y. S., Alnajar, N. K., Alkhanani, J. J., Alghamdi, A., & Abdellatif, R. (2023). Implications of early diagnosis and intervention in the management of neurodevelopmental delay (NDD) in children: A systematic review and meta-analysis. *Cureus*, *15*(5), e38745. DOI: 10.7759/cureus.38745 PMID: 37303321

Alfonso, S., Tsafoulia, L., & Obeid, I. (2023). Frameworks for Dynamic Environments and Neurodiversity. Soft, a Deployable, Stress-Relief and, Adaptive Safe Space. In Mostafa, M., Baumeister, R., Thomsen, M. R., & Tamke, M. (Eds.), *Design for Inclusivity. UIA 2023. Sustainable Development Goals Series*. Springer. DOI: 10.1007/978-3-031-36302-3_31

Ali, S. A., Nisar, H., Ab Aziz, N. A., Fadzil, N. A., Zaber, N. S. M., & Ismail, L. I. (2025). Understanding dyslexia and the potential of artificial intelligence in detecting neurocognitive impairment in dyslexia. In *Artificial Intelligence and Multimodal Signal Processing in Human-Machine Interaction* (pp. 151–170). Academic Press. DOI: 10.1016/B978-0-443-29150-0.00017-2

Allegra, A., Giarratana, R. M., Scola, L., & Balistreri, C. R. (2021). The close link between the fetal programming imprinting and neurodegeneration in adulthood: The key role of "hemogenic endothelium" programming. *Mechanisms of Ageing and Development, 195*, 111461. DOI: 10.1016/j.mad.2021.111461 PMID: 33600833

All-Party Parliamentary Group for Dyslexia and Other Specific Learning Difficulties. (2019). *Educational cost of dyslexia, 2019.* All Party Parliamentary Group on Specific Learning Difficulties. British Dyslexia Association.

Al-Raeei, M. (2025). The smart future for sustainable development: Artificial intelligence solutions for sustainable urbanization. *Sustainable Development, 33*(1), 508–517. DOI: 10.1002/sd.3131

Alzahrani, S., & Algahtani, F. (2025). Implications and Identification of Specific Learning Disability Using Weighted Ensemble Learning Model. *Child: Care, Health and Development, 51*(1), e70026. DOI: 10.1111/cch.70026

Amend, E. R., & Peters, D. B. (2021). The importance of accurate assessment of gifted students: Issues with misdiagnosis, missed diagnoses, and twice-exceptionality. In *Handbook for counselors serving students with gifts and talents* (pp. 713–731). Routledge., DOI: 10.4324/9781003235415-44

American Psychiatric Association. (2013). *Diagnostic and statistical manual of mental disorders* (5th ed.). Author.

American Psychiatric Association. (2015). *Neurodevelopmental disorders: DSM-5® selections*. American Psychiatric Publishing.

Andersen, A. C., Thomsen, P. H., Lydersen, S., Sund, A. M., Haugan, A. L. J., & Nøvik, T. S. (2025). Behaviour ratings of executive functions in adolescents with ADHD: Correlation with core symptoms and functional impairment measured by multiple informants. *Nordic Journal of Psychiatry*, *79*(5), 347–353. DOI: 10.1080/08039488.2025.2506558 PMID: 40401509

André, I., Carmo, A., Abreu, A., Estevens, A., & Malheiros, J. (2012). Learning for and from the city: The role of education in urban social cohesion. *Belgeo. Revue Belge de Geographie*, (4). Advance online publication. DOI: 10.4000/belgeo.8587

Andreou, G., Lymperopoulou, V., & Aslanoglou, V. (2025, September). *Structural language in neurodevelopmental disorders: Comparison between Autism Spectrum Disorder (ASD) and Developmental Language Disorder (DLD).* In *Frontiers in Education* (Vol. 10, p. 1641303). Frontiers. DOI: 10.3389/feduc.2025.1641303

Andreou, E., Didaskalou, E., & Vlachou, A. (2015). Bully/victim problems among Greek pupils with special educational needs: Associations with loneliness and self-efficacy for peer interactions. *Journal of Research in Special Educational Needs*, *15*(4), 235–246. DOI: 10.1111/1471-3802.12028

Andreou, G., & Argatzopoulou, S. (2025). Digital assessment tools for bilingual students with neurodevelopmental disorders. *International Journal of Learning Disabilities*, *58*(2), 145–160. DOI: 10.1002/ijl.441

Andreou, G., & Raxioni, K. (2022). Language development, reading and word learning in Autism Spectrum Disorder (ASD): A review on eye tracking studies. *International Journal of Developmental Disabilities*, •••, 1–18. DOI: 10.1080/20473869.2021.2024404

Angelides, P., Savva, K., & Hajisoteriou, C. (2012). Leading inclusion: Special teachers as leaders in the development of inclusive education. *International Studies in Educational Administration (Commonwealth Council for Educational Administration & Management (CCEAM)), 40*(1). https://search.informit.org/doi/10.3316/aeipt.190896

Antolini, G., & Colizzi, M. (2023, April). *Where do neurodevelopmental disorders go? Casting the eye away from childhood towards adulthood.* In *Healthcare* (Vol. 11, No. 7, p. 1015). MDPI. DOI: 10.3390/healthcare11071015

Aravamuthan, B., Turnacioglu, S. O., & Gropman, A. L. (2024). *Neurodevelopmental Disabilities: Conceptual Framework.* In *Swaiman's Pediatric Neurology* (pp. 636–641). Elsevier. DOI: 10.1016/B978-0-443-10944-7.00068-8

Aravamuthan, B., Turnacioglu, S. O., & Gropman, A. L. (2024). Neurodevelopmental Disabilities: Conceptual Framework. In *Swaiman's Pediatric Neurology* (pp. 636-641). Elsevier. https://doi.org/10.1016/B978-0-443-10944-7.00068-8

Åsberg Johnels, J., Fernell, E., Kjellmer, L., Gillberg, C., & Norrelgen, F. (2022). Language/cognitive predictors of literacy skills in 12-year-old children on the autism spectrum. *Logopedics, Phoniatrics, Vocology*, *47*(3), 166–170. DOI: 10.1080/14015439.2021.1884897

Astle, D. E., Holmes, J., Kievit, R., & Gathercole, S. E. (2022). Annual Research Review: The transdiagnostic revolution in neurodevelopmental disorders. *Journal of Child Psychology and Psychiatry, and Allied Disciplines*, *63*(4), 397–417. DOI: 10.1111/jcpp.13481

Atasever, F., Şahin, N. Ö., & Işık, C. M. (2025). Altered expression of MeCP2 and PTEN genes in the molecular basis of specific learning disorder. *Journal of Molecular Neuroscience*, *75*(2), 74. Advance online publication. DOI: 10.1007/s12031-025-02370-3

Augustine, L., Lygnegård, F., & Granlund, M. (2022). Trajectories of participation, mental health, and mental health problems in adolescents with self-reported neurodevelopmental disorders. *Disability and Rehabilitation*, *44*(9), 1595–1608. DOI: 10.1080/09638288.2021.1955304 PMID: 34353177

Austin, R. D., & Pisano, G. P. (2017). Neurodiversity as a competitive advantage. *Harvard Business Review*, *95*(3), 96–103. Retrieved April 1, 2024, from https://www.hbs.edu/faculty/Pages/item.aspx?num=52624&=undefined

Baksh, R. A., Pape, S. E., Chan, L. F., Aslam, A. A., Gulliford, M. C., & Strydom, A. (2023). Multiple morbidity across the lifespan in people with Down syndrome or intellectual disabilities: A population-based cohort study using electronic health records. *The Lancet. Public Health*, *8*(6), e453–e462. DOI: 10.1016/S2468-2667(23)00057-9

Baldwin, L., Baum, S., Pereles, D., & Hughes, C. (2015). Twice-exceptional learners: The journey toward a shared vision. *Gifted Child Today*, *38*(4), 206–214. DOI: 10.1177/1076217515597277

Bale, T. L., Baram, T. Z., Brown, A. S., Goldstein, J. M., Insel, T. R., McCarthy, M. M., Nemeroff, C. B., Reyes, T. M., Simerly, R. B., Susser, E. S., & Nestler, E. J. (2010). Early life programming and neurodevelopmental disorders. *Biological Psychiatry*, *68*(4), 314–319. DOI: 10.1016/j.biopsych.2010.05.028 PMID: 20674602

Barnes, M. A. (2023). Executive functions: What are they good for? A perspective from intervention research. *Mind, Brain and Education : the Official Journal of the International Mind, Brain, and Education Society*, *17*(4), 238–245. DOI: 10.1111/mbe.12371

Basak, D., Kaul, V., Ananthanarayan, S., Kaul, A., & Chacko, N. K. (2025). Building 6 AI-Enabled for Smart Cities Children with Autism Spectrum Disorder. *Handbook of Artificial Intelligence for Smart City Development: Management Systems and Technology Challenges*, 106. https://doi.org/10.1201/9781003225317

Baughan, N., Pell, J. P., Mackay, D. F., Clark, D., King, A., & Fleming, M. (2023). Educational outcomes in childhood cancer survivors: A Scotland-wide record-linkage study of 766,217 schoolchildren. *PLoS One*, *18*(7), e0286840. DOI: 10.1371/journal.pone.0286840 PMID: 37494295

Bazen, L., de Bree, E. H., van den Boer, M., & de Jong, P. F. (2023). Perceived negative consequences of dyslexia: The influence of person and environmental factors. *Annals of Dyslexia*, *73*(2), 214–234. DOI: 10.1007/s11881-022-00274-0 PMID: 36449221

Beauchamp, J., Choo, T. M., Nayeb, L., & Vender, E. (2022). Linguistically responsive assessment of bilingual students with neurodevelopmental disorders. *Frontiers in Education*, *7*, 908231. DOI: 10.3389/feduc.2022.908231

Beauchamp, M. L., Amorim, K., Wunderlich, S. N., Lai, J., Scorah, J., & Elsabbagh, M. (2022). Barriers to access and utilization of healthcare services for minority-language speakers with neurodevelopmental disorders: A scoping review. *Frontiers in Psychiatry*, *13*, 915999. DOI: 10.3389/fpsyt.2022.915999 PMID: 36090362

Beckman, L., Janson, S., & von Kobyletzki, L. (2016). Associations between neurodevelopmental disorders and factors related to school, health, and social interaction in schoolchildren: Results from a Swedish population-based survey. *Disability and Health Journal*, *9*(4), 663–672. DOI: 10.1016/j.dhjo.2016.05.002

Bednarz, H. M., Maximo, J. O., Murdaugh, D. L., O'Kelley, S., & Kana, R. K. (2017). "Decoding versus comprehension": Brain responses underlying reading comprehension in children with autism. *Brain and Language*, *169*, 39–47. DOI: 10.1016/j.bandl.2017.01.002

Benassi, E., Camia, M., Giovagnoli, S., & Scorza, M. (2022). Impaired school well-being in children with specific learning disorder and its relationship to psychopathological symptoms. *European Journal of Special Needs Education*, *37*(1), 74–88. DOI: 10.1080/08856257.2020.1842975

Bench-Capon, T., & Ashley, K. D. (2017). Argumentation in artificial intelligence and law. *The Knowledge Engineering Review*, *32*, e10.

Benitez, P., & Domeniconi, C. (2023). Equivalence-based instruction to teaching reading by families and teachers' students with autism and/or intellectual disabilities. *Behavioral Interventions*, *38*(3), 861–880. Advance online publication. DOI: 10.1002/bin.1932

Ben-Naim, S., Laslo-Roth, R., Einav, M., Biran, H., & Margalit, M. (2019). Academic self-efficacy, sense of coherence, hope and tiredness among college students with learning disabilities. In *Postsecondary educational opportunities for students with special education needs* (pp. 18–34). Routledge. DOI: 10.4324/9781351107570-3

Berchiatti, M., Ferrer, A., Badenes-Ribera, L., & Longobardi, C. (2022). School Adjustments in Children with Attention Deficit Hyperactivity Disorder (ADHD): Peer Relationships, the Quality of the Student-Teacher Relationship, and Children's Academic and Behavioral Competencies. *Journal of Applied School Psychology*, *38*(3), 241–261. DOI: 10.1080/15377903.2021.1941471

Berglund Melendez, A., Malmsten, M., Einberg, E. L., Clausson, E. K., & Garmy, P. (2020). Supporting students with neurodevelopmental disorders in school health care—School nurses' experiences. *International Journal of Environmental Research and Public Health*, *17*(16), 5752. DOI: 10.3390/ijerph17165752

Berninger, V. W. (2019). *Reading and writing acquisition: A developmental neuropsychological perspective*. Routledge., DOI: 10.4324/9780429303821

Bernstein Ratner, N., & Brundage, S. B. (2024). Advances in understanding stuttering as a disorder of language encoding. *Annual Review of Linguistics*, *10*(1), 127–143. DOI: 10.1146/annurev-linguistics-030521-044754

Berry-Kravis, E., Filipink, R. A., Frye, R. E., Golla, S., Morris, S. M., Andrews, H., Choo, T.-H., & Kaufmann, W. E. (2021). Seizures in fragile X syndrome: Associations and longitudinal analysis of a large clinic-based cohort. *Frontiers in Pediatrics*, *9*, 736255. DOI: 10.3389/fped.2021.736255

Bertollo, A. G., Puntel, C. F., da Silva, B. V., Martins, M., Bagatini, M. D., & Ignácio, Z. M. (2025). Neurobiological relationships between neurodevelopmental disorders and mood disorders. *Brain Sciences*, *15*(3), 307. DOI: 10.3390/brainsci15030307

Bettcher Erickson, N. D. (2025). *The impact of educators' perceptions of neurodivergence on their self-efficacy in teaching middle and high school students with ADHD and autism in rural Minnesota* (Publication No. 1022) [Master's thesis, Minnesota State University Moorhead]. *Dissertations, Theses, and Projects.* https://red.mnstate.edu/thesis/1022

Bhakta, N., Liu, Q., Ness, K. K., Baassiri, M., Eissa, H., Yeo, F., Chemaitilly, W., Ehrhardt, M. J., Bass, J., Bishop, M. W., Shelton, K., Lu, L., Huang, S., Li, Z., Caron, E., Lanctot, J., Howell, C., Folse, T., Joshi, V., & Robison, L. L. (2017). The cumulative burden of surviving childhood cancer: An initial report from the St Jude Lifetime Cohort Study (SJLIFE). *Lancet, 390*(10112), 2569–2582. DOI: 10.1016/S0140-6736(17)31610-0 PMID: 28890157

Billman, E., Smith, S. M., & Jain, S. L. (2023). Perspectives of childhood cancer survivors as young adults: A qualitative study of illness education resources and unmet information needs. *Journal of Cancer Education, 38*(4), 1140–1148. DOI: 10.1007/s13187-022-02240-1 PMID: 36456891

Bishop, D. V. M., Snowling, M. J., Thompson, P. A., Greenhalgh, T., & the CATALISE-2 Consortium. (2017). CATALISE: A multinational and multidisciplinary Delphi consensus study of problems with language development. Phase 2: Terminology. *PeerJ* Preprints, *5*, e2484v2. DOI: 10.7287/peerj.preprints.2484v2

Bishop, D. V. M., Snowling, M. J., Thompson, P. A., & Greenhalgh, T.CATALISE Consortium. (2016). CATALISE: A multinational and multidisciplinary Delphi consensus study, Identifying language impairments in children. *PLoS One, 11*(7), e0158753. DOI: 10.1371/journal.pone.0158753

Blázquez, A., Rodriguez-Revenga, L., Alvarez-Mora, M. I., & Calvo, R. (2025). Clinical and genetic findings in autism spectrum disorders analyzed using exome sequencing. *Frontiers in Psychiatry, 16*, 1515793. DOI: 10.3389/fpsyt.2025.1515793

Bölte, S., Leifler, E., Berggren, S., & Borg, A. (2021). Inclusive practice for students with neurodevelopmental disorders in Sweden. *Scandinavian Journal of Child and Adolescent Psychiatry and Psychology, 9*(9), 15. DOI: 10.21307/sjcapp-2021-002

Bombonato, C., Del Lucchese, B., Ruffini, C., Di Lieto, M. C., Brovedani, P., Sgandurra, G., Cioni, G., & Pecini, C. (2024). Far transfer effects of trainings on executive functions in neurodevelopmental disorders: A systematic review and meta-analysis. *Neuropsychology Review, 34*(1), 98–133. DOI: 10.1007/s11065-022-09574-z PMID: 36633797

Bonifacci, P., Montuschi, M., Lami, L., & Snowling, M. J. (2014). Parents of children with dyslexia: Cognitive, emotional and behavioural profile. *Dyslexia (Chichester, England)*, *20*(2), 175–190. DOI: 10.1002/dys.1469 PMID: 24265066

Bonti, E., Zerva, I. K., Koundourou, C., & Sofologi, M. (2024). The High Rates of Comorbidity among Neurodevelopmental Disorders: Reconsidering the Clinical Utility of Distinct Diagnostic Categories. *Journal of Personalized Medicine*, *14*(3), 300. DOI: 10.3390/jpm14030300

Bonuck, K., Peristeri, E., & Tsimpli, I. M. (2022). Developmental language disorder and literacy acquisition in bilingual children. *Journal of Child Language*, *49*(6), 1137–1156. DOI: 10.1017/S0305000921000651

Bonuck, K., Shafer, V., Battino, R., Valicenti-McDermott, R. M., Sussman, E. S., & McGrath, K. (2022). Language disorders research on bilingualism, school-age, and related difficulties: A scoping review of descriptive studies. *Academic Pediatrics*, *22*(4), 518–525. DOI: 10.1016/j.acap.2021.12.002 PMID: 34896271

Borasio, F., Mascheretti, S., & Lorusso, M. L. (2023). New digital technologies for diagnosis and rehabilitation of neurodevelopmental disorders. *Applied Sciences (Basel, Switzerland)*, *13*(22), 12165. DOI: 10.3390/app132212165

Borasio, G., & Morales, J., & TM, S. (2023). Remote interventions and bilingual neurodevelopmental support post-COVID-19. *International Journal of School & Educational Psychology*, *11*(3), 211–229.

Bosch, R., Pagerols, M., Rivas, C., Sixto, L., Bricollé, L., Español-Martín, G., Prat, R., Ramos-Quiroga, J. A., & Casas, M. (2022). Neurodevelopmental disorders among Spanish school-age children: Prevalence and sociodemographic correlates. *Psychological Medicine*, *52*(14), 3062–3072. DOI: 10.1017/S0033291720005115 PMID: 33436129

Bosia, M., ... (2022). Adult consequences of neurodevelopmental disorders. In R. Cavallaro & C. Colombo (Eds.), *Fundamentals of psychiatry for health care professionals* (pp. 113–130). Springer. DOI: 10.1007/978-3-031-07715-9_7

Bostrom, N. (2014). *Superintelligence: Paths, dangers, strategies*. Oxford University Press.

Bőthe, B., Koós, M., Tóth-Király, I., Orosz, G., & Demetrovics, Z. (2019). Investigating the Associations Of Adult ADHD Symptoms, Hypersexuality, and Problematic Pornography Use Among Men and Women on a Largescale, Non-Clinical Sample. *Journal of Sexual Medicine*, *16*(4), 489–499. DOI: 10.1016/j.jsxm.2019.01.312

Boulton, K. A., Hodge, M. A., Jewell, A., Ong, N., Silove, N., & Guastella, A. J. (2023). Diagnostic delay in children with neurodevelopmental conditions attending a publicly funded developmental assessment service: Findings from the Sydney Child Neurodevelopment Research Registry. *BMJ Open*, *13*(2), e069500. DOI: 10.1136/bmjopen-2022-069500 PMID: 36725093

Bowers, J. S. (2016). The practical and principled problems with educational neuroscience. *Psychological Review*, *123*(5), 600–612. https://psycnet.apa.org/buy/2016-10827-001. DOI: 10.1037/rev0000025 PMID: 26938449

Brachman, R. J., & Levesque, H. J. (2004). *Knowledge representation and reasoning*. Morgan Kaufmann.

Brænden, A., Coldevin, M., Zeiner, P., Stubberud, J., & Melinder, A. (2023). Executive function in children with disruptive mood dysregulation disorder compared to attention-deficit/hyperactivity disorder and oppositional defiant disorder, and in children with different irritability levels. *European Child & Adolescent Psychiatry*, •••, 1–11. DOI: 10.1007/s00787-023-02143-6 PMID: 36680626

Breaux, R. P., & Harvey, E. A. (2019). A longitudinal study of the relation between family functioning and preschool ADHD symptoms. *Journal of Clinical Child and Adolescent Psychology*, *48*(5), 749–764. DOI: 10.1080/15374416.2018.1437737 PMID: 29578799

Brèthes, H., Cavalli, E., Denis-Noël, A., Melmi, J. B., El Ahmadi, A., Bianco, M., & Colé, P. (2022). Text reading fluency and text reading comprehension do not rely on the same abilities in university students with and without dyslexia. *Frontiers in Psychology*, *13*, 866543. DOI: 10.3389/fpsyg.2022.866543 PMID: 35615197

Brikell, I., Burton, C., Mota, N. R., & Martin, J. (2021). Insights into attention-deficit/hyperactivity disorder from recent genetic studies. *Psychological Medicine*, *51*(13), 2274–2286. DOI: 10.1017/S0033291721000982

Bruinsma, G. I., Wijnen, F., & Gerrits, E. (2024). Communication in daily life of children with developmental language disorder: Parents' and teachers' perspectives. *Language, Speech, and Hearing Services in Schools*, *55*(1), 105–129. DOI: 10.1044/2023_LSHSS-23-00051

Brunswick, N., & Bargary, S. (2022). Self-concept, creativity, and developmental dyslexia in university students: Effects of age of assessment. *Dyslexia (Chichester, England)*, *28*(3), 293–308. DOI: 10.1002/dys.1722 PMID: 35818173

Bubeck, S., Chandrasekaran, V., Eldan, R., . . . Zhang, Y. (2023). Sparks of artificial general intelligence: Early experiments with GPT-4. *arXiv preprint arXiv:2303.12712*.

Buchanan, B. G., & Shortliffe, E. H. (1984). *Rule-based expert systems: The MYCIN experiments of the Stanford heuristic programming project.* Addison-Wesley. https://doi.org//arXiv.2303.12712DOI: 10.48550

Bukva, A., & Memisevic, H. (2025). The relationship between executive functions and mathematics achievements in early-grade elementary students. *Child Neuropsychology, 31*(4), 635–648. DOI: 10.1080/09297049.2024.2421222 PMID: 39466137

Bullen, J. C., Zajic, M. C., McIntyre, N., Solari, E., & Mundy, P. (2022). Patterns of math and reading achievement in children and adolescents with autism spectrum disorder. *Research in Autism Spectrum Disorders, 92*, 101933. DOI: 10.1016/j.rasd.2022.101933

Burnley, A., St Clair, M., Dack, C., Thompson, H., & Wren, Y. (2024). Exploring the psychosocial experiences of individuals with developmental language disorder during childhood: A qualitative investigation. *Journal of Autism and Developmental Disorders, 54*(8), 3008–3027. DOI: 10.1007/s10803-023-05946-3

Bussing, R., Zima, B. T., Gary, F. A., Mason, D. M., Leon, C. E., Sinha, K., & Garvan, C. W. (2003). Social networks, caregiver strain, and utilization of mental health services among elementary school students at high risk for ADHD. *Journal of the American Academy of Child and Adolescent Psychiatry, 42*(7), 842–850. DOI: 10.1097/01.CHI.0000046876.27264.BF PMID: 12819444

Bussmann, N., Giudici, P., Marinelli, D., & Papenbrock, J. (2021). Explainable machine learning in credit risk management: A fintech perspective. *Frontiers in Artificial Intelligence, 3*, 26. Advance online publication. DOI: 10.3389/frai.2020.00026 PMID: 33733145

Busso, D. S., & Pollack, C. (2015). No brain left behind: Consequences of neuroscience discourse for education. *Learning, Media and Technology, 40*(2), 168–186. DOI: 10.1080/17439884.2014.908908

Butterfuss, R., & Kendeou, P. (2018). The role of executive functions in reading comprehension. *Educational Psychology Review, 30*(3), 801–826. DOI: 10.1007/s10648-017-9422-6

Cameron, H. E. (2016). Beyond cognitive deficit: The everyday lived experience of dyslexic students at university. *Disability & Society, 31*(2), 223–239. DOI: 10.1080/09687599.2016.1152951

Canu, D., Ioannou, C., Müller, K., Martin, B., Fleischhaker, C., Biscaldi, M., Beauducel, A., Smyrnis, N., van Elst, L. T., & Klein, C. (2022a). Evidence towards a continuum of impairment across neurodevelopmental disorders from basic ocular-motor tasks. *Scientific Reports, 12*(1), 16521. DOI: 10.1038/s41598-022-19661-z PMID: 36192503

Canu, D., Ioannou, C., Müller, K., Martin, B., Fleischhaker, C., Biscaldi, M., Beauducel, A., Smyrnis, N., van Elst, L. T., & Klein, C. (2022b). Visual search in neurodevelopmental disorders: Evidence towards a continuum of impairment. *European Child & Adolescent Psychiatry, 31*(8), 1–18. DOI: 10.1007/s00787-021-01756-z PMID: 33751240

Capobianco, M., Puzzo, C., Di Matteo, C., Costa, A., & Adriani, W. (2025). Current virtual reality-based rehabilitation interventions in neuro-developmental disorders at developmental ages. *Frontiers in Behavioral Neuroscience, 18*, 1441615. DOI: 10.3389/fnbeh.2024.1441615 PMID: 39882439

Capodieci, A., Ruffini, C., Frascari, A., Rivella, C., Bombonato, C., Giaccherini, S., Scali, V., Luccherino, L., Viterbori, P., Traverso, L., Usai, M. C., Marzocchi, G. M., & Pecini, C. (2023). Executive functions in children with specific learning disorders: Shedding light on a complex profile through teleassessment. *Research in Developmental Disabilities, 142*, 104621. DOI: 10.1016/j.ridd.2023.104621

Carlsson, T., Molander, F., Taylor, M. J., Jonsson, U., & Bölte, S. (2021). Early environmental risk factors for neurodevelopmental disorders: A systematic review of twin and sibling studies. *Development and Psychopathology, 33*(4), 1448–1495. DOI: 10.1017/S0954579420000620 PMID: 32703331

Carrick, A., & Hamilton, C. J. (2023). Heated behaviour in the classroom for children with FASD: The relationship between characteristics associated with ADHD, ODD and ASD, hot executive function and classroom-based reward systems. *Children (Basel, Switzerland), 10*(4), 685. DOI: 10.3390/children10040685 PMID: 37189934

Carter, C., & Sellman, E. (2013). A view of dyslexia in context: Implications for understanding differences in essay writing experience amongst higher education students identified as dyslexic. *Dyslexia (Chichester, England), 19*(3), 149–164. DOI: 10.1002/dys.1457 PMID: 23780811

Cavalli, E., Colé, P., Brèthes, H., Lefevre, E., Lascombe, S., & Velay, J. L. (2019). E-book reading hinders aspects of long-text comprehension for adults with dyslexia. *Annals of Dyslexia, 69*(2), 243–259. DOI: 10.1007/s11881-019-00182-w PMID: 31313046

Cha, J. H., Ryu, S., Park, M., Lim, B. C., Kim, Y. J., & Moon, J. H. (2025). Developmental screening of neurodevelopmental disorders before age 6: A nationwide health screening program. *Pediatric Research, 97*(5), 1566–1574. DOI: 10.1038/s41390-024-03516-6 PMID: 39261660

Chalmpe, M., & Vlachos, F. (2025). Are there distinct subtypes of developmental dyslexia? *Frontiers in Behavioral Neuroscience, 18*, 1512892. DOI: 10.3389/fnbeh.2024.1512892

Chan, E. R. L. (2018). Neurodivergent themed neighbourhoods as A strategy to enhance the liveability of cities: The blueprint of an autism village, its benefits to neurotypical environments. *Urban Science (Basel, Switzerland), 2*(2), 42. DOI: 10.3390/urbansci2020042

Cheek, C. L., Garcia, J. L., Mehta, P. D., Francis, D. J., & Grigorenko, E. L. (2023). The exceptionality of twice-exceptionality: Examining combined prevalence of giftedness and disability using multivariate statistical simulation. *Exceptional Children, 90*(1), 43–56. DOI: 10.1177/00144029221150929

Chen, H., Gu, X. H., Zhou, Y., Ge, Z., Wang, B., Siok, W. T., Wang, G., Huen, M., Jiang, Y., Tan, L.-H., & Sun, Y. (2017). A genome-wide association study identifies genetic variants associated with mathematics ability. *Scientific Reports, 7*(1), 40365. DOI: 10.1038/srep40365

Chen, L., Chen, P., & Lin, Z. (2020). Artificial intelligence in education: A review. *IEEE Access : Practical Innovations, Open Solutions, 8*, 75264–75278. DOI: 10.1109/ACCESS.2020.2988510

Chen, Y., & Jahromi, L. B. (2025). Self-Regulation and Academic Learning in Preschoolers with Autism Spectrum Disorder: Links to School Engagement and Levels of Autism Characteristics. *Journal of Autism and Developmental Disorders, 55*(4), 1286–1301. DOI: 10.1007/s10803-024-06288-4

Cherkasova, M., Sulla, E. M., Dalena, K. L., Pondé, M. P., & Hechtman, L. (2013). Developmental course of attention deficit hyperactivity disorder and its predictors. *Journal of the Canadian Academy of Child and Adolescent Psychiatry*, 47, *22*(1), 47-54. PMID: 23390433; PMCID: PMC3565715.

Cheung, Y. T., Brinkman, T. M., Li, C., Mzayek, Y., Srivastava, D., Ness, K. K., Patel, S. K., Howell, R. M., Oeffinger, K. C., Robison, L. L., Armstrong, G. T., & Krull, K. R. (2018). Chronic health conditions and neurocognitive function in aging survivors of childhood cancer: A report from the Childhood Cancer Survivor Study. *Journal of the National Cancer Institute, 110*(4), 411–419. DOI: 10.1093/jnci/djx224 PMID: 29088360

Cho, E., Ju, U., Kim, E. H., Lee, M., Lee, G., & Compton, D. L. (2023). Relations among motivation, executive functions, and reading comprehension: Do they differ for students with and without reading difficulties? *Scientific Studies of Reading*, *27*(4), 289–310. DOI: 10.1080/10888438.2022.2127357

Choo, A. L., Smith, S., Pratt, A., & Leon Guerrero, S. (2023). Multilingualism and neurodevelopmental disorders. *Frontiers in Psychology*, *14*, 1267023. DOI: 10.3389/fpsyg.2023.1267023 PMID: 37655193

Choo, T. M., Yuen, M., & Leung, K. (2023). Multilingualism and neurodevelopmental disorders: Implications for assessment and intervention. *Journal of Child Psychology and Psychiatry, and Allied Disciplines*, *64*(8), 953–968. DOI: 10.1111/jcpp.13745

Cioni, G., Inguaggiato, E., & Sgandurra, G. (2016). Early intervention in neurodevelopmental disorders: Underlying neural mechanisms. *Developmental Medicine and Child Neurology*, *58*(S4), 61–66. DOI: 10.1111/dmcn.13050 PMID: 27027609

Cirino, P. T. (2023). Framing executive function as a construct and its relation to academic achievement. *Mind, Brain and Education : the Official Journal of the International Mind, Brain, and Education Society*, *17*(1), 5–24. DOI: 10.1111/mbe.12360 PMID: 38093790

Clark, M., & Adams, D. (2022). Resilience in autism and intellectual disability: A systematic review. *Review Journal of Autism and Developmental Disorders*, *9*(1), 39–53. DOI: 10.1007/s40489-021-00239-w

Clouder, L., Karakus, M., Cinotti, A., Ferreyra, M. V., Fierros, G. A., & Rojo, P. (2020). Neurodiversity in higher education: A narrative synthesis. *Higher Education*, *80*(4), 757–778. DOI: 10.1007/s10734-020-00513-6

Conner, C., Allor, J. H., Al Otaiba, S., Yovanoff, P., & LeJeune, L. (2022). Early Reading Outcomes in Response to a Comprehensive Reading Curriculum for Students With Autism Spectrum Disorder and Intellectual Disability. *Focus on Autism and Other Developmental Disabilities*, *10883576221137905*. Advance online publication. DOI: 10.1177/10883576221137905

Cook, B., Van Bockstaele, S., Crow, S. B., Sainsbury, D., Butterworth, S., & Filson, S. (2025). Neurodevelopmental disorders in children with cleft lip and palate: A systematic review. *European Child & Adolescent Psychiatry*, *34*(6), 1731–1738. DOI: 10.1007/s00787-024-02636-y PMID: 39724178

Cornoldi, C., Giofrè, D., & Toffalini, E. (2023). Cognitive characteristics of intellectually gifted children with a diagnosis of ADHD. *Intelligence*, *97*, 101736. DOI: 10.1016/j.intell.2023.101736

Cortés Pascual, A., Moyano Muñoz, N., & Quílez Robres, A. (2019). The relationship between executive functions and academic performance in primary education: Review and meta-analysis. *Frontiers in Psychology, 10*, 449759. DOI: 10.3389/fpsyg.2019.01582 PMID: 31354585

Cortese, S., Solmi, M., Michelini, G., Bellato, A., Blanner, C., Canozzi, A., Eudave, L., Farhat, L. C., Højlund, M., Köhler-Forsberg, O., Leffa, D. T., Rohde, C., de Pablo, G. S., Vita, G., Wesselhoeft, R., Martin, J., Baumeister, S., Bozhilova, N. S., Carlisi, C. O., & Correll, C. U. (2023). Candidate diagnostic biomarkers for neurodevelopmental disorders in children and adolescents: A systematic review. *World Psychiatry; Official Journal of the World Psychiatric Association (WPA), 22*(1), 129–149. DOI: 10.1002/wps.21037 PMID: 36640395

Cortiella, C., & Horowitz, S. H. (2014). *The state of learning disabilities: Facts, trends and emerging issues*. National Center for Learning Disabilities.

Costa, L. J. C., Green, J. A., & Hooper, S. R. (2020). The relations among the development of written language and executive functions for children in elementary school. In Talinn, P. (Ed.), *Writing across the lifespan* (pp. 201–221). Parlor Press., DOI: 10.37514/PER-B.2020.1053.2.10

Costescu, C., Ro an, A., David, C., Cozma, L., & Calota, A. (2023). The relation between cognitive and emotional processes in children and adolescents with neurodevelopmental disorders—A meta-analysis. *European Journal of Investigation in Health, Psychology and Education, 13*(12), 2811–2826. DOI: 10.3390/ejihpe13120194 PMID: 38131893

Coughtrey, A., Millington, A., Bennett, S., Christie, D., Hough, R., Su, M. T., Constantinou, M. P., & Shafran, R. (2018). The effectiveness of psychosocial interventions for psychological outcomes in pediatric oncology: A systematic review. *Journal of Pain and Symptom Management, 55*(3), 1004–1017. DOI: 10.1016/j.jpainsymman.2017.09.022 PMID: 28962919

Cowan, T. M., Galley, K. O., & Miller, C. A. (2023). Clinical bilingual evaluations: Exploring ways to overcome barriers in the direct assessment of both languages when there is a suspicion of language disorder. *Perspectives of the ASHA Special Interest Groups, 8*(3), 607–617. DOI: 10.1044/2023_PERSP-22-00097

Craig, F., Operto, F. F., De Giacomo, A., Margari, L., Frolli, A., Conson, M., Ivagnes, S., Monaco, M., & Margari, F. (2016). Parenting stress among parents of children with neurodevelopmental disorders. *Psychiatry Research, 242*, 121–129. DOI: 10.1016/j.psychres.2016.05.016 PMID: 27280521

Crisci, G., Caviola, S., Cardillo, R., & Mammarella, I. C. (2021). Executive functions in neurodevelopmental disorders: Comorbidity overlaps between attention deficit and hyperactivity disorder and specific learning disorders. *Frontiers in Human Neuroscience, 15*, 594234. DOI: 10.3389/fnhum.2021.594234

Cristofani, C., Sesso, G., Cristofani, P., Fantozzi, P., Inguaggiato, E., Muratori, P., Narzisi, A., Pfanner, C., Pisano, S., Polidori, L., Ruglioni, L., Valente, E., Masi, G., & Milone, A. (2020). The role of executive functions in the development of empathy and its association with externalizing behaviors in children with neurodevelopmental disorders and other psychiatric comorbidities. *Brain Sciences, 10*(8), 489. DOI: 10.3390/brainsci10080489 PMID: 32731515

Cuba Bustinza, C., Adams, R. E., Claussen, A. H., Vitucci, D., Danielson, M. L., Holbrook, J. R., Charania, S. N., Yamamoto, K., Nidey, N., & Froehlich, T. E. (2022). Factors associated with bullying victimization and bullying perpetration in children and adolescents with ADHD: 2016 to 2017 national survey of children's health. *Journal of Attention Disorders, 26*(12), 1535–1548. DOI: 10.1177/10870547221085502

Cuffe, S. P., Visser, S. N., Holbrook, J. R., Danielson, M. L., Geryk, L. L., Wolraich, M. L., & McKeown, R. E. (2020). ADHD and Psychiatric Comorbidity: Functional Outcomes in a School-Based Sample of Children. *Journal of Attention Disorders, 24*(9), 1345–1354. DOI: 10.1177/1087054715613437

Cummings, K., Watkins, A., Jones, C., Dias, R., & Welham, A. (2022). Behavioural and psychological features of PTEN mutations: A systematic review of the literature and meta-analysis of the prevalence of autism spectrum disorder characteristics. *Journal of Neurodevelopmental Disorders, 14*(1), 1. DOI: 10.1186/s11689-021-09406-w

Da Silva, B. S., Grevet, E. H., Silva, L. C. F., Ramos, J. K. N., Rovaris, D. L., & Bau, C. H. D. (2023). An overview on neurobiology and therapeutics of attention-deficit/hyperactivity disorder. *Discover Mental Health, 3*(1), 2. DOI: 10.1007/s44192-022-00030-1

da Silva, E. J. C., & Eisenstein, E. (2026). Transition to adolescence in neurodevelopmental disorders: Challenges and clinical perspectives. *Jornal de Pediatria, 102*, 101463. DOI: 10.1016/j.jped.2025.101463 PMID: 41120046

Dafsari, H. S., Deneubourg, C., Singh, K., Maroofian, R., Suprenant, Z., Kho, A. L., Ingham, N. J., Steel, K. P., Sheshadri, P., Baur, F., Hentrich, L., Gerisch, B., Zamani, M., Alves, C., Siddiqui, A., Dafsari, H. S., Salari, M., Lang, A. E., Harris, M., & Jungbluth, H. (2025). Mutations in the key autophagy tethering factor EPG5 link neurodevelopmental and neurodegenerative disorders including early-onset parkinsonism. *Annals of Neurology*, *98*(5), 932–950. Advance online publication. DOI: 10.1002/ana.78013 PMID: 41053928

Dahl-Leonard, K., Hall, C., Capin, P., Solari, E. J., Demchak, A., & Therrien, W. J. (2023). Examining fidelity reporting within studies of foundational reading interventions for elementary students with or at risk for dyslexia. *Annals of Dyslexia*, *73*(2), 1–26. DOI: 10.1007/s11881-023-00279-3 PMID: 36701045

Dahl, V., Ramakrishnan, A., Spears, A. P., Jorge, A., Lu, J., Bigio, N. A., & Chacko, A. (2020). Psychoeducation interventions for parents and teachers of children and adolescents with ADHD: A systematic review of the literature. *Journal of Developmental and Physical Disabilities*, *32*(2), 257–292. DOI: 10.1007/s10882-019-09691-3

Dahmane, W. M., Ouchani, S., & Bouarfa, H. (2025). Smart cities services and solutions: A systematic review. *Data and Information Management*, *9*(2), 100087. DOI: 10.1016/j.dim.2024.100087

Dai, X., Cheung, Y. C. H., Zhuang, X., Ng, E. F. W., & Wong, D. F. K. (2024). Context-specific resilience through a cultural lens: Social-ecological factors among Chinese families of children with neurodevelopmental disabilities. *Journal of Autism and Developmental Disorders*, •••, 1–11. DOI: 10.1007/s10803-024-06605-x PMID: 39436517

Davidson, M. M., & Ellis Weismer, S. (2017). Reading comprehension of ambiguous sentences by school-age children with autism spectrum disorder. *Autism Research*, *10*(12), 2002–2022. DOI: 10.1002/aur.1850

de Bree, E. H., van den Boer, M., Toering, B. M., & de Jong, P. F. (2022). A stitch in time…: Comparing late-identified, late-emerging and early identified dyslexia. *Dyslexia (Chichester, England)*, *28*(3), 276–292. DOI: 10.1002/dys.1712 PMID: 35586881

De Felice, A., Ricceri, L., Venerosi, A., Chiarotti, F., & Calamandrei, G. (2015). Multifactorial origin of neurodevelopmental disorders: Approaches to understanding complex etiologies. *Toxics*, *3*(1), 89–129. DOI: 10.3390/toxics3010089 PMID: 29056653

Deacon, L., Macdonald, S. J., & Donaghue, J. (2022). "What's wrong with you, are you stupid?" Listening to the biographical narratives of adults with dyslexia in an age of 'inclusive' and 'anti-discriminatory' practice. *Disability & Society*, *37*(3), 406–426. DOI: 10.1080/09687599.2020.1815522

Deb, S. S., Roy, M., Bachmann, C., & Bertelli, M. O. (2022). Specific learning disorders, motor disorders, and communication disorders. In *Textbook of Psychiatry for Intellectual Disability and Autism Spectrum Disorder* (pp. 483–511). Springer. DOI: 10.1007/978-3-319-95720-3_18

Dell'Osso, L., Lorenzi, P., & Carpita, B. (2019). The neurodevelopmental continuum towards a neurodevelopmental gradient hypothesis. *Journal of Psychopathology*, *25*(4), 179–182. https://hdl.handle.net/11568/1172285

Denhart, H. (2008). Deconstructing barriers: Perceptions of students labeled with learning disabilities in higher education. *Journal of Learning Disabilities*, *41*(6), 483–497. DOI: 10.1177/0022219408321151 PMID: 18931016

Deserno, M. K., Bathelt, J., Groenman, A. P., & Geurts, H. M. (2023). Probing the overarching continuum theory: Data-driven phenotypic clustering of children with ASD or ADHD. *European Child & Adolescent Psychiatry*, *32*(10), 1909–1923. DOI: 10.1007/s00787-022-01986-9 PMID: 35687205

Desvaux, T., Danna, J., Velay, J. L., & Frey, A. (2024). From gifted to high potential and twice-exceptional: A state-of-the-art meta-review. *Applied Neuropsychology. Child*, *13*(2), 165–179. DOI: 10.1080/21622965.2023.2252950 PMID: 37665678

Dewey, D. (2018). What is comorbidity and why does it matter in neurodevelopmental disorders? *Current Developmental Disorders Reports*, *5*(4), 235–242. DOI: 10.1007/s40474-018-0152-3

Diamond, A. (2013). Executive functions. *Annual Review of Psychology*, *64*(1), 135–168. DOI: 10.1146/annurev-psych-113011-143750 PMID: 23020641

Doikou-Avlidou, M. (2015). The educational, social and emotional experiences of students with dyslexia: The perspective of postsecondary education students. *International Journal of Special Education*, *30*(1), 132–145.

Duncan, A. F., & Matthews, M. A. (2018). Neurodevelopmental outcomes in early childhood. *Clinics in Perinatology*, *45*(3), 377–392. DOI: 10.1016/j.clp.2018.05.001 PMID: 30144844

DuPaul, G. J., Gormley, M. J., & Laracy, S. D. (2013). Comorbidity of LD and ADHD: Implications of DSM-5 for assessment and treatment. *Journal of Learning Disabilities*, *46*(1), 43–51. DOI: 10.1177/0022219412464351 PMID: 23144063

Durrett, C. (2025). *Neuro-Inclusive Community Design: Lessons from Sólheimar Ecovillage*. Taylor & Francis. DOI: 10.4324/9781003642336

Earey, A. (2013). Parental experiences of support for pupils with dyslexia: Ignoring the effect on parents. *Support for Learning*, *28*(1), 35–40. DOI: 10.1111/1467-9604.12013

Eberhard, D., Billstedt, E., & Gillberg, C. (2022). Neurodevelopmental disorders and comorbidity in young adults attending a psychiatric outpatient clinic. *Psychiatry Research*, *313*, 114638. DOI: 10.1016/j.psychres.2022.114638 PMID: 35597136

ECDPM. (2016). *Implementation of the 2030 Agenda in the European Union*. Retrieved March 6, 2024, from https://ecdpm.org/download_file/429/1387

Edin, J. (2025). Segregated special educational solutions designated for pupils diagnosed with neurodevelopmental disorders: School leaders' understandings. *Education Inquiry*, *16*(2), 139–156. DOI: 10.1080/20004508.2023.2173120

Egset, K. S., Stubberud, J., Ruud, E., Hjort, M. A., Eilertsen, M. E. B., Sund, A. M., Hjemdal, O., Weider, S., & Reinfjell, T. (2024). Neurocognitive outcome and associated factors in long-term adult survivors of childhood acute lymphoblastic leukemia, treated without cranial radiation therapy. *Journal of the International Neuropsychological Society*, *30*(2), 233–243. DOI: 10.1017/S1355617724000080 PMID: 38465668

Eguia, K. F., Ng, S. S., Capio, C. M., & Wong, T. W. (2025). Applications of motor learning strategies in children with neurodevelopmental disorders: A scoping review. *Review Journal of Autism and Developmental Disorders*, •••, 1–16. DOI: 10.1007/s40489-025-00500-6

Eiser, C., Hill, J. J., & Vance, Y. H. (2000). Examining the psychological consequences of surviving childhood cancer: Systematic review as a research method in pediatric psychology. *Journal of Pediatric Psychology*, *25*(6), 449–460. DOI: 10.1093/jpepsy/25.6.449 PMID: 10980049

Elliott, J. G., & Grigorenko, E. L. (2024). *The dyslexia debate revisited*. Cambridge University Press. DOI: 10.1017/9781009083140

Emmers, E., Baeyens, D., & Petry, K. (2020). Attitudes and self-efficacy of teachers towards inclusion in higher education. *European Journal of Special Needs Education*, *35*(2), 139–153. DOI: 10.1080/08856257.2019.1628337

Erdmann, F., Frederiksen, L. E., Bonaventure, A., Mader, L., Hasle, H., Robison, L. L., & Winther, J. F. (2021). Childhood cancer: Survival, treatment modalities, late effects, and improvements over time. *Cancer Epidemiology*, *71*, 101733. DOI: 10.1016/j.canep.2020.101733 PMID: 32461035

Espinas, D. R., Ramos, M. R., & Fernández, P. (2025). Interventions for children and adolescents with specific learning disorders: A review and future directions. *Pediatric Research*. Advance online publication. DOI: 10.1038/s41390-025-04261-0

Espin, L., García, I., del Pino Sánchez, M., Román, F., & Salvador, A. (2019). Effects of psychosocial stress on the hormonal and affective response in children with dyslexia. *Trends in Neuroscience and Education*, *15*, 1–9. DOI: 10.1016/j.tine.2019.03.001 PMID: 31176466

Eubanks, V. (2018). *Automating inequality: How high-tech tools profile, police, and punish the poor*. St. Martin's Press.

European Agency of Special Needs Education and Inclusive Education. (2018). *EASIE. European Agency Statistics on Inclusive Education. Data Set Cross-country Data Report*. Electronic.

European Commission. (2023a). *Artificial intelligence in education: Guidelines for policy and practice*. Publications Office of the EU.

European Commission. (2023b). *Ethical guidelines on the use of artificial intelligence in education*. Publications Office of the EU.

Ewe, L. P., & Aspelin, J. (2025). Teachers' relational competence towards students with neurodevelopmental symptoms: A microscopic relational analysis. *Journal of Research in Special Educational Needs*, *25*(2), 225–236. DOI: 10.1111/1471-3802.12721

Faedda, N., Romani, M., Rossetti, S., Vigliante, M., Pezzuti, L., Cardona, F., & Guidetti, V. (2019). Intellectual functioning and executive functions in children and adolescents with attention deficit hyperactivity disorder (ADHD) and specific learning disorder (SLD). *Scandinavian Journal of Psychology*, *60*(5), 440–446. DOI: 10.1111/sjop.12562 PMID: 31242533

Faísca, L., Reis, A., & Araújo, S. (2023). Cognitive subtyping of university students with dyslexia in a semi-transparent orthography: What can weaknesses and strengths tell us about compensation? *Journal of Cultural Cognitive Science*, *7*(2), 1–16. DOI: 10.1007/s41809-023-00124-4

Faraone, S. V., & Larsson, H. (2019). Genetics of attention deficit hyperactivity disorder. *Molecular Psychiatry*, *24*(4), 562–575. DOI: 10.1038/s41380-018-0070-0

Ferrante, F. J., Escobar Grisales, D., López, M. F., Lopes da Cunha, P., Sterpin, L. F., Vonk, J. M., Chaná Cuevas, P., Estienne, C., Hesse, E., Amoruso, L., Orozco Arroyave, J. R., & García, A. M. (2025). Cognitive phenotyping of Parkinson's disease patients via digital analysis of spoken word properties. *Movement Disorders*, *40*(11), 2354–2366. Advance online publication. DOI: 10.1002/mds.70005

Ferreira, P., David, C., Costescu, C., Vera, L., Herrera, G., Lopes, S., Stilwell, D., Ferreira, A., Domingues, D., Brito, J., Campos, J., Paiva, A. M., Simão, A. M. V., Heldal, I., Stefanut, T., Rosan, A., & Trindade, F. (2025). Neurodevelopmental disorders: Assessing and training working memory. *BMC Psychology*, *13*(1), 1163. DOI: 10.1186/s40359-025-02912-9

Filipe, M. (2021). How do executive functions issues affect writing in students with neurodevelopmental disorders. In *Executive functions and writing* (p. 160). Oxford University Press. DOI: 10.1093/oso/9780198863564.003.0007

Fine, J. R., Bafna, T., Griffith, S. C., Ransdell, J. M., Barredo, J. C., Isrow, D. M., & Isrow, D. (2023). Effects of cancer treatment on the education and cognition of South Florida pediatric cancer survivors. *Cureus*, *15*(4), e37001. DOI: 10.7759/cureus.37001 PMID: 37139042

Finkelstein, S., Sharma, U., & Furlonger, B. (2021). The inclusive practices of classroom teachers: A scoping review and thematic analysis. *International Journal of Inclusive Education*, *25*(6), 735–762. DOI: 10.1080/13603116.2019.1572232

Finlay-Jones, A., Varcin, K., Leonard, H., Bosco, A., Alvares, G., & Downs, J. (2019). Very early identification and intervention for infants at risk of neurodevelopmental disorders: A transdiagnostic approach. *Child Development Perspectives*, *13*(2), 97–103. DOI: 10.1111/cdep.12319

Fletcher, J. M., Francis, D. J., Foorman, B. R., & Schatschneider, C. (2021). Early detection of dyslexia risk: Development of brief, teacher-administered screens. *Learning Disability Quarterly*, *44*(3), 145–157. DOI: 10.1177/0731948720931870 PMID: 34584341

Flink, T., & Kaldewey, D. (2018). The language of science policy in the twenty-first century. *Basic and Applied Research: The Language of Science Policy in the Twentieth Century*, *4*, 251–284. DOI: 10.2307/j.ctv8bt0z7.15

Floridi, L. (2019). *The logic of information: A theory of philosophy as conceptual design*. Oxford University Press. DOI: 10.1093/oso/9780198833635.001.0001

Floridi, L., & Chiriatti, M. (2020). GPT-3: Its nature, scope, limits, and consequences. *Minds and Machines*, *30*(4), 681–694. DOI: 10.1007/s11023-020-09548-1

Fogel, Y. (2022). Cognitive strategies: Moderating the relationship between executive functions and daily functioning. *International Journal of Environmental Research and Public Health, 19*(24), 16845. DOI: 10.3390/ijerph192416845 PMID: 36554722

Foley-Nicpon, M., & Teriba, A. (2022). Policy considerations for twice-exceptional students. *Gifted Child Today, 45*(4), 212–219. DOI: 10.1177/10762175221110943

Francés, L., Quintero, J., Fernández, A., Ruiz, A., Caules, J., Fillon, G., Hervás, A., & Soler, C. V. (2022). Current state of knowledge on the prevalence of neurodevelopmental disorders in childhood according to the DSM-5: A systematic review in accordance with the PRISMA criteria. *Child and Adolescent Psychiatry and Mental Health, 16*(1), 27. DOI: 10.1186/s13034-022-00462-1 PMID: 35361232

Francés, L., Ruiz, A., Soler, C. V., Francés, J., Caules, J., Hervás, A., Carretero, C., Cardona, B., Quezada, E., Fernández, A., & Quintero, J. (2023). Prevalence, comorbidities, and profiles of neurodevelopmental disorders according to the DSM-5-TR in children aged 6 years old in a European region. *Frontiers in Psychiatry, 14*, 1260747. DOI: 10.3389/fpsyt.2023.1260747 PMID: 38025459

Franke, B., Michelini, G., Asherson, P., Banaschewski, T., Bilbow, A., Buitelaar, J. K., Cormand, B., Faraone, S. V., Ginsberg, Y., Haavik, J., Kuntsi, J., Larsson, H., Lesch, K.-P., Ramos-Quiroga, J. A., Réthelyi, J. M., Ribases, M., & Reif, A. (2018). Live fast, die young? A review on the developmental trajectories of ADHD across the lifespan. *European Neuropsychopharmacology, 28*(10), 1059–1088. DOI: 10.1016/j.euroneuro.2018.08.001

Frazier, T. W., Crowley, E., Shih, A., Vasudevan, V., Karpur, A., Uljarevic, M., & Cai, R. Y. (2022). Associations between executive functioning, challenging behavior, and quality of life in children and adolescents with and without neurodevelopmental conditions. *Frontiers in Psychology, 13*, 1022700. DOI: 10.3389/fpsyg.2022.1022700 PMID: 36337537

Freedman, L., Zivan, M., Farah, R., & Horowitz-Kraus, T. (2020). Greater functional connectivity within the cingulo-opercular and ventral attention networks is related to better fluent reading: A resting-state functional connectivity study. *NeuroImage. Clinical, 26*, 102214. DOI: 10.1016/j.nicl.2020.102214 PMID: 32092682

Frei-Landau, R., Grobgeld, E., & Guberman, R. (2023). Implementing digital neuroscience in special-needs-teacher education: Exploring student-teachers' multifaceted learning outcomes related to teaching children with neurodevelopmental disorders. *Frontiers in Psychology, 14*, 1232315. DOI: 10.3389/fpsyg.2023.1232315

Friedman, N. P., & Robbins, T. W. (2022). The role of prefrontal cortex in cognitive control and executive function. *Neuropsychopharmacology : Official Publication of the American College of Neuropsychopharmacology, 47*(1), 72–89. DOI: 10.1038/s41386-021-01132-0 PMID: 34408280

Gachechiladze, M. A., & Dougherty, J. D. (2024). Neurodevelopmental genetic associations across the translational space-time continuum. *Biological Psychiatry, 95*(9), 825–827. DOI: 10.1016/j.biopsych.2024.02.1010 PMID: 38599709

Gaines, K., & Rodrigues, R. (2024). Smart Design and AI for Developmental Disabilities: Characteristics of People with Autism Spectrum Disorders and Other Developmental Disabilities. In *The Routledge Companion to Smart Design Thinking in Architecture & Urbanism for a Sustainable, Living Planet* (pp. 333-344). Routledge. DOI: 10.4324/9781003384113

Gallego-Méndez, J., Perez-Gomez, J., Calzada-Rodríguez, J. I., Denche-Zamorano, Á. M., Mendoza-Muñoz, M., Carlos-Vivas, J., Garcia-Gordillo, M. Á., & Adsuar, J. C. (2020). Relationship between health-related quality of life and physical activity in children with hyperactivity. *International Journal of Environmental Research and Public Health, 17*(8), 2804. DOI: 10.3390/ijerph17082804 PMID: 32325730

Georgitsi, M., Dermitzakis, I., Soumelidou, E., & Bonti, E. (2021). The polygenic nature and complex genetic architecture of specific learning disorder. *Brain Sciences, 11*(5), 631. DOI: 10.3390/brainsci11050631

Germanò, E., Gagliano, A., & Curatolo, P. (2010). Comorbidity of ADHD and dyslexia. *Developmental Neuropsychology, 35*(5), 475–493. DOI: 10.1080/87565641.2010.494748 PMID: 20721770

Gewirtz, S. (2006). Towards a contextualized analysis of social justice in education. *Educational Philosophy and Theory, 38*(1), 69–81. DOI: 10.1111/j.1469-5812.2006.00175.x

Gialluisi, A., Andlauer, T. F. M., Mirza-Schreiber, N., Moll, K., Becker, J., Hoffmann, P., Ludwig, K. U., Czamara, D., Pourcain, B. S., Honbolygó, F., Tóth, D., Csépe, V., Huguet, G., Chaix, Y., Iannuzzi, S., Demonet, J.-F., Morris, A. P., Hulslander, J., Willcutt, E. G., & Schulte-Körne, G. (2021). Genome-wide association study reveals new insights into the heritability and genetic correlates of developmental dyslexia. *Molecular Psychiatry, 26*(7), 3004–3017. DOI: 10.1038/s41380-020-00898-x

Giannoulis, S. V., Müller, D., Kennedy, J. L., & Gonçalves, V. (2024). Systematic review of mitochondrial genetic variation in attention-deficit/hyperactivity disorder. *European Child & Adolescent Psychiatry, 33*(6), 1675–1685. DOI: 10.1007/s00787-022-02030-6

Gibson, S. (2012). Narrative accounts of university education: Socio-cultural perspectives of students with disabilities. *Disability & Society*, *27*(3), 353–369. DOI: 10.1080/09687599.2012.654987

Gilhuber, C. S., Raulston, T. J., & Galley, K. (2023). Language and communication skills in multilingual children on the autism spectrum: A systematic review. *Autism*, *13623613221147780*(6), 1516–1531. Advance online publication. DOI: 10.1177/13623613221147780

Glasson, E. J., Buckley, N., Chen, W., Leonard, H., Epstein, A., Skoss, R., Jacoby, P., Blackmore, A. M., Bourke, J., & Downs, J. (2020). Systematic review and meta-analysis: Mental health in children with neurogenetic disorders associated with intellectual disability. *Journal of the American Academy of Child and Adolescent Psychiatry*, *59*(9), 1036–1048. DOI: 10.1016/j.jaac.2020.01.006

Glazzard, J., & Dale, K. (2013). Trainee teachers with dyslexia: Personal narratives of resilience. *Journal of Research in Special Educational Needs*, *13*(1), 26–37. DOI: 10.1111/j.1471-3802.2012.01254.x

Gnanavel, S., & Hussain, S. (2018). Audit of physical health monitoring in children and adolescents receiving antipsychotics in neurodevelopmental clinics in Northumberland. *World Journal of Psychiatry*, *8*(1), 27–32. DOI: 10.5498/wjp.v8.i1.27 PMID: 29568729

Gnanavel, S., Sharma, P., Kaushal, P., & Hussain, S. (2019). Attention deficit hyperactivity disorder and comorbidity: A review of literature. *World Journal of Clinical Cases*, *7*(17), 2420–2426. DOI: 10.12998/wjcc.v7.i17.2420

Goldstein, S., & Naglieri, J. A. (2014). *Executive functioning*. Springer., DOI: 10.1007/978-1-4614-8106-5

Gooch, D., Snowling, M., & Hulme, C. (2011). Time perception, phonological skills and executive function in children with dyslexia and/or ADHD symptoms. *Journal of Child Psychology and Psychiatry, and Allied Disciplines*, *52*(2), 195–203. DOI: 10.1111/j.1469-7610.2010.02312.x PMID: 20860755

Goodall, J. S. (2016). Technology and school-home communication. *International Journal of Pedagogies and Learning*, *11*(2), 118–131. DOI: 10.1080/22040552.2016.1227252

Goodfellow, I., Bengio, Y., & Courville, A. (2016). *Deep learning*. MIT Press.

Goodman, P. B., Palka, J. M., Nguyen, A., Carroll, B., & Holland, A. A. (2024). Effects of screen time on emotional and behavioral functioning in pediatric cancer survivors: The potential mediating role of sleep and fatigue. *Journal of Pediatric Neuropsychology*, *10*(3), 202–212. Advance online publication. DOI: 10.1037/h0101949

Goren, H., & Yemini, M. (2017). Global citizenship education redefined – A systematic review of empirical studies on global citizenship education. *International Journal of Educational Research*, *82*, 170–183. DOI: 10.1016/j.ijer.2017.02.004

Graff, M. (2024). Cognitive characteristics of "pure" Tourette Syndrome and what we can learn from neurological study. *Journal of Cognitive Psychology*, *36*(6), 673–687. DOI: 10.1080/20445911.2024.2361958

Gregory, D., & Murray, F. (2026). Neurodiversity in Higher Education: Reflections From Lived Experience. In *Neurodivergent Education and Lifelong Learning* (pp. 49-72). IGI Global Scientific Publishing.

Grigorenko, E. L., Compton, D. L., Fuchs, L. S., Wagner, R. K., Willcutt, E. G., Grigorenko, E. L., Klin, A., & Volkmar, F. (2003). Annotation: Hyperlexia: disability or superability? *Journal of Child Psychology and Psychiatry, and Allied Disciplines*, *44*(8), 1079–1091. DOI: 10.1111/1469-7610.00193

Gudmundsdottir, E., Frigge, H. M., Saemundsen, E., Jensinudottir, K. B., & Njardvik, U. (2025). Rate differences in referrals and diagnostic outcomes of neurodevelopmental disorders between children with native and migrant backgrounds: A retrospective cohort study. *European Child & Adolescent Psychiatry*, *34*(10), 1–13. DOI: 10.1007/s00787-025-02744-3 PMID: 40423709

Güneş Özler, N., & Yıldız, G. (2025). Privacy education for young children with neurodevelopmental disabilities: Teacher and parent experiences. *Sex Education*, •••, 1–16. DOI: 10.1080/14681811.2025.2476147

Gur, R. E. (2025). Pathways to psychosis: Factors associated with risk for psychosis spectrum disorders. *The American Journal of Psychiatry*, *182*(11), 984–990. DOI: 10.1176/appi.ajp.20250950 PMID: 41174893

Gutiérrez-Ruiz, K., & Santoya Montes, Y. (2025). Early detection of neurodevelopmental disorders as a strategy for educational inclusion in early childhood education. *International Journal of Disability Development and Education*, *72*(3), 523–539. DOI: 10.1080/1034912X.2024.2370799

Guzzardo, M. T., Khosla, N., Adams, A. L., Bussmann, J. D., Engelman, A., Ingraham, N., Gamba, R., Jones-Bey, A., Moore, M. D., Toosi, N. R., & Taylor, S. (2021). "The ones that care make all the difference": Perspectives on student-faculty relationships. *Innovative Higher Education, 46*(1), 41–58. DOI: 10.1007/s10755-020-09522-w PMID: 33012971

Hadders-Algra, M. (2021). Early diagnostics and early intervention in neurodevelopmental disorders: Age-dependent challenges and opportunities. *Journal of Clinical Medicine, 10*(4), 861. DOI: 10.3390/jcm10040861 PMID: 33669727

Hadders-Algra, M. (2022). The developing brain: Challenges and opportunities to promote school readiness in young children at risk of neurodevelopmental disorders in low- and middle-income countries. *Frontiers in Pediatrics, 10*, 989518. DOI: 10.3389/fped.2022.989518

Haenlein, M., & Kaplan, A. (2019). A Brief History of Artificial Intelligence: On the Past, Present, and Future of Artificial Intelligence. *California Management Review, 61*(4), 5–14. DOI: 10.1177/0008125619864925

Hamilton, A., Tallberg, P., Ilahi, F., Hoff, E., Ahmadi, B., & Daukantaitė, D. (2025). Behavioral manifestations of executive functioning in Swedish youth with ADHD, autism, and psychiatric comorbidity: A comparative analysis with community controls. *Child Neuropsychology, 31*(5), 771–790. DOI: 10.1080/09297049.2024.2434736 PMID: 39655647

Hannigan, L. J., Askeland, R. B., Ask, H., Tesli, M., Corfield, E., Ayorech, Z., Magnus, P., Njølstad, P. R., Øyen, A.-S., Stoltenberg, C., Andreassen, O. A., Ronald, A., Smith, G. D., Reichborn-Kjennerud, T., & Havdahl, A. (2023). Developmental milestones in early childhood and genetic liability to neurodevelopmental disorders. *Psychological Medicine, 53*(5), 1750–1758. DOI: 10.1017/S0033291721003330 PMID: 37310338

Hansen, B. H., Oerbeck, B., Skirbekk, B., Petrovski, B. É., & Kristensen, H. (2018). Neurodevelopmental disorders: Prevalence and comorbidity in children referred to mental health services. *Nordic Journal of Psychiatry, 72*(4), 285–291. DOI: 10.1080/08039488.2018.1444087 PMID: 29488416

Hansen, K. D., & Dawson, D. L. (2020). "We can do better": Community college faculty preparedness for teaching students with learning disabilities. *Journal of Diversity in Higher Education, 13*(4), 309–319. DOI: 10.1037/dhe0000142

Harkness, K., Bray, S., Durber, C. M., Dewey, D., & Murias, K. (2025). Assessing the contribution of measures of attention and executive function to diagnosis of ADHD or autism. *Journal of Autism and Developmental Disorders, 55*(4), 1353–1364. DOI: 10.1007/s10803-024-06275-9 PMID: 38478161

Hebbecker, K., Förster, N., & Souvignier, E. (2019). Reciprocal effects between reading achievement and intrinsic and extrinsic reading motivation. *Scientific Studies of Reading, 23*(5), 419–436. DOI: 10.1080/10888438.2019.1598413

He, F., Huang, H., Ye, L., Wen, X., & Cheng, A. S. (2022). Meta-analysis of neurocognitive rehabilitation for cognitive dysfunction among pediatric cancer survivors. *Journal of Cancer Research and Therapeutics, 18*(7), 2058–2065. DOI: 10.4103/jcrt.jcrt_1429_22 PMID: 36647970

Heiervang, A. L., Stevenson, J., & Hugdahl, K. (2001). Behavior problems in children with dyslexia. *Nordic Journal of Psychiatry, 55*(4), 251–256. DOI: 10.1080/080394801681019101 PMID: 11839115

Henderson, L. M., Clarke, P. J., & Snowling, M. J. (2014). Reading comprehension impairments in autism spectrum disorders. *L'Année Psychologique, 114*(4), 779–797.

Herreras, E. B. (2025). Executive functioning profiles in children with neurodevelopmental disorders. *Behavioral Sciences (Basel, Switzerland), 15*(9), 1256. DOI: 10.3390/bs15091256 PMID: 41009286

Heselton, G. A. (2023). Childhood adversity, resilience, and autism: A critical review of the literature. *Disability & Society, 38*(7), 1251–1270. DOI: 10.1080/09687599.2021.1983416

Holmes, W., Bialik, M., & Fadel, C. (2022). *Artificial intelligence in education: Promises and implications for teaching and learning* (2nd ed.). Center for Curriculum Redesign.

Holt-Damant, K., Guaralda, M., Taylor Gomez, M., & Nicollet, C. (2013). Urban jungles: Making cities healthy places for australians with neurodiversity. In *Proceedings of the 6th Making Cities Liveable Conference in conjunction with Sustainable Transformation Conference* (pp. 116–132). AST Management Pty Ltd.

Horowitz-Kraus, T. (2023). The role of executive functions in fluent reading: Lessons from reading acquisition and remediation. *Mind, Brain and Education : the Official Journal of the International Mind, Brain, and Education Society, 17*(2), 89–102. DOI: 10.1111/mbe.12357

Hoza, B., & Shoulberg, E. K. (2024). Transdiagnostic considerations are critical to understanding childhood neurodevelopmental disorders. *Frontiers in Human Neuroscience, 18*, 1385873. DOI: 10.3389/fnhum.2024.1385873 PMID: 38774129

Huang, Y., He, M., Li, A., Lin, Y., Zhang, X., & Wu, K. (2020). Personality, behavior characteristics, and life quality impact of children with dyslexia. *International Journal of Environmental Research and Public Health, 17*(4), 1415. DOI: 10.3390/ijerph17041415 PMID: 32098297

Hunter, W., Rath, S., Barnes, K., Hilliard, L., White, C. L., & McGiffert-Sandoval, D. (2024). "Take me to the bridge": Transitional support for minoritized twice-exceptional learners. *Teaching Exceptional Children, 57*(6), 448–457. DOI: 10.1177/00400599241239332

Hustyi, K. M., Logue, J. J., & Hall, S. S. (2025). Skill-based treatment for challenging behavior in autism spectrum disorder: A scoping review of treatment characteristics and outcomes. *Research in Autism Spectrum Disorders, 119*, 102523. DOI: 10.1016/j.rasd.2024.102523

Hutson, J., & Hutson, P. (2023). *Neuroinclusive workplaces and biophilic design: Strategies for promoting occupational health and sustainability in smart cities*. Glob Health Econ Sustain. DOI: 10.36922/ghes.0549

Iannotta, F., La Monica, I., Di Iorio, M. R., Freda, V., Sica, A., de Bartolomeis, A., Pastore, L., Iasevoli, F., & Lombardo, B. (2025). A paradigmatic case of genetic overlap between neurodevelopment disorders and schizophrenia aligning with the neurodevelopmental continuum hypothesis. *International Journal of Molecular Sciences, 26*(9), 3970. DOI: 10.3390/ijms26093970 PMID: 40362210

Ilaria, B., Marcella, B., Valentina, P., Juri, T., Tony, U., Andrea, Z., & Calabrò, G. (2022). Clustering analysis of factors affecting academic career of university students with dyslexia in Italy. *Scientific Reports, 12*(1), 9010. DOI: 10.1038/s41598-022-12985-w PMID: 35637234

Indelicato, E., Zech, M., Amprosi, M., & Boesch, S. (2022). Untangling neurodevelopmental disorders in adulthood: A movement disorder is the clue. *Orphanet Journal of Rare Diseases, 17*(1), 55. DOI: 10.1186/s13023-022-02218-8 PMID: 35172867

Indelicato, E., Zech, M., Eberl, A., & Boesch, S. (2025). Insights on the shared genetic landscape of neurodevelopmental and movement disorders. *Current Neurology and Neuroscience Reports, 25*(1), 24. DOI: 10.1007/s11910-025-01414-w

Inguaggiato, E., Sgandurra, G., & Cioni, G. (2017). Brain plasticity and early development: Implications for early intervention in neurodevelopmental disorders. *Neuropsychiatrie de l'Enfance et de l'Adolescence*, *65*(5), 299–306. DOI: 10.1016/j.neurenf.2017.03.009

Irkey, T., & Wiater, M. (2023). Availability of Municipal Services for Neurodiverse People. *Journal of Psychosocial Rehabilitation and Mental Health*, *10*(4), 473–493. DOI: 10.1007/s40737-023-00336-5

Irungu, , S. KAli, , N. AMbugua, , G. (2025). Self-reported teachers 'self-efficacy in supporting children with neurodevelopmental disorders in early-year education within Kiambu county, Kenya. *African Journal of Emerging Issues*, *7*(8), 1-14. https://ajoeijournal.org/sys/index.php/ajoei/article/view/819

Isaksson, J., Eklund, F., Remnélius, K. L., Black, M. H., & Bölte, S. (2025). Neurodevelopmental conditions and adaptive functioning–a co-twin control study. *Journal of Child Psychology and Psychiatry, and Allied Disciplines*, jcpp.70073. Advance online publication. DOI: 10.1111/jcpp.70073 PMID: 41188694

Jack, C., Crane, L., Kenny, A., Blaisdell, C., & Davis, R. (2025). "There's only so much the school can change about itself… Before you need to change something about yourself"—A qualitative analysis of the experiences of neurodivergent student teachers. *Autism in Adulthood: Challenges and Management*, *7*(4), 435–446. DOI: 10.1089/aut.2024.0047

Jacobs, L., Parke, A., Ziegler, F., Headleand, C., & De Angeli, A. (2022). Learning at school through to university: The educational experiences of students with dyslexia at one UK higher education institution. *Disability & Society*, *37*(4), 662–683. DOI: 10.1080/09687599.2020.1829553

Jain, D., Multani, K. S., Dodiya, A., Benani, U., & Iyer, A. (2025). Adaptive behavior and its differences between children with autism spectrum disorder and social communication disorder. *Autism*, *29*(4), 829–837. DOI: 10.1177/13623613251317787

Jansen, S., Vissers, L. E., & de Vries, B. B. (2023). The genetics of intellectual disability. *Brain Sciences*, *13*(2), 231. DOI: 10.3390/brainsci13020231

Johnsen, S. K. (2021). Definitions, models, and characteristics of gifted students. In *Identifying gifted students* (pp. 1–32). Routledge., DOI: 10.4324/9781003235682-1

Johnson, C., Rossiter, H., Cartmell, B., Domingos, M., & Svanaes, S. (2019). *Evaluation of disabled students' allowances.* London: DfE.

Jordan, M. I., & Mitchell, T. M. (2015). Machine learning: Trends, perspectives, and prospects. *Science*, *349*(6245), 255–260. DOI: 10.1126/science.aaa8415 PMID: 26185243

Jurafsky, D., & Martin, J. H. (2023). *Speech and language processing* (3rd ed.). Stanford University.

Justus, S. A., Pogue, E. M., & Simanovich, V. (2025). Pre-K–12 teachers' views on ASD + ADHD: Prevalence estimates and teaching preparedness. *Children (Basel, Switzerland)*, *12*(3), 342. DOI: 10.3390/children12030342

Kamio, Y., Robins, D., Kelley, E., Swainson, B., & Fein, D. (2007). Atypical lexical/ semantic processing in high-functioning autism spectrum disorders without early language delay. *Journal of Autism and Developmental Disorders*, *37*(6), 1116–1122. DOI: 10.1007/s10803-006-0254-3

Kamran, M., Siddiqui, S., & Adil, M. S. (2023). Breaking barriers: The influence of teachers' attitudes on inclusive education for students with mild learning disabilities (MLDs). *Education Sciences*, *13*(6), 606. DOI: 10.3390/educsci13060606

Kargiotidis, A., & Manolitsis, G. (2023). Are children with early literacy difficulties at risk for anxiety disorders in late childhood? *Annals of Dyslexia*, •••, 1–15. DOI: 10.1007/s11881-023-00291-7 PMID: 37935991

Karlson, C. W., Sarver, D. E., Raiker, J. S., Espil, F. M., Cox, A. D., Elkin, T. D., & Annett, R. D. (2020). The contribution of neurocognitive functions to academic and psychological outcomes in pediatric cancer: A latent profile analysis. *Child Neuropsychology*, *26*(7), 881–899. DOI: 10.1080/09297049.2020.1734553 PMID: 32157961

Kattari, S. K., & Erickson, D. (2025). "If you're asking for my diversity, you're asking for all of me": Challenging experiences of neurodivergent higher education faculty in the United States. *Journal of Diversity in Higher Education*. Advance online publication. DOI: 10.1037/dhe0000639

Ke, L., Su, X., Yang, S., Du, Z., Huang, S., & Wang, Y. (2023). New trends in developmental coordination disorder: Multivariate, multidimensional and multimodal. *Frontiers in Psychiatry*, *14*, 1116369. DOI: 10.3389/fpsyt.2023.1116369

Kelleher, E., & Corvin, A. (2015). Overlapping etiology of neurodevelopmental disorders. In *The genetics of neurodevelopmental disorders* (pp. 29–48). Wiley. DOI: 10.1002/9781118524947.ch2

Kelley, E., Paul, J. J., Fein, D., & Naigles, L. R. (2006). Residual language deficits in optimal outcome children with a history of autism. *Journal of Autism and Developmental Disorders*, *36*(6), 807–828. DOI: 10.1007/s10803-006-0111-4

Kelly, D. R., & Erwin, V. M. (2022). Specific learning difficulty tutors: Direct supports for navigating disabilities and the university environment. *Disability & Society*, •••, 1–24. DOI: 10.1080/09687599.2022.2114882

Kenna, T. (2022). Cities of neurodiversity: New directions for an urban geography of neurodiversity. *Area*, *54*(4), 646–654. DOI: 10.1111/area.12803

Kenna, T. (2023). Neurodiversity in the city: Exploring the complex geographies of belonging and exclusion in urban space. *The Geographical Journal*, *189*(2), 370–382. DOI: 10.1111/geoj.12512

Kennecke, A. K., Wessel, D., & Heine, M. (2022). Dyslexia and accessibility guidelines–how to avoid barriers to access in public services. *i-com, 21*(1), 139-155. DOI: 10.1515/icom-2021-0040

Kern, L., Mathur, S. R., Albrecht, S. F., Poland, S., Rozalski, M., & Skiba, R. (2017). Recommendations for school-based mental health services. *Behavioral Disorders*, *43*, 223–226. DOI: 10.1177/0198742917728355

Kern, M., Johnson, K., & Peverly, S. (2023). The need for integrated supports in schools and the role of school psychologists. In *Learning disorders across the lifespan: A mental health framework* (pp. 25–42). Springer International Publishing. DOI: 10.1007/978-3-031-21772-2_4

Khan, F., Amatya, B., Galea, M. P., Gonzenbach, R., & Kesselring, J. (2017). Neurorehabilitation: Applied neuroplasticity. *Journal of Neurology*, *264*(3), 603–615. DOI: 10.1007/s00415-016-8307-9 PMID: 27778158

Khoirunnisa, A. N., & Dewi, L. (2023). Design and Prototype Development of Augmented Reality in Reading Learning for Autism. *Computers*, *12*(3), 55. DOI: 10.3390/computers12030055

King, C. (2025). Commentary: Air pollution and neurodevelopmental disorders: a cause for concern in an urbanising world. *Global Public Health: An International Journal for Research, Policy and Practice*, *20*(1), 2492234. DOI: 10.1080/17441692.2025.2492234 PMID: 40249163

King-Dowling, S., Proudfoot, N. A., & Obeid, J. (2019). Comorbidity among chronic physical health conditions and neurodevelopmental disorders in childhood. *Current Developmental Disorders Reports*, *6*(4), 248–258. DOI: 10.1007/s40474-019-00173-x

King, S. (2022). The education context for twice-exceptional students: An overview of issues in special and gifted education. *Neurobiology of Learning and Memory*, *193*, 107659. DOI: 10.1016/j.nlm.2022.107659

Kjelgaard, M. M., & Tager-Flusberg, H. (2001). An investigation of language impairment in autism: Implications for genetic subgroups. *Language and Cognitive Processes*, *16*(2-3), 287–308. DOI: 10.1080/01690960042000058

Knight, C. (2018). What is dyslexia? An exploration of the relationship between teachers' understandings of dyslexia and their training experiences. *Dyslexia (Chichester, England)*, *24*(3), 207–219. DOI: 10.1002/dys.1593 PMID: 30019501

Kofler, M. J., Irwin, L. N., Soto, E. F., Groves, N. B., Harmon, S. L., & Sarver, D. E. (2019). Executive functioning heterogeneity in pediatric ADHD. *Journal of Abnormal Child Psychology*, *47*(2), 273–286. DOI: 10.1007/s10802-018-0438-2

Koyuncu, A., Ayan, T., İnce Guliyev, E., Erbilgin, S., & Deveci, E. (2022). ADHD and anxiety disorder comorbidity in children and adults: Diagnostic and therapeutic challenges. *Current Psychiatry Reports*, *24*(2), 129–140. DOI: 10.1007/s11920-022-01324-5 PMID: 35076887

Kranz, A. E., Serry, T. A., & Snow, P. C. (2024). Twice-exceptionality unmasked: A systematic narrative review of the literature on identifying dyslexia in the gifted child. *Dyslexia (Chichester, England)*, *30*(1), e1763. DOI: 10.1002/dys.1763 PMID: 38232949

Krull, K. R., Hardy, K. K., Kahalley, L. S., Schuitema, I., & Kesler, S. R. (2018). Neurocognitive outcomes and interventions in long-term survivors of childhood cancer. *Journal of Clinical Oncology*, *36*(21), 2181–2189. DOI: 10.1200/JCO.2017.76.4696 PMID: 29874137

Krull, K. R., Okcu, M. F., Potter, B., Jain, N., Dreyer, Z., Kamdar, K., & Brouwers, P. (2008). Screening for neurocognitive impairment in pediatric cancer long-term survivors. *Journal of Clinical Oncology*, *26*(25), 4138–4143. DOI: 10.1200/JCO.2008.16.8864 PMID: 18757327

Kuja-Halkola, R., Lichtenstein, P., D'Onofrio, B. M., & Larsson, H. (2015). Codevelopment of ADHD and externalizing behaviour from childhood to adulthood. *Journal of Child Psychology and Psychiatry, and Allied Disciplines*, *56*(6), 640–647. DOI: 10.1111/jcpp.12340

Küry, L., & Fischer, C. (2025). The self-perceptions of twice-exceptional children: A systematic review. *Education Sciences*, *15*(1), 44. DOI: 10.3390/educsci15010044

Kuwahara, Y., Yonezawa, M., Miya, H., Moroto, M., & Iehara, T. (2025). Behavioral problems in adolescents and young adults with neurodevelopmental disorders. *Pediatrics International*, *67*(1), e70115. DOI: 10.1111/ped.70115 PMID: 40536041

Lambrechts, A. A., Cavallaro, M., & Lepori, B. (2024). The European Universities initiative: Between status hierarchies and inclusion. *Higher Education*, *88*(4), 1–21. DOI: 10.1007/s10734-023-01167-w PMID: 39391862

Lamont, D., Kenyon, S., & Lyons, G. (2013). Dyslexia and mobility-related social exclusion: The role of travel information provision. *Journal of Transport Geography*, *26*, 147–157. DOI: 10.1016/j.jtrangeo.2012.08.013

Lampri, S., Peristeri, E., Marinis, T., & Andreou, M. (2024). Figurative language processing in autism spectrum disorders: A review. *Autism Research*, *17*(4), 674–689. DOI: 10.1002/aur.3069

Lamsal, R., Finlay, B., Whitehurst, D. G. T., & Zwicker, J. D. (2020). Generic preference-based health-related quality of life in children with neurodevelopmental disorders: A scoping review. *Developmental Medicine and Child Neurology*, *62*(2), 169–177. DOI: 10.1111/dmcn.14301 PMID: 31225644

Landgren, V. (2022). *Neurodevelopmental disorders (ESSENCE): Early detection and outcome in adulthood.* [Doctoral dissertation, Lund University]. ISBN 978-91-8009-861-8 (Print); 978-91-8009-862-5 (PDF).

Lane, J. M., Liu, S. H., Midya, V., Alcala, C. S., Eggers, S., Svensson, K., Martinez-Medina, S., Horton, M. K., White, R. F., Téllez-Rojo, M. M., & Wright, R. O. (2025). Childhood Pb-induced cognitive dysfunction: Structural equation modeling of hot and cold executive functions. *Journal of Exposure Science & Environmental Epidemiology*, *35*(5), 1–10. DOI: 10.1038/s41370-025-00761-7 PMID: 40033031

Larsson, H., Asherson, P., Chang, Z., Ljung, T., Friedrichs, B., Larsson, J.-O., & Lichtenstein, P. (2013). Genetic and environmental influences on adult attention deficit hyperactivity disorder symptoms: A large Swedish population-based study of twins. *Psychological Medicine*, *43*(1), 197–207. DOI: 10.1017/S0033291712001067

Le Cunff, A. L., Ross, F., Westwood, S. J., Koya, S., Caldwell, D. M., Russell, A. E., & Dommett, E. J. (2025). Key research questions to support neurodiversity in higher education: A participatory priority setting exercise. *Education Sciences*, *15*(7), 839. DOI: 10.3390/educsci15070839

Leclerc, J. B., O'Connor, K. P., Gauthier, B., Singer, I., Woods, D. W., Blanchet, P., & Lavoie, M. E. (2024). Comparison of cognitive-behavioral treatments for tics and Tourette syndrome in youth and adults: A randomized controlled trial. *Journal of Behavioral and Cognitive Therapy*, *100491*(2), 100491. Advance online publication. DOI: 10.1016/j.jbct.2024.100491

LeCun, Y., Bengio, Y., & Hinton, G. (2015). Deep learning. *Nature*, *521*(7553), 436–444. DOI: 10.1038/nature14539 PMID: 26017442

Lee, H. K., Chan, W. S., & Tong, S. X. (2023). The heterogeneity and interrelationships among theory of mind, executive function, and reading comprehension deficits in Hong Kong Chinese children with Autism. *Read Writ 36*, 1539–1565. https://doi.org/10.1007/s11145-022-10298-y

Lee, D., Boulton, K. A., Sun, C., Phillips, N. L., Munro, M., Kumfor, F., Demetriou, E. A., & Guastella, A. J. (2025). Attention and executive delays in early childhood: A meta-analysis of neurodevelopmental conditions. *Molecular Psychiatry*, *30*(5), 1906–1914. DOI: 10.1038/s41380-024-02802-3 PMID: 39489868

Lee, K. M., & Olenchak, F. R. (2015). Individuals with a gifted/ADHD diagnosis: Identification, performance, outcomes, and interventions. *Gifted Education International*, *31*(3), 185–199. DOI: 10.1177/0261429414530712

Lee, Y. C., Yang, H. J., Chen, V. C. H., Lee, W. T., Teng, M. J., Lin, C. H., & Gossop, M. (2016). Meta-analysis of quality of life in children and adolescents with ADHD: By both parent proxy-report and child self-report using PedsQL™. *Research in Developmental Disabilities*, *51*, 160–172. DOI: 10.1016/j.ridd.2015.11.009 PMID: 26829402

Lei, W., & Yueyi, L. (2022). Shenzhen, China: The innovation and technology for sustainable development in dyslexia services. In *The Routledge international handbook of dyslexia in education* (pp. 47-54). Routledge. DOI: 10.4324/9781003162520

Leifler, E. (2022). *Educational inclusion for students with neurodevelopmental conditions* [Doctoral dissertation, Karolinska Institutet]. Karolinska Institutet University Press. ISBN 978-91-8016-640-9

Leifler, E. (2022). *Educational inclusion for students with neurodevelopmental conditions.* Karolinska Institutet. https://hdl.handle.net/10616/48104

Leifler, , EBorg, , ABölte, , S. (2024). A multi-perspective study of perceived inclusive education for students with neurodevelopmental disorders. Journal of Autism and Developmental Disorders (italics), 54(4), 1611–1617. https://doi.org/10.1007/s10803-022-05643-7

Leifler, E., Borg, A., & Bölte, S. (2024). A multi-perspective study of perceived inclusive education for students with neurodevelopmental disorders. *Journal of Autism and Developmental Disorders*, *54*(4), 1611–1617. DOI: 10.1007/s10803-022-05643-7

Leitch, S., Sciberras, E., Post, B., Gerner, B., Rinehart, N., Nicholson, J. M., & Evans, S. (2019). Experience of stress in parents of children with ADHD: A qualitative study. *International Journal of Qualitative Studies on Health and Well-being*, *14*(1), 1690091. DOI: 10.1080/17482631.2019.1690091 PMID: 31713473

Lemoine, L., Bernier, T., Peter, L., Noël, Y., & Besançon, M. (2024). Teachers' attitudes toward inclusive education for children with disabilities. *European Journal of Psychology of Education*, *39*(3), 1–34. DOI: 10.1007/s10212-024-00812-x

Leonard, H., Whitehouse, A., Jacoby, P., Benke, T., Demarest, S., Saldaris, J., Wong, K., Reddihough, D., Williams, K., & Downs, J. (2022). Quality of life beyond diagnosis in intellectual disability – latent profiling. *Research in Developmental Disabilities*, *129*, 104322. DOI: 10.1016/j.ridd.2022.104322

Lervåg, A. (2021). Is there a core deficit in specific learning disabilities? *Journal of Child Psychology and Psychiatry, and Allied Disciplines*, *62*(6), 677–679. DOI: 10.1111/jcpp.13434 PMID: 34008198

Leung, K., & Molnar, M. (2025). Clinician biases in evaluating bilingual students with neurodevelopmental disorders. *Journal of Learning Disabilities*, *58*(6), 451–468.

Levi, U., Einav, M., Raskind, I., Ziv, O., & Margalit, M. (2013). Helping students with LD to succeed: The role of teachers' hope, sense of coherence and specific self-efficacy. *European Journal of Special Needs Education*, *28*(4), 427–439. DOI: 10.1080/08856257.2013.820457

Li, X., Chen, Y., & Wang, J. (2023). *Artificial intelligence and inclusive education: Emerging trends and ethical challenges.* ISBN: 978-981-13-8160-7DOI: 10.1007/978-981-13-8161-4

Li'el, , NWilliams, , CKane, , R. (2019). Identifying developmental language disorder in bilingual children from diverse linguistic backgrounds. *International Journal of Speech-Language Pathology, 21*(6), 613–622. https://doi.org/10.1080/17549507.2018.1513073

Lievore, R., Caviola, S., & Mammarella, I. C. (2024). How trait and state mathematics anxiety could affect performance: Evidence from children with and without Specific Learning Disorders. *Learning and Individual Differences*, *112*, 102459. DOI: 10.1016/j.lindif.2024.102459

Limbers, C. A., Ripperger-Suhler, J., Boutton, K., Ransom, D., & Varni, J. W. (2011). A comparative analysis of health-related quality of life and family impact between children with ADHD treated in a general pediatric clinic and a psychiatric clinic utilizing the PedsQL. *Journal of Attention Disorders, 15*(5), 392–402. DOI: 10.1177/1087054709356191 PMID: 20065072

Lisle, K., & Wade, T. J. (2013). Does the presence of a learning disability elicit stigmatization? British *Journal of Education, Society & Behavioral Sciences, 4*(2), 211–225. https://doi.org/10.9734/BJESBS/2014/6379

Liu, J., Jiang, Z., Li, F., Zheng, Y., Cui, Y., Xu, H., & Li, Y. (2025). Prevalence and comorbidity of attention deficit hyperactivity disorder in Chinese school-attending students aged 6–16: A national survey. *Annals of General Psychiatry, 24*(1), 23. DOI: 10.1186/s12991-025-00558-w PMID: 40223102

Liu, J., Wang, Y., & Yi, L. (2023). Heterogeneity and imbalance of reading profiles in Autism Spectrum Disorders. *Research in Autism Spectrum Disorders, 100*, 102088. DOI: 10.1016/j.rasd.2022.102088

Livingston, E. M., Siegel, L. S., & Ribary, U. (2018). Developmental dyslexia: Emotional impact and consequences. *Australian Journal of Learning Difficulties, 23*(2), 107–135. DOI: 10.1080/19404158.2018.1479975

Lonergan, A., Doyle, C., Cassidy, C., MacSweeney Mahon, S., Roche, R. A., Boran, L., & Bramham, J. (2019). A meta-analysis of executive functioning in dyslexia with consideration of the impact of comorbid ADHD. *Journal of Cognitive Psychology, 31*(7), 725–749. DOI: 10.1080/20445911.2019.1669609

López-Rodríguez, S., Coelho, D. R., Renet, C., Vieira, W. F., Tural, Ü., Cassano, P., & Camprodon, J. A. (2025). Noninvasive brain stimulation for neurodevelopmental disorders: A systematic review. *The Journal of Neuropsychiatry and Clinical Neurosciences, 37*(4), 297–312. Advance online publication. DOI: 10.1176/appi.neuropsych.20240127

Lopez-Villatoro, J. M., Galvez-Merlin, A., Diaz-Marsá, M., & Carrasco, J. L. (2025). Social cognition impairment associated to a borderline personality disorder subtype linked to attention deficit hyperactivity disorder. *Journal of Psychiatric Research, 185*, 26–30. DOI: 10.1016/j.jpsychires.2025.03.031 PMID: 40147151

Louveau, C., Ellul, P., Iftimovici, A., Dubreucq, J., Laidi, C., Leyrolle, Q., Purper-Ouakil, D., Jacquemont, S., Lyonnet, S., Barthélémy, C., Krebs, M.-O., Bai, J., Olivier, P., & Chaumette, B. (2023). Neurodevelopmental disorders (NDD) without boundaries: Research and interventions beyond classifications. *Journal of Neural Transmission (Vienna, Austria), 130*(3), 473–479. DOI: 10.1007/s00702-023-02586-w PMID: 36719463

Luckin, R. (2018). *Machine learning and human intelligence: The future of education for the 21st century*. UCL IOE Press.

Luckin, R., Holmes, W., Griffiths, M., & Forcier, L. B. (2016). *Intelligence unleashed: An argument for AI in education*. Pearson Education.

Magaya, M., & Tafirenyika, J. (2025). Early childhood development teachers' coping strategies for managing learners with neurodevelopmental disorders in Zimbabwe's mainstream classes. *Early Child Development and Care, 195*(5-6), 1–13. DOI: 10.1080/03004430.2025.2499455

Magielski, J. H., Ruggiero, S. M., Xian, J., Parthasarathy, S., Galer, P. D., Ganesan, S., Back, A., McKee, J. L., McSalley, I., Gonzalez, A. K., Morgan, A., Donaher, J., & Helbig, I. (2025). The clinical and genetic spectrum of paediatric speech and language disorders. *Brain, 148*(2), 663–674. DOI: 10.1093/brain/awae264

Magnin, E., Ryff, I., & Moulin, T. (2021). Medical teachers' opinions about students with neurodevelopmental disorders and their management. *BMC Medical Education, 21*(1), 16. DOI: 10.1186/s12909-020-02413-w

Makris, G., Eleftheriades, A., & Pervanidou, P. (2023). Early life stress, hormones, and neurodevelopmental disorders. *Hormone Research in Paediatrics, 96*(1), 17–24. DOI: 10.1159/000523942 PMID: 35259742

Malatesta, G., Marzoli, D., Prete, G., & Tommasi, L. (2021). Human lateralization, maternal effects, and neurodevelopmental disorders. *Frontiers in Behavioral Neuroscience, 15*, 668520. DOI: 10.3389/fnbeh.2021.668520 PMID: 33828467

Marchak, J. G., Christen, S., Mulder, R. L., Baust, K., Blom, J. M., Brinkman, T. M., Elens, I., Harju, E., Kadan-Lottick, N. S., Khor, J. W. T., Lemiere, J., Recklitis, C. J., Wakefield, C. E., Wiener, L., Constine, L. S., Hudson, M. M., Kremer, L. C. M., Skinner, R., Vetsch, J., & Michel, G. (2022). Recommendations for the surveillance of mental health problems in childhood, adolescent, and young adult cancer survivors: A report from the International Late Effects of Childhood Cancer Guideline Harmonization Group. *The Lancet. Oncology, 23*(4), e184–e196. DOI: 10.1016/S1470-2045(21)00750-6 PMID: 35358467

Mareva, S., Astle, D., Baker, K., Gathercole, S., Holmes, J., Kievit, R., Manly, T., Akarca, D., Bathelt, J., Bettencourt, M., Bennett, M., Bignardi, G., Bishop, S., Bottacin, E., Bridge, L., Brkic, D., Bryant, A., Butterfield, S., Byrne, E., & Holmes, J.Lead InvestigatorsTeam of Researchers and PhD Students. (2024). Mapping neurodevelopmental diversity in executive function. *Cortex, 172*, 204–221. DOI: 10.1016/j.cortex.2023.11.021 PMID: 38354470

Marini, A., Piccolo, B., Taverna, L., Berginc, M., & Ozbič, M. (2020). The complex relation between executive functions and language in preschoolers with developmental language disorders. *International Journal of Environmental Research and Public Health, 17*(5), 1772. DOI: 10.3390/ijerph17051772 PMID: 32182903

Marocchini, E. (2023). Impairment or difference? The case of Theory of Mind abilities and pragmatic competence in the Autism Spectrum. *Applied Psycholinguistics, 44*(3), 1–19. DOI: 10.1017/S0142716423000024

Marschik, P. B., Pokorny, F. B., Peharz, R., Zhang, D., O'Muircheartaigh, J., Roeyers, H., Bölte, S., Spittle, A. J., Urlesberger, B., Schuller, B., Poustka, L., Ozonoff, S., Pernkopf, F., Pock, T., Tammimies, K., Enzinger, C., Krieber, M., Tomantschger, I., Bartl-Pokorny, K. D., & Kaufmann, W. E.BEE-PRI Study Group. (2017). A novel way to measure and predict development: A heuristic approach to facilitate the early detection of neurodevelopmental disorders. *Current Neurology and Neuroscience Reports, 17*(5), 43. DOI: 10.1007/s11910-017-0748-8 PMID: 28390033

Martinez-Santos, A. E., Fernandez De La Iglesia, J. D. C., Sheaf, G., & Coyne, I. (2021). A systematic review of the educational experiences and needs of children with cancer returning to school. *Journal of Advanced Nursing, 77*(7), 2971–2994. DOI: 10.1111/jan.14784 PMID: 33598984

Marusak, H. A., Iadipaolo, A. S., Harper, F. W., Elrahal, F., Taub, J. W., Goldberg, E., & Rabinak, C. A. (2018). Neurodevelopmental consequences of pediatric cancer and its treatment: Applying an early adversity framework to understanding cognitive, behavioral, and emotional outcomes. *Neuropsychology Review, 28*(2), 123–175. DOI: 10.1007/s11065-017-9365-1 PMID: 29270773

Mazalán, P., & Mazalán, P. (2021). The city as a place prepared for neurodiversity. *Architektúra & Urbanizmus, 55*(1-2), 18-31. ISSN: 0044-8680

Mazur-Palandre, A., Quignard, M., & Witko, A. (2021). Confronting lexical choice and error distribution in written French: New insights into the linguistic insecurity of students with dyslexia. *Brain Sciences, 11*(7), 922. DOI: 10.3390/brainsci11070922 PMID: 34356156

McCallum, R. S., Bell, S. M., Coles, J. T., Miller, K. C., Hopkins, M. B., & Hilton-Prillhart, A. (2013). A model for screening twice-exceptional students (gifted with learning disabilities) within a response to intervention paradigm. *Gifted Child Quarterly*, *57*(4), 209–222. DOI: 10.1177/0016986213500070

McClain, M. B., Yoho, S. E., Drill, R. B., Haverkamp, C. R., Schwartz, S. E., Barker, B. A., & Upton, S. R. (2023). Reading Skills and Background Noise in Autistic and Non-autistic Children: A Pilot Study. *Contemporary School Psychology*, •••, 1–13. DOI: 10.1007/s40688-023-00450-y

McDermott, J. M., Troller-Renfree, S., Vanderwert, R., Nelson, C. A., Zeanah, C. H., & Fox, N. A. (2013). Psychosocial deprivation, executive functions, and the emergence of socio-emotional behavior problems. *Frontiers in Human Neuroscience*, *7*, 167. DOI: 10.3389/fnhum.2013.00167 PMID: 23675333

McDonough, E. M., Flanagan, D. P., Sy, M., & Alfonso, V. C. (2017). Specific Learning Disorder. In Goldstein, S., & DeVries, M. (Eds.), *Handbook of DSM-5 Disorders in Children and Adolescents*. Springer. DOI: 10.1007/978-3-319-57196-6_4

McFall, G. P., Sapkota, S., Thibeau, S., & Dixon, R. A. (2017). Trajectories and modifiers of executive function: Normal aging to neurodegenerative disease. In *Executive Function* (pp. 263–277). Routledge. DOI: 10.4324/9781315160719-18

Melcher, J., Hays, R., & Torous, J. (2020). Digital phenotyping for mental health of college students. *Evidence-Based Mental Health*, *23*(4), 161–166. DOI: 10.1136/ebmental-2020-300180 PMID: 32998937

Mercera, G., Kalis, T., Vervoort-Schel, J., & Moonen, X. (2025). In the eye of the beholder: Visualizing strengths, burdens and desires through the lens of neurodivergent children, their parents, and professionals using the Yucel Method. *Advances in Neurodevelopmental Disorders*, *9*(2), 315–329. DOI: 10.1007/s41252-024-00411-3

Meredith, R. M. (2015). Sensitive and critical periods during neurotypical and aberrant neurodevelopment: A framework for neurodevelopmental disorders. *Neuroscience and Biobehavioral Reviews*, *50*, 180–188. DOI: 10.1016/j.neubiorev.2014.12.001 PMID: 25496903

Meßler, C. F., Holmberg, H. C., & Sperlich, B. (2018). Multimodal therapy involving high-intensity interval training improves the physical fitness, motor skills, social behavior, and quality of life of boys with ADHD: A randomized controlled study. *Journal of Attention Disorders*, *22*(8), 806–812. DOI: 10.1177/1087054716636936 PMID: 27013028

Micai, M., Fulceri, F., Caruso, A., Guzzetta, A., Gila, L., & Scattoni, M. L. (2020). Early behavioral markers for neurodevelopmental disorders in the first 3 years of life: An overview of systematic reviews. *Neuroscience and Biobehavioral Reviews*, *116*, 183–201. DOI: 10.1016/j.neubiorev.2020.06.027 PMID: 32610179

Michael, K. (2024). *An Examination of Methods for Increasing Teachers' Treatment Fidelity for Special Education Students* (Order No. 31295138). Available from ProQuest Dissertations & Theses Global. (3099989987). https://www.proquest.com/dissertations-theses/examination-methods-increasing-teachers-treatment/docview/3099989987/se-2

Michel, G., Brinkman, T. M., Wakefield, C. E., & Grootenhuis, M. (2020). Psychological outcomes, health-related quality of life, and neurocognitive functioning in survivors of childhood cancer and their parents. *Pediatric Clinics of North America*, *67*(6), 1103–1134. DOI: 10.1016/j.pcl.2020.07.005 PMID: 33131537

Michelini, G., Carlisi, C. O., Eaton, N. R., Elison, J. T., Haltigan, J. D., Kotov, R., Krueger, R. F., Latzman, R. D., Li, J. J., Levin-Aspenson, H. F., Salum, G. A., South, S. C., Stanton, K., Waldman, I. D., & Wilson, S. (2024). Where do neurodevelopmental conditions fit in transdiagnostic psychiatric frameworks? Incorporating a new neurodevelopmental spectrum. *World Psychiatry; Official Journal of the World Psychiatric Association (WPA)*, *23*(3), 333–357. DOI: 10.1002/wps.21225 PMID: 39279404

Miezah, D., Porter, M., Rossi, A., Kazzi, C., Batchelor, J., & Reeve, J. (2021). Cognitive profile of young children with Williams syndrome. *Journal of Intellectual Disability Research*, *65*(8), 784–794. DOI: 10.1111/jir.12860

Mintz, J., Holmes, W., Liu, L., & Perez-Ortiz, M. (2023). Artificial Intelligence and K-12 education: Possibilities, pedagogies and risks. *Computers in the Schools*, *40*(4), 325–333. DOI: 10.1080/07380569.2023.2279870

Miranda, A., Tárraga, R., Fernández, M. I., Colomer, C., & Pastor, G. (2015). Parenting stress in families of children with autism spectrum disorder and ADHD. *Exceptional Children*, *82*(1), 81–95. DOI: 10.1177/0014402915585479

Miranda-Ojeda, R., Wickramasinghe, A., Ntolkeras, G., Castanho, I., & Yassin, W. (2025). The neurodiversity framework in medicine: On the spectrum. *Developmental Neurobiology*, *85*(1), e22960. DOI: 10.1002/dneu.22960 PMID: 39874176

Mirkovic, B., Chagraoui, A., Gerardin, P., & Cohen, D. (2020). Epigenetics and attention-deficit/hyperactivity disorder: New perspectives? *Frontiers in Psychiatry*, *11*, 579. DOI: 10.3389/fpsyt.2020.00579

Mirzoev, T., Tull, K. I., Winn, N., Mir, G., King, N. V., Wright, J. M., & Gong, Y. Y. (2022). Systematic review of the role of social inclusion within sustainable urban developments. *International Journal of Sustainable Development and World Ecology*, *29*(1), 3–17. DOI: 10.1080/13504509.2021.1918793

Mishra, A., Lin, H., Singla, R., Le, N., Oraebosi, M., Liu, D., & Cao, R. (2024). Circadian desynchrony in early life leads to enduring autistic-like behavioral changes in adulthood. *Communications Biology*, *7*(1), 1485. DOI: 10.1038/s42003-024-07131-3 PMID: 39528720

Moffitt, T. E. (2018). Male antisocial behavior in adolescence and beyond. *Nature Human Behaviour*, *2*(3), 177–186. DOI: 10.1038/s41562-018-0309-4

Mogavero, M. P., Bruni, O., DelRosso, L. M., & Ferri, R. (2020). Neurodevelopmental consequences of pediatric cancer and its treatment: The role of sleep. *Brain Sciences*, *10*(7), 411. DOI: 10.3390/brainsci10070411 PMID: 32630162

Moll, K., Göbel, S. M., Gooch, D., Landerl, K., & Snowling, M. J. (2016). Cognitive risk factors for specific learning disorder: Processing speed, temporal processing, and working memory. *Journal of Learning Disabilities*, *49*(3), 272–281. DOI: 10.1177/0022219414547221 PMID: 25124507

Moll, K., Snowling, M. J., & Hulme, C. (2020). Introduction to the special issue "Comorbidities between reading disorders and other developmental disorders.". *Scientific Studies of Reading*, *24*(1), 1–6. DOI: 10.1080/10888438.2019.1702045 PMID: 35283618

Montirosso, R., Mascheroni, E., Guida, E., Piazza, C., Sali, M. E., Molteni, M., & Reni, G. (2021). Stress symptoms and resilience factors in children with neurodevelopmental disabilities and their parents during the COVID-19 pandemic. *Health Psychology*, *40*(7), 428–438. DOI: 10.1037/hea0000966 PMID: 34435794

Moore, B. D.III. (2005). Neurocognitive outcomes in survivors of childhood cancer. *Journal of Pediatric Psychology*, *30*(1), 51–63. DOI: 10.1093/jpepsy/jsi016 PMID: 15610985

Morales, J., Calvo, A., & Bialystok, E. (2013). Working memory development in monolingual and bilingual children. *Journal of Experimental Child Psychology*, *114*(2), 187–202. DOI: 10.1016/j.jecp.2012.09.002 PMID: 23059128

Morosini, G., Cuder, A., De Vita, C., Passolunghi, M. C., & Pellizzoni, S. (2025). The relationship between executive functions and early numerical skills: A study on three- and four-year-old preschoolers. *Cognitive Development*, *74*, 101568. DOI: 10.1016/j.cogdev.2025.101568

Morris, I., Michelini, G., & Wilson, S. (2025). Moving toward transdiagnostic dimensional models of neurodiversity and mental health (and away from models of psychopathology). *Journal of Psychopathology and Clinical Science*, *134*(5), 483–485. DOI: 10.1037/abn0001007 PMID: 40310186

Morris-Rosendahl, D. J., & Crocq, M. A. (2020). Neurodevelopmental disorders—The history and future of a diagnostic concept. *Dialogues in Clinical Neuroscience*, *22*(1), 65–72. DOI: 10.31887/DCNS.2020.22.1/macrocq PMID: 32699506

Mostafa, M., Baumeister, R., Thomsen, M. R., & Tamke, M. (Eds.). (2023). *Design for inclusivity*. Springer. DOI: 10.1007/978-3-031-36302-3

Mouzaki, A., Spantidakis, I., Vamvoukas, M., Liva, E., & Ralli, M. (2007). Development of cognitive and academic skills in children with learning difficulties in early primary school grades. Paper presented at the Conference of the Hellenic Association of Speech-Language Pathologists, *Language Difficulties and Written Expression in the Context of School Learning*, Athens, Greece, 23–25.

Mugnaini, D., Lassi, S., La Malfa, G., & Albertini, G. (2009). Internalizing correlates of dyslexia. *World Journal of Pediatrics*, *5*(4), 255–264. DOI: 10.1007/s12519-009-0049-7 PMID: 19911139

Mulder, H., & Cragg, L. (2014). Executive functions and academic achievement: Current research and future directions. *Infant and Child Development*, *23*(1), 1–3. DOI: 10.1002/icd.1836

Müller, B., Schaadt, G., Boltze, J., Emmrich, F., Skeide, M. A., Neef, N. E., Kraft, I., Brauer, J., Friederici, A. D., Kirsten, H., & Wilcke, A.LEGASCREEN Consortium. (2017). ATP 2C2 and DYX 1C1 are putative modulators of dyslexia-related MMR. *Brain and Behavior*, *7*(11), e00851. DOI: 10.1002/brb3.851

Murdaugh, D. L., Maximo, J. O., Cordes, C. E., O'Kelley, S. E., & Kana, R. K. (2017). From word reading to multisentence comprehension: Improvements in brain activity in children with autism after reading intervention. *NeuroImage. Clinical*, *16*, 303–312. DOI: 10.1016/j.nicl.2017.08.012

Murphy, J. M., Guzmán, J., McCarthy, A. E., Squicciarini, A. M., George, M., Canenguez, K. M., Dunn, E. C., Baer, L., Simonsohn, A., Smoller, J. W., & Jellinek, M. S. (2015). Mental health predicts better academic outcomes: A longitudinal study of elementary school students in Chile. *Child Psychiatry and Human Development*, *46*(2), 245–256. DOI: 10.1007/s10578-014-0464-4 PMID: 24771270

Murray, F. (2025). Neurodiversity in higher education: Reflections from lived experience. Neurodivergent Education and Lifelong Learning, 49. ISBN 9798369380468

Murray, C., & Greenberg, M. T. (2001). Relationships with teachers and bonds with school: Social-emotional adjustment correlates for children with and without disabilities. *Psychology in the Schools, 38*(1), 25–41. DOI: 10.1002/1520-6807(200101)38:1<25::AID-PITS4>3.0.CO;2-C

Nadeau, M. F., Massé, L., Argumedes, M., & Verret, C. (2020). Education for students with neurodevelopmental disabilities—Resources and educational adjustments. In *Handbook of Clinical Neurology* (Vol. 174, pp. 369–378). Elsevier. DOI: 10.1016/B978-0-444-64148-9.00027-2

Nalavany, B. A., Logan, J. M., & Carawan, L. W. (2018). The relationship between emotional experience with dyslexia and work self-efficacy among adults with dyslexia. *Dyslexia (Chichester, England), 24*(1), 17–32. DOI: 10.1002/dys.1575 PMID: 29230916

National Institute for Health and Care Excellence. (2018). Attention deficit hyperactivity disorder: diagnosis and management (update), NICE guideline: methods. www.nice.org.uk/guidance/ng87/evidence (Retrieved: March 6th 2024).

Ndou-Chikwena, N. N., & Sefotho, M. M. (2025). Stakeholders' perspectives on the inclusion of neurodiverse learners in the mainstream curricula. *South African Journal of Childhood Education, 15*(1), 1622. https://hdl.handle.net/10520/ejc-sajce_v15_n1_a1622. DOI: 10.4102/sajce.v15i1.1622

Neef, N. E., & Chang, S. E. (2024). Knowns and unknowns about the neurobiology of stuttering. *PLoS Biology, 22*(2), e3002492. DOI: 10.1371/journal.pbio.3002492

Nelson, J. M., & Gregg, N. (2012). Depression and anxiety among transitioning adolescents and college students with ADHD, dyslexia, or comorbid ADHD/dyslexia. *Journal of Attention Disorders, 16*(3), 244–254. DOI: 10.1177/1087054710385783 PMID: 20978271

Nevill, T. P., & Forsey, M. (2022). We are all thrown into one basket": Dyslexia, schools and the (non) enactment of policies of inclusion. *Disability Studies Quarterly, 42*(1). Advance online publication. DOI: 10.18061/dsq.v42i1.7649

Nevill, T., & Forsey, M. (2022). The social impact of schooling on students with dyslexia: A systematic review of the qualitative research on the primary and secondary education of dyslexic students. *Educational Research Review, 36*, 100507. DOI: 10.1016/j.edurev.2022.100507

Newbury, D. F., & Monaco, A. P. (2010). Genetic advances in the study of speech and language disorders. *Neuron, 68*(2), 309–320. DOI: 10.1016/j.neuron.2010.10.001

Nigg, J. T., Sibley, M. H., Thapar, A., & Karalunas, S. L. (2020). Development of ADHD: Etiology, heterogeneity, and early life course. *Annual Review of Developmental Psychology, 2*(1), 559–583. DOI: 10.1146/annurev-devpsych-060320-093413

Nisiotou, I., & Vlachos, F. (2014). Neurodevelopmental disorders: Is there a shared biological basis? [Translated from Greek]. *Child and Adolescent Psychiatry, 2*, 31–41.

Njardvik, U., Wergeland, G. J., Riise, E. N., Hannesdottir, D. K., & Öst, L. G. (2025). Psychiatric comorbidity in children and adolescents with ADHD: A systematic review and meta-analysis. *Clinical Psychology Review, 102571*, 102571. Advance online publication. DOI: 10.1016/j.cpr.2025.102571 PMID: 40245462

Norbury, C. F., & Sparks, A. (2013). Difference or disorder? Understanding variations in bilingual children. *Child Language Teaching and Therapy, 29*(2), 131–144. DOI: 10.1177/0265659012450432

Novita, S. (2016). Secondary symptoms of dyslexia: A comparison of self-esteem and anxiety profiles of children with and without dyslexia. *European Journal of Special Needs Education, 31*(2), 279–288. DOI: 10.1080/08856257.2015.1125694

Obermeyer, Z., & Emanuel, E. J. (2016). Predicting the future—Big data, machine learning, and clinical medicine. *The New England Journal of Medicine, 375*(13), 1216–1219. DOI: 10.1056/NEJMp1606181 PMID: 27682033

Obiweluozo, P. E., Dike, I. C., Ogba, F. N., Elom, C. O., Orabueze, F. O., Okoye-Ugwu, S., Ani, C. K. C., Onu, A. O., Ukaogo, V., Obayi, L. N., Abonyi, S. E., Onu, J., Omenma, Z. O., Okoro, I. D., Eze, A., Igu, N. C. N., Onuigbo, L. N., Umeano, E. C., & Onyishi, C. N. (2021). Stress in teachers of children with neurodevelopmental disorders: Effect of blended rational emotive behavioral therapy. *Science Progress, 104*(4), 00368504211050278. DOI: 10.1177/00368504211050278

OECD. (2023). *Education at a glance.* OECD. Retrieved April 13, 2024, from https://www.oecd.org/education/education-at-a-glance/

OECD. (2024). *AI and the futures of education: Systems, skills and strategies.* OECD Publishing.

Otterman, D. L., Koopman-Verhoeff, M. E., White, T. J., Tiemeier, H., Bolhuis, K., & Jansen, P. W. (2019). Executive functioning and neurodevelopmental disorders in early childhood: A prospective population-based study. *Child and Adolescent Psychiatry and Mental Health, 13*(1), 38. DOI: 10.1186/s13034-019-0299-7 PMID: 31649749

Ottosen, H. F., Bønnerup, K. H., Weed, E., & Parrila, R. (2022). Identifying dyslexia at the university: Assessing phonological coding is not enough. *Annals of Dyslexia*, *72*(3), 566–567. DOI: 10.1007/s11881-022-00259-z PMID: 35672600

Ouellette, J., & Lacoste, B. (2021). From neurodevelopmental to neurodegenerative disorders: The vascular continuum. *Frontiers in Aging Neuroscience*, *13*, 749026. DOI: 10.3389/fnagi.2021.749026 PMID: 34744690

Oyarzábal, A., Musokhranova, U., Barros, L. F., & García-Cazorla, A. (2021). Energy metabolism in childhood neurodevelopmental disorders. *EBioMedicine*, *69*, 103474. DOI: 10.1016/j.ebiom.2021.103474 PMID: 34256347

Özkaynak, B., Aras, N., Daloğlu Çetinkaya, İ., Ersoy, C., Durmaz İncel, Ö., Koca, M., Nalça, İ., Onay, T. T., Öncü, S., Ülger Vatansever, B., Yücesoy, E., & Yücesoy, C. A. (2024). Neurochallenges in smart cities: State-of-the-art, perspectives, and research directions. *Frontiers in Neuroscience*, *18*, 1279668. DOI: 10.3389/fnins.2024.1279668 PMID: 39744330

Ozturk, B., & Tan, S. (2025). Identifying twice-exceptional students: A systematic research review. *Journal of Advanced Academics*, 1932202X251387416. Advance online publication. DOI: 10.1177/1932202X251387416

Pagán, A. F., & Loveland, K. A. (2025). Launching! to adulthood: An updated group therapy program for young adults with neurodevelopmental differences transitioning to adulthood. *Advances in Neurodevelopmental Disorders*, *9*(1), 115–129. DOI: 10.1007/s41252-024-00402-4 PMID: 41445725

Palagan, C. A., Joe, S. S. A., Mary, S. J., & Jijo, E. E. (2025). Predictive analysis-based sustainable waste management in smart cities using IoT edge computing and blockchain technology. *Computers in Industry*, *166*, 104234. DOI: 10.1016/j.compind.2024.104234

Pallanti, S., & Salerno, L. (2023). Neurodevelopmental disorders (NDDs): Beyond the clinical definition and translational approach. *Children (Basel, Switzerland)*, *10*(1), 99. DOI: 10.3390/children10010099 PMID: 36670649

Pan, D. J., & Lin, D. (2023). How do executive functions explain early Chinese reading and writing? *Reading and Writing*, *36*(3), 625–647. DOI: 10.1007/s11145-022-10314-1

Papadatou-Pastou, M., Panagiotidou, D. A., Abbondanza, F., Fischer, U., Paracchini, S., & Karagiannakis, G. (2021). Hand preference and Mathematical Learning Difficulties: New data from Greece, the United Kingdom, and Germany and two meta-analyses of the literature. *Laterality*, *26*(5), 485–538. DOI: 10.1080/1357650X.2021.1906693

Papadopoulos, T. C., Csépe, V., Aro, M., Caravolas, M., Diakidoy, I. A., & Olive, T. (2021). Methodological issues in literacy research across languages: Evidence from alphabetic orthographies. *Reading Research Quarterly*, *56*(S1), S351–S370. https://ila.onlinelibrary.wiley.com/doi/epdf/10.1002/rrq.407. DOI: 10.1002/rrq.407

Papadopoulos, T. C., Kendeou, P., & Spanoudis, G. (2012). Investigating the factor structure and measurement invariance of phonological abilities in a sufficiently transparent language. *Journal of Educational Psychology*, *104*(2), 321–336. https://psycnet.apa.org/record/2012-00883-001. DOI: 10.1037/a0026446

Papadopoulos, T. C., Spanoudis, G., & Chatzoudi, D. (2020). A longitudinal investigation of the double dissociation between reading and spelling deficits: The role of linguistic and executive function skills. *Reading and Writing*, *33*(4), 1075–1104. https://link.springer.com/article/10.1007/s11145-020-10029-1. DOI: 10.1007/s11145-020-10029-1

Papadopoulos, T. C., Spanoudis, G., Ktisti, C., & Fella, A. (2021). Precocious readers: A cognitive or a linguistic advantage? *European Journal of Psychology of Education*, *36*(1), 63–90. https://pure.unic.ac.cy/en/publications/precocious-readers-a-cognitive-or-a-linguistic-advantage. DOI: 10.1007/s10212-020-00470-9

Papantoniou, M., & Kambourologou, M. (n.d.). *Development and disorders of communication and language in autism*. EPEAEK: Access for All. Institute of Educational Policy. Retrieved March 20, 2023, from https://prosvasimo.iep.edu.gr/docs/pdf/epimorfwtiko-uliko-autismos/Autismos1.pdf

Pappas, M. A., Papoutsi, C., & Drigas, A. S. (2018). Policies, practices, and attitudes toward inclusive education: The case of Greece. *Social Sciences (Basel, Switzerland)*, *7*(6), 90. DOI: 10.3390/socsci7060090

Parrila, R. K., Georgiou, G. K., & Papadopoulos, T. C. (2020). Dyslexia in a consistent orthography: Evidence from reading-level match design. *Dyslexia (Chichester, England)*, *26*(4), 343–358. https://onlinelibrary.wiley.com/doi/epdf/10.1002/dys.1650. DOI: 10.1002/dys.1650

Patel, D. R., Cabral, M. D., Ho, A., & Merrick, J. (2020). A clinical primer on intellectual disability. *Translational Pediatrics*, *9*(S1, Suppl 1), S23–S35. DOI: 10.21037/tp.2020.02.02

Patel, T., Johar, P., Kanisetti, V., Talacheru, S., Avinash, V., Das, A., Sahu, S., Goyal, A., Szobody, M. W., Sayers, T., Gullapalli, S., Yallapu, M. M., Shaikh, M. H., Anand, N., Potter-Baker, K., & Gadad, B. S. (2025). Long-term neurocognitive and behavioral outcomes in survivors of pediatric brain tumors: A systematic review. *Frontiers in Neuroscience*, *19*, 1587059. DOI: 10.3389/fnins.2025.1587059 PMID: 40520501

Peall, K. J., Owen, M. J., & Hall, J. (2024). Rare genetic brain disorders with overlapping neurological and psychiatric phenotypes. *Nature Reviews. Neurology, 20*(1), 7–21. DOI: 10.1038/s41582-023-00896-x

Peng, P., Barnes, M., Wang, C., Wang, W., Li, S., Swanson, H. L., Dardick, W., & Tao, S. (2018). A meta-analysis on the relation between reading and working memory. *Psychological Bulletin, 144*(1), 48–76. DOI: 10.1037/bul0000124 PMID: 29083201

Peristeri, E., & Tsimpli, I. M. (2023). Spelling and syntactic difficulties in bilingual children with developmental language disorder. *Applied Psycholinguistics, 44*(1), 87–110. DOI: 10.1017/S0142716422000427

Perry, N., Sun, C., Munro, M., Boulton, K. A., & Guastella, A. J. (2024). AI technology to support adaptive functioning in neurodevelopmental conditions in everyday environments: A systematic review. *NPJ Digital Medicine, 7*(1), 370. DOI: 10.1038/s41746-024-01355-7 PMID: 39702672

Perry, V., Ellis, K., Moss, J., Beck, S. R., Singla, G., Crawford, H., Waite, J., Richards, C., & Oliver, C. (2022). Executive function, repetitive behaviour and restricted interests in neurodevelopmental disorders. *Research in Developmental Disabilities, 122*, 104166. DOI: 10.1016/j.ridd.2021.104166 PMID: 35016127

Perugini, A., Fontanillas, P., Gordon, S. D., Fisher, S. E., Martin, N. G., Bates, T. C., & Luciano, M. (2024). Dyslexia polygenic scores show heightened prediction of verbal working memory and arithmetic. *Scientific Studies of Reading, 28*(5), 549–563. DOI: 10.1080/10888438.2024.2365697

Petersen, A. (2015). Perspectives of Special Education Teachers on General Education Curriculum Access: Preliminary Results: Preliminary Results. *Research and Practice for Persons with Severe Disabilities : the Journal of TASH, 41*(1), 19–35. DOI: 10.1177/1540796915604835

Peters, L., & Ansari, D. (2019). Are specific learning disorders truly specific, and are they disorders? *Trends in Neuroscience and Education, 17*, 100115. DOI: 10.1016/j.tine.2019.100115

Pimenta, R. A., Fuchs, C., Fears, N. E., Mariano, M., & Tamplain, P. (2023). Distinct mental health profiles in children with developmental coordination disorder: A latent class analysis and associations. *Research in Developmental Disabilities, 132*, 104377. DOI: 10.1016/j.ridd.2022.104377

Pino, M., & Mortari, L. (2014). The inclusion of students with dyslexia in higher education: A systematic review using narrative synthesis. *Dyslexia (Chichester, England), 20*(4), 346–369. DOI: 10.1002/dys.1484 PMID: 25293652

Polydoros, G., & Antoniou, A. (2025). Optimization of fraction learning for students with learning difficulties in mathematics: Computer-assisted educational environments. *International Journal of Educational Methodology, 11*(2), 127–141. DOI: 10.12973/ijem.11.2.127

Polydoros, G., & Antoniou, A. S. (2025b). Empowering students with learning disabilities: Examining serious digital games' potential. *Behavioral Sciences (Basel, Switzerland), 15*(3), 282–296. DOI: 10.3390/bs15030282 PMID: 40150177

Polydoros, G., Galitskaya, V., Antoniou, A. S., & Drigas, A. (2025). AI technology integration in elementary geometry. *Scientific Electronic Archives, 18*(2), 1–11. DOI: 10.36560/18220252050

Popp, C., Stoeger, H., & Ziegler, A. (2025). Robust and semi-robust findings from giftedness and talent research: Results of a secondary-order scoping review on meta-analyses. *Journal for the Education of the Gifted, 48*(4), 334–363. Advance online publication. DOI: 10.1177/01623532251372590

Posner, J., Polanczyk, G. V., & Sonuga-Barke, E. (2020). Attention-deficit hyperactivity disorder. *Lancet, 395*(10222), 450–462. DOI: 10.1016/S0140-6736(19)33004-1 PMID: 31982036

Protopapas, A., & Parrila, R. (2018). Is dyslexia a brain disorder? *Brain Sciences, 8*(4), 61. DOI: 10.3390/brainsci8040061 PMID: 29621138

Purpura, G., Annunziata, S., Biancardi, S., Brivio, M., Caporali, C., Mantegazza, G., Piazza, E., Restelli, A., & Cavallini, A. (2025). Clinical and research insights from pre-emptive early intervention for neurodevelopmental disorders: A case series. *Children (Basel, Switzerland), 12*(11), 1489. DOI: 10.3390/children12111489 PMID: 41300607

Quigley, E., & Gallagher, T. (2025). Neurodiversity and higher education: Double masking by neurodivergent students. *European Journal of Special Needs Education*, •••, 1–17. DOI: 10.1080/08856257.2025.2511369

Ramey, D. M., & Freelin, B. N. (2023). Exploring the relationships between school suspension, ADHD diagnoses, and delinquency across different school punitive and special education climates. *Children and Youth Services Review, 148*, 106849. DOI: 10.1016/j.childyouth.2023.106849

Ramteke, A., & Lamture, Y. (2022). Tics and Tourette Syndrome: A literature review of etiological, clinical, and pathophysiological aspects. *Cureus, 14*(8), e28575. DOI: 10.7759/cureus.28575

Randi, J., Newman, T., & Grigorenko, E. L. (2010). Teaching children with autism to read for meaning: Challenges and possibilities. *Journal of Autism and Developmental Disorders, 40*(7), 890–902. DOI: 10.1007/s10803-010-0938-6

Ranieri, J. M., Neil, N., Sadowski, M., & Azzam, M. (2024). Supporting inclusion in informal education settings for children with neurodevelopmental disorders: A scoping review. *Journal of Developmental and Physical Disabilities, 36*(6), 955–993. DOI: 10.1007/s10882-024-09970-8

Regan, S. L., Williams, M. T., & Vorhees, C. V. (2022). Review of rodent models of attention deficit hyperactivity disorder. *Neuroscience and Biobehavioral Reviews, 132*, 621–637. DOI: 10.1016/j.neubiorev.2021.11.041

Reider Lewis, L. (2021). Twice-exceptionality: Maximizing academic & psychosocial success in youth. *Journal of Health Service Psychology, 47*(4), 191–196. DOI: 10.1007/s42843-021-00046-1

Reid, G., & Kirk, J. (2001). *Dyslexia in adults: Education and employment*. Wiley.

Reis, A., Araújo, S., Morais, I. S., & Faísca, L. (2020). Reading and reading-related skills in adults with dyslexia from different orthographic systems: A review and meta-analysis. *Annals of Dyslexia, 70*(3), 339–368. DOI: 10.1007/s11881-020-00205-x PMID: 32918699

Retz, W., Ginsberg, Y., Turner, D., Barra, S., Retz-Junginger, P., Larsson, H., & Asherson, P. (2021). Attention-Deficit/Hyperactivity Disorder (ADHD), antisociality and delinquent behavior over the lifespan. *Neuroscience and Biobehavioral Reviews, 120*, 236–248. DOI: 10.1016/j.neubiorev.2020.11.025

Reyes, J. M., Gentry, J. E., Atchley, S., & Phillips, J. (2024). Special education directors and special education teachers' self-efficacy for serving students with English learner and special education needs. *International Journal of Disability Development and Education, 71*(4), 573–589. DOI: 10.1080/1034912X.2022.2146068

Ricketts, J., Jones, C. R., Happé, F., & Charman, T. (2013). Reading comprehension in autism spectrum disorders: The role of oral language and social functioning. *Journal of Autism and Developmental Disorders, 43*(4), 807–816. DOI: 10.1007/s10803-012-1619-4

Riddick, B. (2013). An examination of the relationship between labeling and stigmatization with special reference to dyslexia. In *Addressing difficulties in literacy development* (pp. 317–333). Routledge.

Riesco, K., Ukawa, Y., Alfreh, P., & Abdelfattah, E. (2025, March). *Analyzing Autism Spectrum Disorder in Children. In 2025 Northeast Section Conference.*

Rinn, A. N. (2024). A critique on the current state of research on the social and emotional experiences of gifted individuals and a framework for moving the field forward. *Gifted Child Quarterly*, *68*(1), 34–48. DOI: 10.1177/00169862231197780

Riva, A., Musetti, A., Bomba, M., Milani, L., Montrasi, V., & Nacinovich, R. (2021). Language-related skills in bilingual children with specific learning disorders. *Frontiers in Psychology*, *11*, 564047. DOI: 10.3389/fpsyg.2020.564047

Roccella, M., & Vetri, L. (2021). Embracing the complexity of neurodevelopmental disorders. *Brain Sciences*, *11*(11), 1385. DOI: 10.3390/brainsci11111385 PMID: 34827384

Roch, M., Cain, K., & Jarrold, C. (2021). Reading for comprehension in individuals with Down syndrome, autism spectrum disorder and typical development: Similar or different patterns of ability? *Brain Sciences*, *11*(7), 828. DOI: 10.3390/brainsci11070828

Rodan, L. H., Stoler, J., Chen, E., Geleske, T., Vergano, S., Holm, I. A., Jones, K., Kalish, J., Monteil, D. C., Pritchard, A. B., Rasmussen, S. A., Russell, B., Santoro, S. L., Trapane, P., & Weaver, K. N. (2025). Genetic evaluation of the child with intellectual disability or global developmental delay: Clinical report. *Pediatrics*, *156*(1), e2025072219. DOI: 10.1542/peds.2025-072219

Rodríguez, C., Areces, D., García, T., Cueli, M., & Gonzalez-Castro, P. (2021). Neurodevelopmental disorders: An innovative perspective via the response to intervention model. *World Journal of Psychiatry*, *11*(11), 1017–1026. DOI: 10.5498/wjp.v11.i11.1017 PMID: 34888170

Romano, S., Esposito, D., Aricò, M., Arigliani, E., Cavalli, G., Vigliante, M., Penge, R., Sogos, C., Pisani, F., & Romani, M. (2024). Giftedness and twice-exceptionality in children suspected of ADHD or specific learning disorders: A retrospective study. *Sci*, *6*(2), 23. DOI: 10.3390/sci6020023

Ronksley-Pavia, M. (2015). A model of twice-exceptionality: Explaining and defining the apparent paradoxical combination of disability and giftedness in childhood. *Journal for the Education of the Gifted*, *38*(3), 318–340. DOI: 10.1177/0162353215592499

Rosenblum, S., Sharfi, K., Elimelech, O. C., Regev, S., & Grinblat, N. (2025). Predicting emotional consequences of disorganization among adults with neurodevelopmental disorders: A structural equation model. *Disability and Rehabilitation*, *47*(8), 1954–1964. DOI: 10.1080/09638288.2024.2385733 PMID: 39119890

Rose, Z. R. (1998). *A study of college graduates with dyslexia* (Vol. 9839593). UMI.

Rovira, P., Demontis, D., Sánchez-Mora, C., Zayats, T., Klein, M., Mota, N. R., Weber, H., Garcia-Martínez, I., Pagerols, M., Vilar-Ribó, L., Arribas, L., Richarte, V., Corrales, M., Fadeuilhe, C., Bosch, R., Martin, G. E., Almos, P., Doyle, A. E., Grevet, E. H., & Ribasés, M. (2020). Shared genetic background between children and adults with attention deficit/hyperactivity disorder. *Neuropsychopharmacology : Official Publication of the American College of Neuropsychopharmacology*, *45*(10), 1617–1626. DOI: 10.1038/s41386-020-0664-5 PMID: 32279069

Ruan, Y., Ye, Y., & McBride, C. (2023). Effectiveness of parent coaching on the literacy skills of Hong Kong Chinese children with and without dyslexia. *Reading and Writing*, *36*, 1–22. DOI: 10.1007/s11145-022-10408-w PMID: 36687998

Ruble, K. J., Paré-Blagoev, E. J., Cooper, S. L., & Jacobson, L. A. (2020a). Assessment of online resources for returning to school during and after treatment of childhood cancer. *Journal of Cancer Education*, *35*(5), 876–884. DOI: 10.1007/s13187-019-01537-y PMID: 31049872

Ruble, K., Paré-Blagoev, J., Cooper, S., & Jacobson, L. A. (2020b). Pediatric oncology provider perspectives and practices: Supporting patients and families in schooling after cancer diagnosis. *Pediatric Blood & Cancer*, *67*(4), e28166. DOI: 10.1002/pbc.28166 PMID: 31930685

Ruffini, C., Osmani, F., Martini, C., Giera, W. K., & Pecini, C. (2024). The relationship between executive functions and writing in children: A systematic review. *Child Neuropsychology*, *30*(1), 105–163. DOI: 10.1080/09297049.2023.2170998 PMID: 36748722

Rungsattatharm, L., Tasingha, P., Trairatvorakul, P., & Chonchaiya, W. (2025). Longitudinal associations between executive function and positive parenting during early childhood and resilience, self-regulation, and behavioral problems in school-age children. *Child and Adolescent Psychiatry and Mental Health*, *19*(1), 19. DOI: 10.1186/s13034-025-00875-8 PMID: 40057732

Russell, S. J., & Norvig, P. (2021). *Artificial intelligence: A modern approach* (4th ed.). Pearson., https://library.giadinh.edu.vn/handle/GDU/2783

Rutherford, M., Maciver, D., Johnston, L., Prior, S., & Forsyth, K. (2021). Development of a pathway for multidisciplinary neurodevelopmental assessment and diagnosis in children and young people. *Children (Basel, Switzerland)*, *8*(11), 1033. DOI: 10.3390/children8111033

Sadozai, A. K., Sun, C., Demetriou, E. A., Lampit, A., Munro, M., Perry, N., Boulton, K. A., & Guastella, A. J. (2024). Executive function in children with neurodevelopmental conditions: A systematic review and meta-analysis. *Nature Human Behaviour*, *8*(12), 2357–2366. DOI: 10.1038/s41562-024-02000-9 PMID: 39424962

Salminen, J., Laasanen, M., Leinonen, L., Karukivi, M., Vornanen, R., Alin, M., & Kyttälä, M. (2025). Professional collaboration for children and adolescents with neurodevelopmental disorders: A scoping review. *European Journal of Special Needs Education*, *40*(1), 16–34. DOI: 10.1080/08856257.2024.2314422

Saloviita, T. (2020). Teacher attitudes towards the inclusion of students with support needs. *Journal of Research in Special Educational Needs*, *20*(1), 64–73. DOI: 10.1111/1471-3802.12466

Sansosti, F. J., Was, C., Rawson, K. A., & Remaklus, B. L. (2013). Eye movements during processing of text requiring bridging inferences in adolescents with higher functioning autism spectrum disorders: A preliminary investigation. *Research in Autism Spectrum Disorders*, *7*(12), 1535–1542. DOI: 10.1016/j.rasd.2013.09.001

Santoro, J. D., Pagarkar, D., Chu, D. T., Rosso, M., Paulsen, K. C., Levitt, P., & Rafii, M. S. (2021). Neurologic complications of Down syndrome: A systematic review. *Journal of Neurology*, *268*(12), 4495–4509. DOI: 10.1007/s00415-020-10179-w

Saß, S., Köller, O., & Zimmermann, F. (2025). Smart but maladapted? Differences in the psychological functioning of intellectually gifted students compared with average-ability students. *Gifted Child Quarterly*, *69*(3), 219–236. Advance online publication. DOI: 10.1177/00169862241310871

Sathyanesan, A., Zhou, J., Scafidi, J., Heck, D. H., Sillitoe, R. V., & Gallo, V. (2019). Emerging connections between cerebellar development, behaviour and complex brain disorders. *Nature Reviews. Neuroscience*, *20*(5), 298–313. DOI: 10.1038/s41583-019-0152-2

Scattolin, M. A. D. A., Resegue, R. M., & Rosário, M. C. D. (2022). The impact of the environment on neurodevelopmental disorders in early childhood. *Jornal de Pediatria*, *98*, 66–72. DOI: 10.1016/j.jped.2021.11.002 PMID: 34914896

Schatz, N. K., Fabiano, G. A., Raiker, J. S., Hayes, T. B., & Pelham, W. E.Jr. (2021). Twenty-year trends in elementary teachers' beliefs about best practices for students with ADHD. *School Psychology*, *36*(4), 203–213. DOI: 10.1037/spq0000442

Schiltz, H., Gohari, D., Park, J., & Lord, C. (2024). A longitudinal study of loneliness in autism and other neurodevelopmental disabilities: Coping with loneliness from childhood through adulthood. *Autism*, *28*(6), 1471–1486. DOI: 10.1177/13623613231217337 PMID: 38153097

Schwab, S., Resch, K., & Alnahdi, G. (2024). Inclusion does not solely apply to students with disabilities: Pre-service teachers' attitudes towards inclusive schooling of all students. *International Journal of Inclusive Education*, *28*(2), 214–230. DOI: 10.1080/13603116.2021.1938712

Segundo-Marcos, R., Carrillo, A. M., Fernández, V. L., & González, M. T. D. (2022). Development of executive functions in late childhood and the mediating role of cooperative learning: A longitudinal study. *Cognitive Development*, *63*, 101219. DOI: 10.1016/j.cogdev.2022.101219

Seiradakis, E. V. (2023). Giving students a voice: Dyslexia and language learning experiences from childhood to adulthood in Greece. *European Journal of Education and Pedagogy*, *4*(1), 43–48. DOI: 10.24018/ejedu.2023.4.1.561

Senter, R., Chow, J. C., & Willis, E. C. (2023). Speech-language pathology interventions for children with executive function deficits: A systematic literature review. *Language, Speech, and Hearing Services in Schools*, *54*(1), 336–354. DOI: 10.1044/2022_LSHSS-22-00013 PMID: 36306507

Serry, T. A., Liamputtong, P., Rose, M. L., & Bretherton, L. (2016). Parents' experiences of their children's supplementary reading intervention: A qualitative exploration. *Australian Journal of Learning Difficulties*, *21*(1), 3–20. DOI: 10.1080/19404158.2015.1128452

Sharkey, R. J., & Nickl-Jockschat, T. (2023). The neurobiology of autism spectrum disorder as it relates to twice-exceptionality. *Neurobiology of Learning and Memory*, *200*, 107740. DOI: 10.1016/j.nlm.2023.107740 PMID: 36894126

Shaywitz, B. A., & Shaywitz, S. E. (2020). The American experience: Towards a 21st century definition of dyslexia. *Oxford Review of Education*, *46*(4), 454–471. DOI: 10.1080/03054985.2020.1793545

Shessel, I., & Reiff, H. B. (1999). Experiences of adults with learning disabilities: Positive and negative impacts and outcomes. *Learning Disability Quarterly*, *22*(4), 305–316. DOI: 10.2307/1511264

Shiue, I. (2016). People with dyslexia and heart, chest, skin, digestive, musculoskeletal, vision, learning, speech and mental disorders were more dissatisfied with neighbourhoods: Scottish Household Survey, 2007–2008. *Environmental Science and Pollution Research International, 23*(23), 23840–23853. DOI: 10.1007/s11356-016-7585-1 PMID: 27628699

Siegel, L. S. (2020). Early identification and intervention to prevent reading failure: A response to intervention (RTI) initiative. *The Educational and Developmental Psychologist, 37*(2), 140–146. DOI: 10.1017/edp.2020.21

Siegle, D., McCoach, D. B., Boldt, G. T., Hamilton, R., Gubbins, E. J., & Callahan, C. M. (2025). What really happens in gifted and talented education? A portrait of three states. *Gifted Child Quarterly*, 00169862251370673. Advance online publication. DOI: 10.1177/00169862251370673

Silver, D., Huang, A., Maddison, C. J., Guez, A., Sifre, L., van den Driessche, G., Schrittwieser, J., Antonoglou, I., Panneershelvam, V., Lanctot, M., Dieleman, S., Grewe, D., Nham, J., Kalchbrenner, N., Sutskever, I., Lillicrap, T., Leach, M., Kavukcuoglu, K., Graepel, T., & Hassabis, D. (2016). Mastering the game of Go with deep neural networks and tree search. *Nature, 529*(7587), 484–489. DOI: 10.1038/nature16961 PMID: 26819042

Simon, J., Hyde, C., Saravanapandian, V., Wilson, R., Distefano, C., Besterman, A., & Jeste, S. (2022). The diagnostic journey of genetically defined neurodevelopmental disorders. *Journal of Neurodevelopmental Disorders, 14*(1), 27. DOI: 10.1186/s11689-022-09439-9 PMID: 35501684

Singer, B. (2015). *Extended stereotype threat: Parental concern that a child will confirm a negative stereotype associated with dyslexia* [Doctoral dissertation, Trinity College]. https://core.ac.uk/download/pdf/232743371.pdf

Singh, B. D., Moore, D. W., Furlonger, B. E., Anderson, A., Fall, R., & Howorth, S. (2021). Reading comprehension and autism spectrum disorder: A systematic review of interventions involving single-case experimental designs. *Review Journal of Autism and Developmental Disorders, 8*(1), 3–21. DOI: 10.1007/s40489-020-00200-3

Slomowitz, R. F., Narayan, A. J., & McGrath, L. M. (2025). Resilience research in learning disabilities: Guiding principles from developmental psychopathology. *Learning and Instruction, 97*, 102090. DOI: 10.1016/j.learninstruc.2025.102090

Smith Gabig, C. (2010). Phonological awareness and word recognition in reading by children with autism. *Communication Disorders Quarterly, 31*(2), 67–85. DOI: 10.1177/1525740108328410

Smythe, T., Scherer, N., Nanyunja, C., Tann, C. J., & Olusanya, B. O. (2024). Strategies for addressing the needs of children with or at risk of developmental disabilities in early childhood by 2030: A systematic umbrella review. *BMC Medicine*, *22*(1), 51. DOI: 10.1186/s12916-024-03265-7 PMID: 38302917

Snowling, M. J., & Hulme, C. (2025). Risk factors for dyslexia: Addressing oral language deficits. *Mind, Brain and Education : the Official Journal of the International Mind, Brain, and Education Society*, *19*(3), 168–175. DOI: 10.1111/mbe.70009

Snowling, M. J., Hulme, C., & Nation, K. (2020). Defining and understanding dyslexia: Past, present and future. *Oxford Review of Education*, *46*(4), 501–513. DOI: 10.1080/03054985.2020.1765756 PMID: 32939103

Snowling, M., & Hulme, C. (2024). Do we really need a new definition of dyslexia? A commentary. *Annals of Dyslexia*, *74*(3), 1–8. DOI: 10.1007/s11881-024-00305-y PMID: 38526759

Sokolova, E., Oerlemans, A. M., Rommelse, N. N., Groot, P., Hartman, C. A., Glennon, J. C., Claassen, T., Heskes, T., & Buitelaar, J. K. (2017). A causal and mediation analysis of the comorbidity between attention deficit hyperactivity disorder (ADHD) and autism spectrum disorder (ASD). *Journal of Autism and Developmental Disorders*, *47*(6), 1595–1604. DOI: 10.1007/s10803-017-3083-7 PMID: 28255761

Solari, E. J., Grimm, R. P., McIntyre, N. S., Zajic, M., & Mundy, P. C. (2019). Longitudinal stability of reading profiles in individuals with higher functioning autism. *Autism*, *23*(8), 1911–1926. DOI: 10.1177/1362361318812423

Soldevila-Pérez, J., Naranjo, M., & Collet, J. (2022). Global inclusive education: Challenges for the future. In *Global inclusive education: Lessons from Spain* (pp. 157–167). Springer International Publishing. DOI: 10.1007/978-3-031-11476-2_11

Sorkos, G., Hajisoteriou, C., & Angelides, P. (2025). Students' agency for critical interculturalism: From intercultural responsibility to informal leadership. *Oxford Review of Education*, *51*(1), 109–128. DOI: 10.1080/03054985.2023.2296091

Southon, C. (2022). The relationship between executive function, neurodevelopmental disorder traits, and academic achievement in university students. *Frontiers in Psychology*, *13*, 958013. DOI: 10.3389/fpsyg.2022.958013 PMID: 36118426

Stein, B., Hoeft, F., & Richter, C. G. (2024). Stress, resilience, and emotional well-being in children and adolescents with specific learning disabilities. *Current Opinion in Behavioral Sciences*, *58*, 101410. DOI: 10.1016/j.cobeha.2024.101410 PMID: 39157263

Stein, D. J., Szatmari, P., Gaebel, W., Berk, M., Vieta, E., Maj, M., de Vries, Y. A., Roest, A. M., de Jonge, P., Maercker, A., Brewin, C. R., Pike, K. M., Grilo, C. M., Fineberg, N. A., Briken, P., Cohen-Kettenis, P. T., & Reed, G. M. (2020). Mental, behavioral and neurodevelopmental disorders in the ICD-11: An international perspective on key changes and controversies. *BMC Medicine*, *18*(1), 21. DOI: 10.1186/s12916-020-1495-2 PMID: 31983345

Stephens, K. R., & Karnes, F. A. (2000). State definitions for the gifted and talented revisited. *Exceptional Children*, *66*(2), 219–238. DOI: 10.1177/001440290006600206

Stewart, S., Houghton, S. J., Glasgow, K., & Macqueen, L. (2025). Worry and depressive symptoms in adolescents with neurodevelopmental disorders. *International Journal of Environmental Research and Public Health*, *22*(2), 185. DOI: 10.3390/ijerph22020185 PMID: 40003411

Stoeger, H., & Zeidner, M. (2019). Self-regulated learning in gifted, talented, and high-achieving learners. *High Ability Studies*, *30*(1–2), 1–8. DOI: 10.1080/13598139.2019.1601326

Sumner, E., & Connelly, V. (2020). Writing and revision strategies of students with and without dyslexia. *Journal of Learning Disabilities*, *53*(3), 189–198. DOI: 10.1177/0022219419899090 PMID: 31941395

Sung, M. C., Ku, B., Leung, W., & MacDonald, M. (2022). The effect of physical activity interventions on executive function among people with neurodevelopmental disorders: A meta-analysis. *Journal of Autism and Developmental Disorders*, *52*(3), 1030–1050. DOI: 10.1007/s10803-021-05009-5 PMID: 33856619

Sutton, R. S., & Barto, A. G. (2018). Temporal-difference learning. *Reinforcement learning: an introduction*, 131-132.

Tajik-Parvinchi, D., Farmus, L., Tablon Modica, P., Cribbie, R. A., & Weiss, J. A. (2021). The role of cognitive control and emotion regulation in predicting mental health problems in children with neurodevelopmental disorders. *Child: Care, Health and Development*, *47*(5), 608–617. DOI: 10.1111/cch.12868

Tan, B. L., Shi, J., Yang, S., Loh, H., Ng, D., Choo, C., & Medalia, A. (2022). The use of virtual reality and augmented reality in psychosocial rehabilitation for adults with neurodevelopmental disorders: A systematic review. *Frontiers in Psychiatry*, *13*, 1055204. DOI: 10.3389/fpsyt.2022.1055204 PMID: 36590624

Tao, R., Yang, Y., Wilson, M., Chang, J. R., Liu, C., & Sit, C. H. (2025). Comparative effectiveness of physical activity interventions on cognitive functions in children and adolescents with neurodevelopmental disorders: A systematic review and network meta-analysis of randomized controlled trials. *The International Journal of Behavioral Nutrition and Physical Activity*, *22*(1), 6. DOI: 10.1186/s12966-024-01702-7 PMID: 39806448

Tarrano, C., Galléa, C., Delorme, C., McGovern, E. M., Atkinson-Clement, C., Brochard, V., Thobois, S., Tranchant, C., Grabli, D., Degos, B., Corvol, J. C., Pedespan, J.-M., Krystkowiak, P., Houeto, J.-L., Degardin, A., Defebvre, L., Beranger, B., Martino, D., Apartis, E., & Worbe, Y. (2024). Psychiatric phenotype in neurodevelopmental myoclonus-dystonia is underpinned by abnormality of cerebellar modulation on the cerebral cortex. *Scientific Reports*, *14*(1), 22341. DOI: 10.1038/s41598-024-73386-9

Tek, S., Mesite, L., Fein, D., & Naigles, L. (2014). Longitudinal analyses of expressive language development reveal two distinct language profiles among young children with autism spectrum disorders. *Journal of Autism and Developmental Disorders*, *44*(1), 75–89. DOI: 10.1007/s10803-013-1853-4

Telaumbanua, E., Harsono, I., & Soegiarto, I. (2024). Urbanisation in Indonesia: The relationship between income inequality, urban infrastructure, access to education, and population growth with social cohesion, environmental resilience, and housing quality to look at urbanisation in Indonesia. *International Journal of Business, Law, and Education*, *5*(1), 603–614. https://ijble.com/index.php/journal/index. DOI: 10.56442/ijble.v5i1.443

Terras, M. M., Thompson, L. C., & Minnis, H. (2009). Dyslexia and psychosocial functioning: An exploratory study of the role of self-esteem and understanding. *Dyslexia (Chichester, England)*, *15*(4), 304–327. DOI: 10.1002/dys.386 PMID: 19384920

Ter-Stepanian, M., Grizenko, N., Cornish, K., Talwar, V., Mbekou, V., Schmitz, N., & Joober, R. (2017). Attention and executive function in children diagnosed with attention deficit hyperactivity disorder and comorbid disorders. *Journal of the Canadian Academy of Child and Adolescent Psychiatry = Journal de l'Académie Canadienne de Psychiatrie de l'Enfant et de l'Adolescent*, *26*(1), 21–30. PMID: 28331500

Thanalingam, Y., Langridge, F., Gordon, I., Russell, J., Muir, C., & Hamm, L. M. (2022). Strategies and tools to aid the transition between paediatric and adult health services for young adults with neurodevelopmental disorders: A scoping review protocol. *BMJ Open*, *12*(11), e065138. DOI: 10.1136/bmjopen-2022-065138 PMID: 36446454

Thapar, A. (2020). ADHD: Progressing from genetic discoveries to biological insights. *The American Journal of Psychiatry*, *177*(9), 802–804. DOI: 10.1176/appi.ajp.2020.20070961

Thapar, A., Cooper, M., & Rutter, M. (2017). Neurodevelopmental disorders. *The Lancet. Psychiatry*, *4*(4), 339–346. DOI: 10.1016/S2215-0366(16)30376-5

Theodoratou, M., Gkintoni, E., & Farmakopoulou, I. (2023). Executive functions and quality of life in neurodevelopmental spectrum: An outline. *Technium Social Sciences Journal*, *39*, 430–439. DOI: 10.47577/tssj.v39i1.8231

Theule, J., Wiener, J., Tannock, R., & Jenkins, J. M. (2013). Parenting stress in families of children with ADHD: A meta-analysis. *Journal of Emotional and Behavioral Disorders*, *21*(1), 3–17. DOI: 10.1177/1063426610387433

TM, S., Morales, J., & Borasio, G. (2023). Research design considerations for bilingual neurodevelopmental studies. *Frontiers in Psychology*, *14*, 1223456.

Tobia, V., Greco, A., Steca, P., & Marzocchi, G. M. (2019). Children's well-being at school: A multi-dimensional and multi-informant approach. *Journal of Happiness Studies*, *20*(3), 841–861. DOI: 10.1007/s10902-018-9974-2

Tokarek, A., Ufel, W., & Grabkowska, M. NeurodiverCITY: Leveraging the potential for neurodiverse deliberation in urban life (June 11, 2024). Available at http://dx.doi.org/DOI: 10.2139/ssrn.4891374

Tong, S. X., Wong, R. W. Y., Kwan, J. L. Y., & Arciuli, J. (2020). Theory of mind as a mediator of reading comprehension differences between Chinese school-age children with autism and typically developing peers. *Scientific Studies of Reading*, *24*(4), 292–306. DOI: 10.1080/10888438.2019.1666133

Tonorezos, E. S., Cohn, R. J., Glaser, A. W., Lewin, J., Poon, E., Wakefield, C. E., & Oeffinger, K. C. (2022, April 16). Long-term care for people treated for cancer during childhood and adolescence. *Lancet*, *399*(10334), 1561–1572. DOI: 10.1016/S0140-6736(22)00460-3 PMID: 35430023

Topol, E. (2019). *Deep medicine: How artificial intelligence can make healthcare human again*. Basic Books.

Tops, W., Callens, M., Lammertyn, J., Van Hees, V., & Brysbaert, M. (2012). Identifying students with dyslexia in higher education. *Annals of Dyslexia*, *62*(3), 186–203. DOI: 10.1007/s11881-012-0072-6 PMID: 22815104

Torous, J., Bucci, S., Bell, I. H., Kessing, L. V., Faurholt-Jepsen, M., Whelan, P., Carvalho, A. F., Keshavan, M., Linardon, J., & Firth, J. (2021). The growing field of digital psychiatry. *World Psychiatry; Official Journal of the World Psychiatric Association (WPA)*, *3*(3), 318–335. DOI: 10.1002/wps.20883 PMID: 34505369

Totsika, V., Liew, A., Absoud, M., Adnams, C., & Emerson, E. (2022). Mental health problems in children with intellectual disability. *The Lancet. Child & Adolescent Health*, *6*(6), 432–444. DOI: 10.1016/S2352-4642(22)00067-0

Trail, B. A. (2022). *Twice-exceptional gifted children: Understanding, teaching, and counseling gifted students*. Routledge. DOI: 10.4324/9781003261216

Truong, D. T., Adams, A. K., Paniagua, S., Frijters, J. C., Boada, R., Hill, D. E., Lovett, M. W., Mahone, E. M., Willcutt, E. G., Wolf, M., Defries, J. C., Gialluisi, A., Francks, C., Fisher, S. E., Olson, R. K., Pennington, B. F., Smith, S. D., Bosson-Heenan, J., & Gruen, J. R. (2019). Multivariate genome-wide association study of rapid automatised naming and rapid alternating stimulus in Hispanic American and African American youth. *Journal of Medical Genetics*, *56*(8), 557–566. https://jmg.bmj.com/content/56/8/557.info. DOI: 10.1136/jmedgenet-2018-105874

Tsujii, N., Usami, M., Naya, N., Tsuji, T., Mishima, H., Horie, J., Fujiwara, M., & Iida, J. (2021). Efficacy and safety of medication for attention-deficit hyperactivity disorder in children and adolescents with common comorbidities: A systematic review. *Neurology and Therapy*, *10*(2), 499–522. DOI: 10.1007/s40120-021-00249-0 PMID: 34089145

UNESCO. (2018). *Handbook on measuring equity in education*. UNESCO Institute for Statistics.

UNESCO. (2021). *AI and education: Guidance for policymakers*. UNESCO Publishing.

UNESCO/The World Bank. (2021). *Cities, culture, creativity: Leveraging culture and creativity for sustainable urban development and inclusive growth*. Retrieved March 6, 2024, from https://documents1.worldbank.org/curated/en/104121621556036559/pdf/Cities-Culture-Creativity-Leveraging-Culture-and-Creativity-for-Sustainable-Urban-Development-and-Inclusive-Growth.pdf

United Nations Organization. (n.d.). *Promotion and protection of the rights of children; Follow-up to the outcome of the special session on children.* Retrieved March 22, 2024, from https://www.un.org/en/ga/62/plenary/children/bkg.shtml

Vaccaro, A., Kimball, E. W., Moore, A., Newman, B. M., & Troiano, P. F. (2018). Narrating the self: A grounded theory model of emerging purpose for college students with disabilities. *Journal of College Student Development, 59*(1), 37–54. DOI: 10.1353/csd.2018.0003

Van Andel, E., Ten Have, M., Bijlenga, D., Beekman, A. T., De Graaf, R., & Kooij, J. S. (2022). Combined impact of ADHD and insomnia symptoms on quality of life, productivity, and health care use in the general population. *Psychological Medicine, 52*(1), 36–47. DOI: 10.1017/S0033291720001592 PMID: 32597743

van der Plas, E., Darji, H., Srivastava, D. K., Schapiro, M., Jeffe, D., Perkins, S., Howell, R., Leisenring, W., Armstrong, G. T., Oeffinger, K., Krull, K., Edelstein, K., & Hayashi, R. J. (2024). Risk factors for neurocognitive impairment, emotional distress, and poor quality of life in survivors of pediatric rhabdomyosarcoma: A report from the Childhood Cancer Survivor Study. *Cancer, 130*(12), 2224–2236. Advance online publication. DOI: 10.1002/cncr.35236 PMID: 38373075

Van Herwegen, J., Ashworth, M., & Palikara, O. (2019). Views of professionals about the educational needs of children with neurodevelopmental disorders. *Research in Developmental Disabilities, 91*, 103422. DOI: 10.1016/j.ridd.2019.05.001

Vanolo, A. (2023). Autistic cities: Critical urbanism and the politics of neurodiversity. *City (London, England), 27*(1,2), 190–208. DOI: 10.1080/13604813.2023.2178273

Vargo, F. E. (2015). *Neurodevelopmental disorders: A definitive guide for educators.* W. W. Norton & Company.

Vender, E., Beauchamp, J., & Nayeb, L. (2021). Equity and access in neurodevelopmental assessment of bilingual students. *International Journal of Bilingual Education and Bilingualism, 24*(7), 1014–1031.

Verona, I., & Carpenter, J. (2023). Neurodiversity, Sustainability, and Inclusion: A Case for Redundancy in Architecture. In Mostafa, M., Baumeister, R., & Thomsen, M.R. *Design for inclusivity.* Springer.

Vigil-Pérez, A., Blázquez, A., Garcia-Delgar, B., Ortiz, A. E., Borràs, R., Morer, A., Escalona, R. C., & Lázaro, L. (2025). Phenomenology of repetitive and restrictive behaviors and sensory phenomena in neurodevelopmental disorders: An exploratory study. *BMC Psychiatry, 25*(1), 163. DOI: 10.1186/s12888-025-06569-x PMID: 39994553

Vogel, C., Marino, B. S., Cassedy, A., Natarus, M., Parikh, P., Hinkle, C., & Allen, K. Y. (2025). The novel associations between frailty and neurodevelopmental outcomes in CHD. *Cardiology in the Young*, *35*(7), 1427–1435. DOI: 10.1017/S1047951125100863 PMID: 40637129

Vogindroukas, I., Stankova, M., Chelas, E. N., & Proedrou, A. (2022). Language and speech characteristics in Autism. *Neuropsychiatric Disease and Treatment*, *18*, 2367–2377. DOI: 10.2147/NDT.S331987

Walenski, M., Tager-Flusberg, H., & Ullman, M. T. (2006). Language in Autism. In S. O. Moldin & J. L. R. Rubenstein (Eds.), *Understanding autism: From basic neuroscience to treatment* (pp. 175–203). CRC Press/Routledge/Taylor & Francis Group. DOI: 10.1201/9781420004205.ch9

Watson, S. M., Gable, R. A., & Morin, L. L. (2016). The role of executive functions in classroom instruction of students with learning disabilities. *International Journal of School and Cognitive Psychology*, *3*(167). Advance online publication. DOI: 10.4172/2469-9837.1000167

Weber, C. L., Boswell, C., & Behrens, W. A. (2025). Recognizing learning and developmental differences. In *Exploring critical issues in gifted education* (pp. 33–62). Routledge. DOI: 10.4324/9781003380634-4

Weitzman, C., Guevara, J., Curtin, M., Macias, M., Kinwa Poon, J., Joseph Smith, P. J., Christine Augustyn, M., Liu, Y. H., Anopawuia Spinks-Franklin, A. I., Zubler, J. M., Navsaria, D., Glusman, M., Uzoatu Anyigbo, C., Chen, V., Gonzales, J. L., Guevara, J. P., Nobuhide Hashikawa, A., Marie Loria, H., Miller-Fitzwater, A., & Degnon, L. (2025). Promoting optimal development: Screening for mental health, emotional, and behavioral problems: Clinical report. *Pediatrics*, *156*(3), e2025073172. DOI: 10.1542/peds.2025-073172 PMID: 40850690

White, M., Lafaille-Magnan, M. E., Roche, C., Jolicoeur-Martineau, A., Wazana, A., & Connor, K. L. (2023). Developmental origins, differential susceptibility, and resilience in neurodevelopmental disorders. In Eisenstat, D. D., Goldowitz, D., Oberlander, T. F., & Yager, J. Y. (Eds.), *Neurodevelopmental pediatrics* (pp. 1–20). Springer. DOI: 10.1007/978-3-031-20792-1_45

Wiater, M., & Firszt, D. (2025). Innovations in the public sector in response to the challenges of neurodiversity–findings from Krakow. *Scientific Papers of Silesian University of Technology. Organization & Management/Zeszyty Naukowe Politechniki Slaskiej. Seria Organizacji i Zarzadzanie*, (218). DOI: 10.29119/1641-3466.2025.218.38

Wiener, J., & Mak, M. (2009). Peer victimization in children with attention-deficit/hyperactivity disorder. *Psychology in the Schools*, *46*(2), 116–131. DOI: 10.1002/pits.20358

Williams, N. M., Zaharieva, I., Martin, A., Langley, K., Mantripragada, K., Fosdal, R., Stefansson, H., Stefansson, K., Magnusson, P., Gudmundsson, O. O., Gustafsson, O., Holmans, P., Owen, M. J., O'Donovan, M., & Thapar, A. (2010). Rare chromosomal deletions and duplications in attention-deficit hyperactivity disorder: A genome-wide analysis. *Lancet*, *376*(9750), 1401–1408. DOI: 10.1016/S0140-6736(10)61109-9

Wilmot, A., Pizzey, H., Leitão, S., Hasking, P., & Boyes, M. (2023). Growing up with dyslexia: Child and parent perspectives on school struggles, self-esteem, and mental health. *Dyslexia (Chichester, England)*, *29*(1), 40–54. DOI: 10.1002/dys.1729 PMID: 36349866

Wilson, P., Ruddock, S., Rahimi-Golkhandan, S., Piek, J., Sugden, D., Green, D., & Steenbergen, B. (2020). Cognitive and motor function in developmental coordination disorder. *Developmental Medicine and Child Neurology*, *62*(11), 1317–1323. DOI: 10.1111/dmcn.14646

Witko, A. (2019). La dyslexie à l'âge adulte: Comment révéler son potentiel pour dépasser son trouble. *L'Orthophoniste, 391,* 18–20. https://hal.archives-ouvertes.fr/hal-02264404

Wolf, M., Gotlieb, R. J., Kim, S. A., Pedroza, V., Rhinehart, L. V., Tempini, M. L. G., & Sears, S. (2024). Towards a dynamic, comprehensive conceptualization of dyslexia. *Annals of Dyslexia*, *74*(3), 1–22. DOI: 10.1007/s11881-023-00297-1 PMID: 38217783

Wolf, T., Debut, L., Sanh, V., Chaumond, J., Delangue, C., Moi, A., & Rush, A. M. (2020, October). Transformers: State-of-the-art natural language processing. In *Proceedings of the 2020 conference on empirical methods in natural language processing: system demonstrations* (pp. 38-45). DOI:DOI: 10.18653/v1/2020.emnlp-demos.6

Wong, C. H., Chan, C. K., & Wong, P. W. (2025). Family outcomes of a parent-implemented early intervention for neurodevelopmental disorders: An exploratory study of family dynamics. *Frontiers in Public Health*, *13*, 1614089. DOI: 10.3389/fpubh.2025.1614089 PMID: 40697837

Wong, M. M., Kampen, R. A., Braden, R. O., Alagöz, G., Hildebrand, M. S., Dingemans, A. J., Corbally, J., den Hoed, J., Mendoza, E., Claassen, W. J. J., Barnett, C., Barnett, M., Brusco, A., Carli, D., de Vries, B. B. A., Elmslie, F., Ferrero, G. B., Jansen, N. A., van de Laar, I. M. B. H., & Fisher, S. E. (2025). SETBP1 variants outside the degron disrupt DNA-binding, transcription and neuronal differentiation capacity to cause a heterogeneous neurodevelopmental disorder. *Nature Communications*, *16*(1), 9021. DOI: 10.1038/s41467-025-64074-x PMID: 41073373

Yeung, M. K. (2022). A systematic review and meta-analysis of facial emotion recognition in autism spectrum disorder: The specificity of deficits and the role of task characteristics. *Neuroscience and Biobehavioral Reviews*, *133*, 104518. DOI: 10.1016/j.neubiorev.2021.104518

Yoshida, K., & Van der Walt, J. L. (2018). The policy-implementation-results linkage for education development and aid effectiveness in the Education 2030 era. *Compare: A Journal of Comparative Education*, *48*(1), 39–55. DOI: 10.1080/03057925.2017.1283583

Young, S., Sedgwick, O., Fridman, M., Gudjonsson, G., Hodgkins, P., Lantigua, M., & González, R. (2015). Co-morbid psychiatric disorders among incarcerated ADHD populations: A meta-analysis. *Psychological Medicine*, *45*(12), 2499–2510. DOI: 10.1017/S0033291715000598

Yuen, H. K., Ge, H., Zhang, C., Wong, Y. T., Chan, E. Y., Tsang, W. W., & Capio, C. M. (2025). Effects of bilingualism on executive function of children with neurodevelopmental disorders: A scoping review. *Children (Basel, Switzerland)*, *12*(9), 1247. DOI: 10.3390/children12091247 PMID: 41007112

Yuen, M., Choo, T. M., & Leung, K. (2025). Cognitive and linguistic interplay in bilingual children with neurodevelopmental disorders. *Journal of Cognition and Development*, *26*(3), 411–430.

Zapata, M. A., & Worrell, F. C. (2023). Disability self-worth and personal meaning relate to psychosocial functioning among employed U.S. adults with LD and ADHD. *Rehabilitation Psychology*, *68*(2), 184–195. DOI: 10.1037/rep0000484 PMID: 36951738

Zawacki-Richter, O., Marín, V. I., Bond, M., & Gouverneur, F. (2019). Systematic review of research on AI applications in higher education. *International Journal of Educational Technology in Higher Education*, *16*(1), 1–27. DOI: 10.1186/s41239-019-0171-0

Zelazo, P. D. (2015). Executive function: Reflection, iterative reprocessing, complexity, and the developing brain. *Developmental Review*, *38*, 55–68. DOI: 10.1016/j.dr.2015.07.001

Zelazo, P. D., & Carlson, S. M. (2020). The neurodevelopment of executive function skills: Implications for academic achievement gaps. *Psychology & Neuroscience*, *13*(3), 273–298. DOI: 10.1037/pne0000208

Zerva, I. K. (2023c). Early Signs of Reading Difficulties and Phonological Awareness in Early School Age. In *Perspectives of Cognitive, Psychosocial, and Learning Difficulties From Childhood to Adulthood: Practical Counseling Strategies* (pp. 21-36). IGI Global. DOI: 10.4018/978-1-6684-8203-2.ch002

Zerva, I. K., & Koundourou, C. (2023). Neurodevelopmental Language Disorders in School Context: Specific Learning Disorders in Reading and Writing. In *Developmental Language Disorders in Childhood and Adolescence* (pp. 195-206). IGI Global. DOI: 10.4018/979-8-3693-0644-4.ch015

Zerva, I. K. (2023a). Metalinguistic abilities: The contribution to writing. *Social Sciences & Humanities Open*, *8*(1), 100609. DOI: 10.1016/j.ssaho.2023.100609

Zerva, K. I. (2023b). The imprint of specific learning difficulties and ADHD on individual and environmental mental health. In Katsarou, D., Koundourou, Ch., & Reppa, G. (Eds.), *New trends and approaches in psychopathology and mental health of children and adults* (pp. 185–204). Gutenberg. (In Greek)

Zhang, H., Yu, J., Tian, Y., Ma, H., Pan, Y., Wang, C., & Zhou, C. (2024). Effects of short-term and long-term physical activity intervention on attention in children with attention deficit hyperactivity disorder: Evidence from event-related potentials. *Frontiers in Psychology*, *14*, 1301740. DOI: 10.3389/fpsyg.2023.1301740

Zhang, J., Yuan, M., Liu, Y., Zhong, X., Wu, J., & Chen, W. (2025). Bisphenol A exposure and neurodevelopmental disorders and problems in children under 12 years of age: A systematic review and meta-analysis. *Journal of Hazardous Materials*, *137731*, 137731. Advance online publication. DOI: 10.1016/j.jhazmat.2025.137731 PMID: 40054188

Zhang, M., Liu, Z., Ma, H., & Smith, D. M. (2020). Chronic physical activity for attention deficit hyperactivity disorder and/or autism spectrum disorder in children: A meta-analysis of randomized controlled trials. *Frontiers in Behavioral Neuroscience*, *14*, 564886. DOI: 10.3389/fnbeh.2020.564886 PMID: 33192364

Zhou, S., Loiacono, E. T., & Kordzadeh, N. (2024). Smart cities for people with disabilities: A systematic literature review and future research directions. *European Journal of Information Systems, 33*(6), 845–862. DOI: 10.1080/0960085X.2023.2297974

Zhou, X., Jia, F., Bambling, M., Edirippulige, S., Whiteford, H., & Lin, J. (2025). Associations between multiple neurodevelopmental disorders and mental health in children. *Journal of Affective Disorders, 120397*. Advance online publication. DOI: 10.1016/j.jad.2025.120397 PMID: 41072876

Zoccante, L., Antolini, G., Passarella, L., Puttini, E., Rizzo, V., & Colizzi, M. (2022). Neurodevelopmental disorders and psychosocial issues later in life. In *Prevention in mental health: From risk management to early intervention* (pp. 125–141). Springer International Publishing. DOI: 10.1007/978-3-030-97906-5_7

Zoccante, L., Ciceri, M. L., Chamitava, L., Di Gennaro, G., Cazzoletti, L., Zanolin, M. E., Darra, F., & Colizzi, M. (2021). Postural control in childhood: Investigating the neurodevelopmental gradient hypothesis. *International Journal of Environmental Research and Public Health, 18*(4), 1693. DOI: 10.3390/ijerph18041693 PMID: 33578752

Zorzi, S., & Berteotti, L. (2025). *Independent living for persons with neurodevelopmental disorders*. IntechOpen. DOI: 10.5772/intechopen.1008380

About the Author

Irini K. Zerva is a Postdoctoral Researcher and Teaching and Research Associate at the Faculty of Medicine of the Aristotle University of Thessaloniki, where her work focuses on neurodevelopmental disorders from a lifespan and interdisciplinary medical perspective. She also serves as a Lecturer at Neapolis University Pafos and as an Adjunct Lecturer at the University of Nicosia in the fields of Psychology and Special Education. Dr. Zerva specializes in Neurodevelopmental Disorders, with particular expertise in Specific Learning Disorders (SLD) and Attention Deficit Hyperactivity Disorder (ADHD). Her research and clinical work are grounded in the clinical continuum model of neurodevelopmental conditions, integrating medical, cognitive, linguistic, and psychosocial parameters in diagnostic assessment and intervention planning. Her scholarly focus includes metalinguistic skills, neurocognitive mechanisms underlying learning, and the interaction between medical history and developmental trajectories. She holds a Ph.D. in Special Pedagogy and Psychology (Metalinguistic Skills) from the National and Kapodistrian University of Athens and currently conducts postdoctoral research within a medical academic environment, strengthening the interface between medicine, psychology, and education. She also holds two Master's degrees (M.Sc.) in Special Education (Speech Therapy and Counseling) and Educational Policy and Administration, as well as two undergraduate degrees in Linguistics and Pedagogy, reflecting a structured and specialized academic trajectory in language, cognition, and educational sciences. With more than 20 years of clinical and educational experience, Dr. Zerva has worked extensively with children and adolescents presenting neurodevelopmental profiles, contributing to comprehensive diagnostic evaluations, individualized intervention programs, and interdisciplinary clinical collaboration. She has authored and co-authored scientific articles and book chapters in the areas of neurodevelopmental disorders, language development, and learning disabilities. She is an elected member of the American Psychological Association (APA).

Index

A

B

C

D

E

T

U

V

W

www.ingramcontent.com/pod-product-compliance
Lightning Source LLC
LaVergne TN
LVHW062219200926
848484LV00006B/1436

* 9 7 9 8 3 6 9 3 6 3 3 1 7 *